Brian Fleming Research & Learning Library
Ministry of Education
Ministry of Training, Colleges & Universities
900 Bay St. 13th Floor, Mowat Block
Toronto, ON M7A 1L2

D1804161

Mathematics Education Library

More information about this series at http://www.springer.com/series/6276

Kay Owens

Visuospatial Reasoning

An Ecocultural Perspective for Space, Geometry and Measurement Education

Kay Owens
School of Teacher Education
Charles Sturt University
Dubbo, NSW, Australia

ISSN 0924-4921 ISSN 2214-983X (electronic)
ISBN 978-3-319-02462-2 ISBN 978-3-319-02463-9 (eBook)
DOI 10.1007/978-3-319-02463-9
Springer Cham Heidelberg New York Dordrecht London

Library of Congress Control Number: 2014945742

© Springer International Publishing Switzerland 2015
This work is subject to copyright. All rights are reserved by the Publisher, whether the whole or part of the material is concerned, specifically the rights of translation, reprinting, reuse of illustrations, recitation, broadcasting, reproduction on microfilms or in any other physical way, and transmission or information storage and retrieval, electronic adaptation, computer software, or by similar or dissimilar methodology now known or hereafter developed. Exempted from this legal reservation are brief excerpts in connection with reviews or scholarly analysis or material supplied specifically for the purpose of being entered and executed on a computer system, for exclusive use by the purchaser of the work. Duplication of this publication or parts thereof is permitted only under the provisions of the Copyright Law of the Publisher's location, in its current version, and permission for use must always be obtained from Springer. Permissions for use may be obtained through RightsLink at the Copyright Clearance Center. Violations are liable to prosecution under the respective Copyright Law.
The use of general descriptive names, registered names, trademarks, service marks, etc. in this publication does not imply, even in the absence of a specific statement, that such names are exempt from the relevant protective laws and regulations and therefore free for general use.
While the advice and information in this book are believed to be true and accurate at the date of publication, neither the authors nor the editors nor the publisher can accept any legal responsibility for any errors or omissions that may be made. The publisher makes no warranty, express or implied, with respect to the material contained herein.

Printed on acid-free paper

Springer is part of Springer Science+Business Media (www.springer.com)

This book is dedicated to the people of Papua New Guinea and the Glen Lean Ethnomathematics Centre at the University of Goroka, PNG.

Acknowledgements

I wish to thank the many people of Papua New Guinea and Aboriginal Australia who have welcomed me into their homes and their knowledges, directly and indirectly. I hope this book illustrates a glimpse of the wealth of your cultures and my respect for you. There are too many to thank individually but I would particularly like to acknowledge Wilfred Kaleva and Rex Matang for their years of sharing and caring. I also want to thank the Dubbo Aboriginal community especially the Elders, my colleagues at Charles Sturt University (too many to name), and Diane McNaboe for teaching me Wiradjuri. I want to thank my husband Chris for his endless patience, support, and editorial work over the many years that the ideas of this book have been recorded. For their comments on drafts, I appreciate and was inspired by Peter Galbraith, Philip Clarkson, and the writing group in my School of Teacher Education at CSU, and for their contributions both in conversations, research and writing to the contributing authors—Kate Highfield, Ravi Jawahir, and Marcos Cherinda.

Contents

1	**Introduction: Visuospatial Reasoning in Context**................................	1
	The Challenge	1
	Historical Overview of Relevant Research on Spatial Abilities and Visual Imagery	2
	Visuospatial Reasoning and School Test Performance	3
	Visuospatial Reasoning in Mathematics Education Research	5
	Defining Visuospatial Reasoning..	6
	Extending the Meaning of Visuospatial Reasoning	9
	Visuospatial Reasoning and Geometry	10
	Bridging from the Psychological to the Ecocultural Perspective	11
	An Ecocultural Perspective on Learning	12
	Ecocultural Identity and Mathematical Identity	14
	Moving Forward	16
2	**Visuospatial Reasoning in Twentieth Century Psychology-Based Studies**	19
	The Challenge	19
	Visuospatial Reasoning and Studies on Spatial Abilities	20
	Visuospatial Reasoning from an Information Processing Perspective......	22
	Studies of Learners with Disabilities	24
	Age and Visuospatial Reasoning	26
	Visuospatial Reasoning on Different Tasks	29
	Personal Approaches to Visuospatial Reasoning.....................................	31
	Training...	33
	Key Study on Children's Visuospatial Reasoning	35
	Visuospatial Reasoning—Getting Inside Children's Heads	40
	Holistic, Concrete, Pictorial Imagery..	43
	Dynamic Visuospatial Reasoning	45
	Visuospatial Action Reasoning	47
	Pattern Visuospatial Reasoning...	51
	Procedural Visuospatial Reasoning...	57

	Visuospatial Reasoning in Concept Development	59
	Visuospatial Reasoning in Problem Solving	62
	Visuospatial Reasoning in Learning	63
	Visuospatial Reasoning, Metaphor, and Metonomy	65
	Context and Visuospatial Reasoning About 3D Shapes	67
	School Learning Experiences and 3D Visuospatial Reasoning	72
	Visuospatial Reasoning in Context	73
	Attention	75
	Unlimited Capacity Model of Attention for Action	76
	Ecology and Visual Perception	77
	Attention and Responsiveness	78
	Developing a Theoretical Framework of Visuospatial Reasoning	82
	Assessment Tasks	84
	Moving Forward	87
3	**Changing Perspective: Sociocultural Elaboration**	91
	The Challenge	91
	Local Context	92
	An Example of an Ecocultural Practice	94
	Values in Education	95
	Signification, Meaning, and Becoming	96
	A Genetic Approach to Ecocultural Perspectives	99
	Visuospatial Reasoning in a Navigating Team's Sociocultural Knowledge	104
	Visuospatial Reasoning in Measurement	106
	Embodiment of Spatial Knowledge	107
	Early Childhood Experiences	107
	Critical Perspective of Place	111
	Moving Forward	114
4	**Place, Culture, Language, and Visuospatial Reasoning**	115
	The Challenge	115
	The Role of Language	116
	Cultural Ways of Thinking Reflected in Language	118
	Locating, Visuospatial Reasoning, and Communicating	120
	Language Patterns in Papua New Guinea	122
	Frames of Reference for Space and Place	123
	Language in Comparing and Measuring space	127
	Implications of Language for Visuospatial Reasoning	131
	Maps as Representations of Visuospatial Reasoning	131
	Shapes	136
	Moving Forward	138
5	**Visuospatial Reasoning in Cultural Activities in Papua New Guinea**	141
	The Challenge	141

Earlier Studies on Spatial Abilities and Visualisation Recognising Ecocultural Contexts	143
Overview of PNG Material Cultures Involving Visuospatial Reasoning	159
Visuospatial Reasoning That Connects Euclidean Geometry and Topology	167
Continuous String Bag: *Bilum*	170
String Figures	172
Weaving Patterns	175
Tattoos	176
Visuospatial Reasoning with Number	177
The Measurement Study	180
Visuospatial Reasoning About Length	181
Visuospatial Reasoning About Area	182
Visuospatial Reasoning About Volume	185
Visuospatial Reasoning About Three-Dimensional Designs	188
Roofs	189
Bridges	190
The Ecocultural Holistic Context	191
The Importance of Ratio or Multiplicative Thinking in Reasoning	192
Architecture Students' Visuospatial Reasoning	193
Ecocultural Visuospatial Reasoning as Architects	194
Responsiveness During Rich Activity	199
Decision-Making	200
Power, Identity, and Relationships in Architecture	200
Visuospatial Reasoning with an Ecocultural Perspective	201
Moving Forward	202

6 Visuospatial Reasoning in Other Cultures ... 205

The Challenge	205
Early Studies of Visuospatial Reasoning with an Ecocultural Perspective	206
Navajo Knowledge	206
Pacific Navigation	208
Australian Aboriginal Astronomy	211
Circle Geometry and Straight-Edged Shapes	212
Creative Designs Across the World	214
Settlement Patterns and Shelters: Place, not Just Position	220
Moving Forward	222

7 The Impact of an Ecocultural Perspective of Visuospatial Reasoning on Mathematics Education ... 223

The Challenge	223
Impact on Mathematics	225
Impact on Understanding Mathematical Learning	230
The Context for the Current School System	231

	Impact on Mathematics Education	233
	Tacit Knowledge in Visuospatial Reasoning	237
	Language and Concepts in Mathematics Education	239
	Impact on Mathematical Identity	240
	Moving Forward	244
8	**The Importance of an Ecocultural Perspective for Indigenous and Transcultural Education**	245
	Kay Owens, Marcos Cherinda and Ravi Jawahir	
	The Challenge	245
	Continuities in Education Between Community and School	246
	Language-Based Activities in Multicultural Classrooms	248
	Visuospatial Reasoning in Metaphors: Selection and Relevance	249
	PNG Secondary Teachers' Ethnomathematics Studies	250
	PNG Elementary School Teacher Education Study	254
	Working with Mental and Physical Visuospatial Representations in Africa	261
	Mental Mapping of the Navajo, USA	263
	Yup'ik Mathematics Education	264
	Thoughts on an Ecocultural Curriculum	265
	Brazilian Initiatives	266
	Language and Inquiry for Visuospatial Reasoning in Geometry: Mauritius	267
	Moving Forward	272
9	**Visuospatial Reasoning in Contexts with Digital Technology**	275
	Kay Owens and Kate Highfield	
	The Challenge	275
	Ecocultural Perspective of Measurement in Changing Worlds	276
	The Role of Digital Media in Developing Self-Regulation for Learning	277
	Visuospatial Reasoning in Geometry and Measurement Learning Through Digital Technology	281
	Visuospatial Reasoning in the Digital Age Taking Account of Ecocultural Contexts	287
	Moving Forward	289
10	**An Ecocultural Perspective on Visuospatial Reasoning in Geometry and Measurement Education**	291
	The Challenge	291
	Mathematics	293
	Theories of Mathematics Education	294
	The Synergy of Research Studies on Visuospatial Reasoning	296
	Visuospatial Reasoning	299
	The Importance of Visuospatial Reasoning	301
	Contexts for Visuospatial Reasoning	302

Mathematics Teacher Education for Visuospatial Reasoning	303
Visuospatial Reasoning in Geometry and Measurement	304
Developing Views of Spatiality	305
The Issue of Equity	306
Challenges Addressed	308

Abbreviations 309

Appendix A   **A Synthesis of Problem-Solving Processes** 311

Appendix B   **Test of Visuospatial Reasoning for Young Children** 313

Appendix C   **Map of PNG and Some Counting System Details** 317

Appendix D   **Making the *Bilum* Pattern 50 Toea or Soccer Ball** 321

Appendix E   **Example of Learning Plan for Cultural Mathematics** 323

Appendix F   **Selection of Curriculum Statements and Issues** 327

Appendix G   **Types of Geometry** 333

References 335

Author Index 365

Subject Index 373

List of Figures

Prologue	Elders at Kopnung village, Jiwaka Province, PNG, demonstrate cultural practices (weaving and trap-making with the children learning visuospatially	xxix
Fig. 1.1	Drawings to illustrate Australian Aboriginal art about relationships	10
Fig. 1.2	Developing identity as a mathematical thinker............................	15
Fig. 2.1	Confidence intervals for means of the difference between scores on two-dimension delayed posttest and pretest for spatial versus number groups...	40
Fig. 2.2	Children attending to parts of shapes. (a) Michael attends to the sides of the shape. (b) Children attend to the angle by marking and turning their finger away from their thumb	45
Fig. 2.3	Examples of dynamic visuospatial reasoning. (a) Sam's hexagon "like a square". (b) Michael's rectangles, and M to W. (c) Peter's trapezium to parallelogram......................	46
Fig. 2.4	Victor, grade 2, recognising equal angles on different shapes and on shapes in different orientations	48
Fig. 2.5	Peter and Victor mentally folding the pentomino shapes to form an open cube	48
Fig. 2.6	Mean order of pentominoes being made and correlated data for adults and children. (a) Mean order of shapes being made by children and adults. (b) Correlations in order of appearance of shapes...	50
Fig. 2.7	Kathy's large triangle and rectangle from the tangram pieces.......	50
Fig. 2.8	Examples of pattern visuospatial reasoning. (a) Peter's hexagon. (b) Sally's 4, 8, and 9 squares with 24 matches. (c) Pattern repeated for both types of rhombus. (d) Right-angled triangle pattern that was also used for equilateral triangles by Sally. (e) Lena's pattern of squares	51

Fig. 2.9	Worksheet for tiling shapes (Form S) and grade 2 child completing the items	55
Fig. 2.10	Examples of children's responses to test items on covering areas with tiles	56
Fig. 2.11	Dominance of similarity and horizontal lines in initial trials before visuospatial reasoning improved	58
Fig. 2.12	Using action imagery to develop procedural imagery	59
Fig. 2.13	Selection of test items: joining 3D shapes together to make other shapes	68
Fig. 2.14	Items for recognising shapes from other perspectives	69
Fig. 2.15	Selected items from test: imagining folding 2D shape paper to make a 3D shape	71
Fig. 2.16	Performance of low attainers from both groups on three-dimensional shape test	73
Fig. 2.17	Aspects of problem solving	79
Fig. 2.18	Tasks for assessment in Count Me Into Space	85
Fig. 3.1	The role of Elders in the education of the child (Owens et al., 2012)	93
Fig. 3.2	The interplay of genetic aspects in cultural contexts (Saxe, 2012, p. 33)	101
Fig. 4.1	The interplay of registers in first language (L1) and second language (L2). (Prediger et al., 2012, p. 6216)	117
Fig. 4.2	Shelters in the camp, Yalata, South Australia (Owens, 1966)	126
Fig. 4.3	Mapping the school (multilingual town school, Goroka, PNG) (Owens 2001a, 2001b)	133
Fig. 4.4	Non-orthogonal maps indicating visuospatial reasoning about place	135
Fig. 5.1	Rock art in Papua New Guinea. (a, b) Tabar Island drawings (Gunn, 1986); (c) Snake River (1975)	144
Fig. 5.2	Objects that become part of visuospatial reasoning and identity connections. Trobriand dancers with models of discs, 1975. Note tapa cloth worn by lead male and "grass" skirts by teenage girls	146
Fig. 5.3	Kambea Rambu string figures and story. (a) The course set for Kambea Rambu. (b) Passing Puri. (c) At Ronga village. (d) Wando Range. (e) 4 m deep well along roadside. (f) Cane bridge crossing Suku River. (g) Rushing imaginary Suku River. (h) Akero rainforest where wild orchids and possums abound. (i) Showing bird of paradise nest along the road side. (j) A bee heap adjacent to bird's nest. (k) Side view of Mount Kambea showing its exciting	

	feature. (l) Showing treeless Peak of Mount Kambea. (m) The targeted point at the top of Kambea Peak (Kagua/Erave teacher, 2007)..	147
Fig. 5.4	Tattoo design drawn by Rea, Motu teacher at Tubusereia and copies by grade 2 children who chose which design to do from her self-made big book. (a) Chest tattoo incorporating tear design. (b) Back tattoo. (c) Hand tattoo. (d) Stomach tattoo. (e) Leg tattoo. (f) Drawing the rectangles for the basic hand tattoo. (g) Next stage, with others' drawings. (h) Proud of the finished drawing. (i) Lack of rulers encouraged hand drawing (2013)	148
Fig. 5.5	Visuospatial reasoning in counting by Tolai community, PNG. (a) Nuts bundled in sets of 6 and 4 more providing words for 10 from this visuospatial arrangement. (b) Marking length or number of shell-money. (c) 10×10×3. (Paraide, 2010)..	148
Fig. 5.6	Triangular pattern used in planting cash and other crops	149
Fig. 5.7	Typical bridges in PNG requiring visuospatial reasoning to build. (a) Cantilever/suspension bridge that I crossed around 1986 across the tributary of the Busu between Boana and Hobu. It had been rebuilt after a flood. (b) Suspension bridge below Boana, one of two crossing the upper reaches of the Busu that I crossed in 1973 and another I crossed in 1983. Bridges have to be replaced or repaired when materials rot	149
Fig. 5.8	PNG architecture students' sculptures. (a) David used pattern, stability, repetition, and measurement. (b) Willie's holistic sculpture illustrates his use of curves and repetition. (c) Ian used three-dimensional shapes in his compact sculpture. (d) TKeps developed a functional idea. (e) Fred incorporates traditional decoration and curves. (f) Taurus used the sea devil and counterpoint balance. (g) Fing developed ideas from modern buildings, repetition and asymmetry. (h) Bell's sculpture shows traditional influences..	150
Fig. 5.9	PNG gardens form area units. (a) Drains divide garden into areas, Kopnung, Jiwaka, 2006. (b) Three stalks in each mound in rows of two or three mounds............................	151
Fig. 5.10	PNG round houses. (a) Large, round house, Kaveve, Eastern Highlands, 2006. (b) Using rope tied to centre post to mark the edge of the round house. The leg at the other end of the rope is dragged to mark the circle, Kaveve. (c) Lower area space in a small round house covered with utensils and nuts drying. Karuku nuts drying over fireplace, Kaveve...	151

Fig. 5.11 Rectangular houses. (a) Men's house for 25 men with four corner sleeping rooms, Kopnung, Jiwaka, 2006. (b) Making a model and showing measuring techniques, Kopnung, Jiwaka, 2006 152

Fig. 5.12 PNG preparing a *mumu* (Kaveve village, Eastern Highlands, 2006). (a) Heating the stones. (b) Adding the food, karuku nuts. (c) Covering with dry grass, green grass, and then soil. (d) Pouring water through a hole onto the mumu stones to make steam to cook food 152

Fig. 5.13 PNG coastal houses. (a) House with woven decorated walls, basic house on nine posts, Malalamai, Madang Province, 2006. (b) Placing morata on the roof of a house, checking spaces, Kela, Morobe, 1997. (c) Weaving walls *blind* with a design, Lolobai, Morobe, 1997. (d) Morata ready for a particular size house, Mis, Madang, 2006. (e) Designing new house styles, and single outrigger canoe, Lolobai, Morobe coast, 1997. (f) Long house for families, Kanganaman, Middle Sepik river, 1983 153

Fig. 5.14 Leaf and cane weaving. (a) Weaving patterns for walls, *diagonal, zigzag, diamond* Kopnung, Jiwaka Province, 2006. (b) Woven basket, internal and base layer wider, pandanus, Kepara village, Central Province, 2004. (c) Basket display—Yalibu, Southern Highlands trays—introduced; Buka, Bougainville lady's handbag and tray; fish traps; Sepik mask. Sepik carved man looking on. (d) Sepik mask , 1986. (e) Woven floor mat, Milne Bay, 1984. (f) Carry bag that men learn to make in the men's house. Note bilum with baby in background. (g) Hat making passed down by men 154

Fig. 5.15 Various carrying and storage objects, PNG. (a) Palm frond temporary basket for carrying food or distributing in exchanges. (b) Limbom basket, Wosera, East Sepik, 1983. (c) Karuka nut basket for trading from highlands to coast when stored nuts are dry, Kaveve, Eastern Highlands, 2006. (d) Mountain design woollen bilum with karuka nuts, Kaveve, Eastern Highlands, 2006 155

Fig. 5.16 PNG continuous string bag, *bilum*, designs (a) Continuous string bags (*bilum*) requires string to be joined by rubbing on thigh. Women learn from each other. Bena Bena, Eastern Highlands, 1975. (b) Rai coast bush string being made and wound onto spindle, Madang Province, 2013. (c) Two-strand twisted wool bilum, Watabung, Simbu Province, 1978 with close-up. (d) Bilum made by Engan woman, PNG highlands with closeup, variation on the rainbow design, 2001. (e) Close-up rainbow design, Kamano-Kafe, Eastern

List of Figures xix

Highlands, 2006. (f) Computer – network design made by Kamano-Kafe woman, Eastern Highlands, Province, PNG, 2006 with close-up. (g) Soccer ball bilum, 2005 with close-up. (h) Open stitch of traditional carrying bilum, 1997. (i) The flower pattern design from tiles used by Caroline from Sepik living in Madang, 1997 (Jondu, 2008). (j) Both sides of a bush string traditional bilum, East Sepik Province, 2001. (k) Bilum in plastic, Tufi, Oro Province, note the use of colours and different stitches for effect and the combination of rectangles in rows to form squares, 1986. (l) Girl at Malalamai, Rai coast, Madang carrying bilum, strengthening neck muscles, being like mother, 2004. (m) Pig toy on cane frame with large stitch bilum and light bilum underneath, looking inside at frame, 1997. (n) Two different loops used for effect or for different purposes, Balob Teachers College student project, 1997. (o) Bilum and two kinds of shell for chest piece and status, 1997 156

Fig. 5.17 PNG stone carvings and bindings. (a) Bindings on stone adze still used for canoe and sago working; stone axe, Highlands 1976; bindings on arrows (jagged one for men) 1975, arm band from Rai coast Madang 2013. (b) Rare stone carvings, Hagen, 1985. See also Fig. 5.20d 159

Fig. 5.18 Diverse wooden carvings from across PNG. (a) Bilum hook, hardwood, Palambi, East Sepik River, 1983. (b) Tray, Kiriwina, Trobriand Islands, 1973; Mat Gulf Province, 1997. (c) Carvings from Tami Island, plates 1978, bowls 1980. (d) Carvings from Trobriands, diversity, kwila and ebony woods, small gourd, 1975. (e) West New Britain—single bird with flat wings, 1975; developments to snake and birds with raised wings, 66 cm, 1982. (f) Trobriands, 1975. (g) Sepik carvings. Males, kundu drum, 128 cm ~1986. (h) Yam storage and display house, Trobriands, 1975. (i) Large seafaring canoe prow, Trobriand Islands, 1975. (j) Model of lakatoi canoe, Tubusereia, Papuan coast, Central Province, 2013. (k) Story board, Kambot, Murik Lakes, Sepik River, East Sepik, 1973. (l) Story board, Kambot, 1988. Note variations from earlier. Village woman with adze making sago. Connections of people with totems, animals, place, myths. (m) Carved poles of Haus Tambarans, middle Sepik River, East Sepik, 1983. (n) Carved logs representing all provinces, Sepik carvers, designed by Department of Architecture and Building Science, PNG University of Technology, for University of Goroka Library, ~5 m 2002 160

Fig. 5.19 Pottery from Sepik River, East Sepik, and Markham Valley, Morobe, PNG. (a) Wosera pot, East Sepik, 1983. (b) "Stove" Chambri Lake, East Sepik for sago pancake plate and for smoke to reduce mosquitoes, 1983. (c) Sepik River pot, 1997. (d) Head only of sago storage pots—humour, made for tourists, Chambri Lakes, 1978. (e) Pot for cooking snake, Zumin, 1984; nearby Adzera village, 2006. (f) Mask: Tortoise shell, cassowary feather, basket, clay, East Sepik, 1985 163

Fig. 5.20 Tapa from across PNG. (a) Section of tapa used as cloak for dancing. 180 cm × 110 cm, Oro Province, 1975. (b) Tapa cloth, 55 cm × 136 cm, Oro Province, 1976. (c) Long strip of tapa worn by youth and men wrapped between legs and around waist. Oro, 1976. (d) Anga tapa used as coat, raincoat, and hiding club, Menyamya, Morobe, 1985 164

Fig. 5.21 Examples of diverse dances, decorations, and displays associated with *singsing,* PNG. (a) Mudmen masks, *Gilitue*, preparing for cultural activity; see Table 8.1 for making of mask (John 2007). (b) Highland warriors with bilum skirts, painted faces, feathered headdresses, Goroka Show, 2005. (c) Wig and bamboo strips indicated pig kills in exchanges, 1975 (see Fig. 8.1b for making wigs from another cultural group. (d) Bena headgear, 1976. (e) Huon Peninsula kundu drums and "sails" headdress, 1978. (f) Henganofi tell a story of war and grief, 1984. (g) Anga dances at their show 1985 165

Fig. 5.22 PNG village activities requiring visuospatial reasoning. (a) Village court case seeking compensation of a pig and money. (b) Pig kill sharing after compensation. (c) Cash crops—coffee drying, tobacco, sugar cane Bena, Eastern Highlands, 1975 166

Fig. 5.23 PNG modern pottery 166

Fig. 5.24 Timor Leste crafts, ~1999. (a) Ikat weaving. (b) Small basket with fitting lid 167

Fig. 6.1 Representations from the Marshall Islands. (a) Swells diagram (Davenport, 1960, p. 22). Stick chart for learning places (Bryan, 1938, p. 13). (b) Representation by Hutchins of the navigating system with the moving boat being the reference point (Hutchins, 1995, p. 89) 209

Fig. 6.2 Visuospatial difference reflected in tiles from around the world. (a) Circle construction lines and resulting tessellations La Alhambra, Spain. (b) Firenze duomo, Italy. (c) La Alhambra ceiling. (d) Tiles in Portuguese palace, 3D illusion. (e) 3D tiles on cheda Thailand 213

List of Figures

Fig. 6.3	Creating designs in the Pacific. (a) One of at least 91 sand drawings recorded in Vanuatu (Deacon & Wedgwood, 1934, p. 148). (b) Rotational and reflective symmetry in creating sand drawing using a base tracing, note the start (s) and finish points (f) (Ascher, 1994, p. 53)	215
Fig. 6.4	Rows of dots used as markers for the curved lines to create the design. (a) Tshokwe (Central Africa) *Sona:* "the marks on the ground left by a chicken when it is chased" (Ascher, 1994, p. 42). (b) Tamil Nadu *kolam*: nose jewel that is embedded in other *kolam* (Ascher, 2002, p. 65)	215
Fig. 6.5	Fijian tapa, 1977	216
Fig. 6.6	Tongan tapa designs. (Finau and Stillman, 1995). (a) Repeated rotated design with symmetries and a "perfect" colouring. (b) Repeated rotated design with symmetries and "imperfect" colouring but still attractive. (c) Various patterns incorporated into the repeated motif design. (d) *Kupesi* board coconut fronds bend to form plant designs	217
Fig. 6.7	The artisan and learners' designs. (Cherinda, 2012). (a) Artisan's basic mat. (b) Highlighting with colour. (c) Another colouring	218
Fig. 6.8	Cherinda's diagram of ethnomathematics in terms of the mathematical learner. (Cherinda, 2012)	218
Fig. 7.1	Visuospatial reasoning about ratios represented by irrational numbers in western mathematics. (a) Extending the house, half as much again. Malalamai, Madang Province, PNG. (b) Two ropes to form a right angle in PNG (Yamu, ~2000)	226
Fig. 7.2	Excerpt from student project on traditional measurement and shape construction (Odobu, 2007)	229
Fig. 7.3	Cultural identity reflected in mathematical identity (Kono, 2007)	242
Fig. 7.4	Examples of built environment linked to school mathematics (Pepeta, 2007)	243
Fig. 8.1	Examples of visuospatial reasoning in ethnomathematics for secondary schools. (a) The vines from the handrail to the walking platform provides a metaphor for sine wave (Yambi, 2004). (b) The wig shape is outlined by the *parabolas* and *dotted lines*. The *head circle* is marked (Piru, 2005). (c) Fishing net ecocultural visuospatial representations (Martin, 2007). (d) Visuospatial arrangements in the five stone game: first line illustrates how stones are to be swept up (singles, twos, three); second line illustrates how rectangle and triangle numbers can be extended to give new patterns (David, 2007)	251

Fig. 8.2	Links between ecocultural mathematics and school mathematics (John, 2007)	253
Fig. 8.3	Tabare weaving shapes from Julius' (2007) project	254
Fig. 8.4	Schools in Papua New Guinea. (a) Typical bush material school, Morobe, PNG, ~ 1984. (b) Group work in Tsigimil Primary School grade 3. (c) Elementary school, Atzera language, Binimap, Morobe, 2006	255
Fig. 8.5	Design of key principles for teacher professional development in *Cultural Mathematics*	256
Fig. 8.6	Design of inquiry method for *Cultural Mathematics* based on Murdoch (1998)	257
Fig. 8.7	Bringing ecocultural experiences into the elementary classroom. (a) Preparing a dictionary of mathematical terms. (b) Trying out the lesson idea of using the body to make shapes following the squares (*diamond*) in the weaving. (c) Simulating using steps to measure when planting rice. (d) Sharing the making of rope for a bilum and making a bilum relating it to mathematics. (e) Sharing fish between families according to need. Recording of children's solutions explaining how they worked from one addition to another with visual supports	259
Fig. 8.8	Weaving in Mozambique to encourage visuospatial reasoning in geometry. (a) Student motivated to learn in Mozambique. (b) Translation experience. (c) Repeated to produce rotational symmetry. (d) Weaving a reflection pattern	262
Fig. 8.9	Children learning from manipulative activities in Mauritius. (a) Noting angle size (Group 4). (b) Different shapes and sizes (Group 3). (c) Different quadrilaterals (Group 4). (d) Halving a rectangle (Group 4)	270
Fig. 9.1	Children's visuospatial reasoning when playing by manipulating objects on computer screens. (a) Block patterning. (b) Enlarged lion. (c) Lengthened seals	281
Fig. 9.2	Drawings by children of the movement of their robot. (a) Pictorial idiosyncratic. (b) Emergent spatial. (c) Symbolic emergent spatial. (d) Symbolic partial-spatial. (e) Symbolic spatial. (f) Integrated spatial	283
Fig. 10.1	Developing identity as a mathematical thinker	295
Fig. 10.2	Simplified design of principles for teaching ecocultural mathematics (Owens, Edmonsd-Wathen, & Bino, 2014)	296

List of Tables

Table 2.1	Visualisation and orientation	21
Table 2.2	Comparison of grade 2 and grade 4 students' scores	39
Table 2.3	Percentage of incidents involving different types of visual imagery	42
Table 2.4	Percentage of responses of children on pretest and on delayed posttest	55
Table 2.5	Percentages of students who were correct on completing prisms and recognising shapes from different perspectives	70
Table 2.6	A framework for geometry based on visuospatial reasoning	83
Table 2.7	Examples of different visuospatial reasoning strategies for a task	86
Table 2.8	Student improvement on assessment tasks	87
Table 3.1	Alternative numbers in Gahuku	95
Table 4.1	Examples of added suffixes for movement in Wiradjuri	124
Table 4.2	Words related to size from a small selection of different languages in PNG	129
Table 5.1	Selection of names for classificatory groups in Kilivila	179
Table 5.2	Seasons in Melpa or Hagen, Western Highlands, PNG	192
Table 8.1	Performance of pupils in items requiring verbal skill	269
Table 9.1	Uses of robots to develop visuospatial reasoning for concepts	286

Table A.1	Problem-solving processes from some research literature	311
Table C.1	Alternative numbers in Gahuku provided by two men from Kaveve, Eastern Highlands, PNG (Table 3.1 from this table)	318
Table C.2	Names for classificatory groups in Kilivila (Table 5.1 is a selection from this table)	319

Prologue

My Story

My story begins when I was in high school and I enjoyed solving geometry problems such as those involving congruent triangles. It was like a game to me. I was solving problems by looking for alternative solutions and applying tactics; but I did not use the name "problem solving". At the University of Sydney, I majored in Mathematics and Psychology but also studied Latin (language, history, and philosophy), Anthropology (much of which came from studies of the people of Papua New Guinea and Africa), and Education. However, my membership of Student Action for Aborigines (Australian Freedom Ride, 1965) began my understanding of Indigenous cultures and struggles from an Indigenous perspective. I had long conversations with Aboriginal friends and with people who respected the Aboriginal and Papua New Guinean (PNG) communities with whom they lived for many years and whose languages they had learnt. I also learnt from personal visits and tutoring at the Foundation for Aboriginal Affairs (set up and run by the Aboriginal community) and "a home" where children, now known as the stolen generation, were sent by the Northern Territory government. I thought deeply about my own Christian faith and considered the underpinnings of numerous political perspectives.

After a teacher education course and high school teaching, I married and lived in Papua New Guinea where the next critical part of the story begins. During this time, I taught mathematics at the PNG University of Technology (Unitech) where the Mathematics Education Centre (MEC) was supporting numerous research studies that took account of culture but were mostly educational psychology studies. Along with other technology degrees, I taught students in architecture and valuation in which visuospatial reasoning and geometry were important components. I taught Education and Health Education at Balob Teachers College (and in the community and Unitech voluntarily) so an interest in cultural influences grew. My colleagues and their spouses among many others influenced my thinking about culture and education and to them I am grateful for truly enriching my life over many years—Glen Lean, with whom I worked closely, Geoff Smith, Jack and Mary Woodward, Dan and Carrie Luke, Rod and Yombu Selden, Neil and Betty (deceased) Roberts, Wilfred and later Roa Kaleva, Ken and Helen Costigan, Chris and Naomi Wilkins, Philip Clarkson, Else and Ron Schardt, Teresa (deceased) and Paul Hamadi, Mea Dobunaba (deceased), Tess (deceased) and Geoff Chan, Deveni and Marion Temu, and Misty Baloiloi; and many students including Ai Wandaki, Migleri, Petrus, and Patrick, and many families from Morobe, Gulf, Central and Milne Bay at Church in the Taraka settlement or from other connections. Special friends were Ron (deceased) and Rosemary Elias and Margaret Peter (deceased). I went to many villages with friends, bushwalking friends, students to visit their families, or for supervising cultural school experiences. My growing interest in cultural visuospatial reasoning was strengthened on return visits to Balob and Unitech with its Architectural Heritage Centre.

In the 1980s, I had the pleasure to work under Professor Kathleen Collard at Unitech with an international group of mathematicians and mathematics educators. The Department, supported by the MEC, had some great resources for teaching just the right mathematics for the range of technologies, with explanations and some worked examples and a short set of exercises building up the concepts and procedures well. It was a modified self-paced mastery learning approach but effective. We also used calculators and computers, being very early innovators in using technology. For one professional learning session Kathleen promoted problem solving which she said was the new approach in the UK. The very word problem raised a negative connotation for me so I was quickly engaged to see what value there was in this approach. Having a go at using this new problem-based learning, I prepared a problem scenario with the building science lecturers. Being a different approach to the highly directed materials, it did not go down too well with students. However, there was clear merit in generating group work and mathematical conversations about mathematics in a relevant context.

Alan Bishop and Ken (MacKenzie) Clements both made visits to the MEC at Unitech resulting in profound influences on research overseas and in PNG and on themselves, Glen Lean, and myself. Alan, Ken, and I have continued to converse on issues of PNG education, and space and geometry, Ken being my doctoral supervisor. Alan and Ken carried out research with Glen Lean on spatial abilities and the impact of experience and culture on these abilities. The MEC had many papers on this topic that sparked my interest to research this area. I enjoyed working with Architecture and Surveying students but also seeing how computer graphics played a role in reasoning and decision-making in statistics and modelling.

On returning to Australia after 15 years in PNG, with Bob Perry and colleagues at the University of Western Sydney, we developed a social constructivist classroom for students requiring additional mathematics to become teachers. This teaching approach results in some valuable insights into students' responsiveness to problem solving (Owens, Perry, Conroy, Geoghegan, & Howe, 1998). There has been a continuing emphasis on problem solving through mathematics associations and curricular documents (see our literature review for one syllabus revision, Owens & Perry, 1998) and the work of many eminent researchers arguing for the importance of problem solving (Appendix A).

Prologue

In the late 1980s, certain space and geometry problems with pentominoes, tangram sets, and pattern blocks had become popular but I wondered what students were learning from these activities. Hence began my enquiry first with adult learners (preservice teachers) and then young school students working through a series of activities. I tackled and synthesised the broad literature from psychology on spatial abilities and visual imagery and then developed a group test appropriate for young children and related to classroom activities to measure visuospatial reasoning. I explored how young children were thinking and using visuospatial reasoning when they undertook ten one-hour geometry tasks. My findings on visuospatial reasoning have been incorporated into Chap. 2 of this book. I continued to pursue this area of research in terms of teacher education and adults' visuospatial reasoning about angle equality. Lynne Outhred has been a constant critiquing and collaborating collegial friend.

During the 25 years since we lived in PNG, I have continued to visit for periods of a couple of weeks to several months to work with PNG colleagues on various projects. Over 40 years, I have visited, revisited, or stayed (one night to 3 weeks) in more than 63 villages in more than 52 language groups, not to mention the many times I have talked with university students, colleagues, and participant researchers about their thinking when carrying out village activities. Our in-depth conversations built on our anthropological studies and my study of how cultural beliefs and practices impact on health and well-being and mathematical understanding. People shared their village technologies and rich cultures with us and thus began my inquiry into the mathematical proficiencies of PNG villagers and how they enrich my own understanding of mathematics.

There are always significant moments in one's life that have directed one's thinking. Some of these occurred for me when observing my PNG students or villagers. In my first year in PNG, a mature-aged, Jimi valley student, nose pierced at initiation, told me that he had spent some money to buy a block of land in Australia by responding to a newspaper advertisement but he had not had a response about it. Although he did not progress with his university studies, I did spend time trying to teach him addition with little success. At the time, I was unaware of his PNG counting system (a two-, five-cycle system) but I was aware of how he, and many students, learnt by taking risks.

Village ethnoscience and technology is indelibly written on my memory. I watched men making canoes with a stone adze among their implements, the curves and thickness and overall shape were well established in their visuospatial memory. I was fascinated to watch women set up a fire at Labu with several layers of limbom palm covered with shells from the saltwater lake, and a fire blown by the wind to start at the front to cook the shells perfectly. They picked up each shell and returned it to the fire if not ready or else they dropped it in the mesh to remove sand and then into the bag for smashing with a stick for lime (*kambang*) chewed with betel nut (*buai*). It was sold in the city of Lae's market reached by sailing canoe across the treacherous mouth of the Markham River or by dinghy with a hole in the bottom where the passengers put their feet on a plastic bag to seal it. We watched at Yombu Seldon's village out of Tufi, and in many other places, the making of sago with scraping the pith with an adze, the sheaths placed at appropriate angles, the fibrous collar for strainer, for the washing and collecting of sago starch. Making *tapa*,[1] string, and *marita*[2] soup were other ethnotechnologies requiring extensive visuospatial thinking.

When we returned in 1997, another significant moment occurred. I visited the Department of Architecture and Building Studies at Unitech and was amazed at the wonderful sculptures that first year students had made from three colours of paper and cardboard without using glue or sticky tape. I carried out a retrospective interview study of their problem solving (Owens, 1999b, see Chap. 5). Many of them used cultural practices to join the paper pieces and make attractive visuospatial arrangements. What stayed with me was their identity. If we are Papua New Guineans, we will be good designers and architects. Would students also say, we are Papua New Guineans, we will be good mathematicians? I had my answer when reading through the ethnomathematics projects written in 1995, 2000–2013 at the University of Goroka (UoG). Students were writing proudly about their amazing parents, Elders and ancestors, their practices and their mathematics. They were incorporating ecocultural identities into their mathematical identities (Owens, 2012a, see Chaps. 5 and 8).

Our recent research projects have extended our knowledge in the areas of space, geometry, and measurement. One of the projects was related to delivery of education and involved an evaluation of an Australian funded project for improving teacher education for primary and secondary schools in PNG (Clarkson, Owens, Toomey, Kaleva, & Hamadi, 2001). (While this research kept me in touch with PNG education, I will not refer to it further in this book.) Another involved the cataloguing and production of electronic materials for a website at the Glen Lean Ethnomathematics Centre at UoG. I read many of the original records of first contact which Glen had collected for his study of counting systems. By organising his work into electronic documents and databases, we became more familiar with his extraordinary study, its strengths and his difficulties to report from diverse data sources that gave different information (see Owens, 2001c and later chapters). Then we began a long-term study on measurement. At UoG, I worked with Rex Matang (deceased) and Wilfred Kaleva who are very special people, Martin Imong, Charly Muke, and from Madang Teachers College Sorongke Sondo, and more recently Vagi Bino, Geori Kravia, and Susie Daino. I am in debt to all of them for their collegiality and personal care of me.

For the measurement project, assisted by PNG colleagues, participant researchers, and research assistants, we have gathered data (from 345 languages from 425 sources) from linguistic records, by open-ended questionnaires, semi-structured interviews, focus groups, story-telling, and place visits. These were supplemented by the ethnomathematics projects. The questionnaires were completed by preservice and inservice teachers and staff who selected to complete them for us from three tertiary institutions in PNG. All consented for the information to be collated

---

[1] Tapa is made from the inner bark of trees, particularly the tulip tree (two leaves sprout at the same point).

[2] Marita is a very long, red nut of a pandanus tree which makes an oily red, flavoured soup.

and shared in research reports and papers. Most questionnaires were completed electronically. We made visits to ten villages specifically to gather data on measurement and space. In each place, we talked with Elders, often in a village group and often more than one group, and to people visiting from other language groups. All these were previously negotiated by participant researchers from the area. In addition, I carried out 15 interviews at Universities and colleges and we held five focus groups (students from four distinct regions and one group of linguists) to discuss emerging summaries of the data. The interviews and visited villages were chosen to include Papuan (Non-Austronesian) highland and lowland language groups and coastal Austronesian language groups from ten Provinces. Most interviews were in Tok Pisin, a few in English. All interviews were tape-recorded and/or videorecorded with consent and transcribed by Tok Pisin speakers (a language which I also speak fluently). There was code switching between languages. Language words were translated during the interview if used. There is continuing evidence of specific cultural practices and also the growing loss of these practices along with the languages. It is urgent that these ecocultural practices be valued and maintained in ways negotiated with the communities. Results from these studies can be found in Owens (2007, 2012b, 2013b), Owens and Kaleva (2008a, 2008b), and in Chaps. 4, 5, and 8 of this book. Returning to Australia, we were able to meet up with Australian Indigenous friends and make new ones through teaching first at University of Western Sydney and then at Charles Sturt University. Both have special programs for Indigenous students. My Indigenous friends in Dubbo made me feel welcome when we came to live here and I have particularly valued the wisdom of the Elders and what I have learnt through partnerships and projects like the bidialectal approach to teaching English (Standard Australian and Aboriginal English), Stronger Smarter Leadership evaluation, rural and Indigenous quality teacher projects, and a forum on continuities in education with the local community.

Elders at Kopnung village, Jiwaka Province, PNG, demonstrate cultural practices (weaving and trap-making with the children learning visuospatially).

I have also been privileged to attend international conferences to share and debate issues around ethnomathematics with researchers in Sweden, Africa, the Americas, and New Zealand. In particular, I mention Norma Presmeg, Gerry Goldin, and Keith Jones who, at different times, discussed extensively with me the notion of visuospatial reasoning in different contexts, and also Ylva Yannok Nutti, Annica Andersson (who accompanied me on one PNG trip), Lisbeth Lindberg, Bill Barton, Uenuku Fairhall, Colleen McMurchy-Pilkington, Tamsin Meaney, Jerry Lipka, Barabara Adams, Rik Pinxten, Maria do Carmino Santos Domite, Maria Cecilia de Castello Branco Fantinato, Marcos Cherinda, Milton Rosa, Daniel Orey, and Cris Edmonds-Wathen but really too many to list them all. In Sweden, Macau, and Yemen, I visited communities to discuss culture and mathematics education. Ravi Jawahir studied geometry in Mauritius with me. In the last 10 years, I taught subjects linked to Human Society and Its Environment and social justice. Consequently, issues of identity, environmental and cultural sustainability, and the importance of education situated in the place of the students (both local and global) have dominated my thinking. All these meetings and experiences have enriched my thinking. I hope I can share this from my life story in this book.

This book takes you through my journey to establish an ecocultural perspective for mathematics education, especially in areas relevant to visuospatial reasoning in space, geometry, and measurement. While the rich PNG and Australian Indigenous cultures form the basis of the argument, this book is relevant to other places as I am sure the following chapters, breadth of references, and other researchers' findings and arguments will testify. I want to acknowledge the endless times my PNG and Australian Indigenous friends have shared with me and whether wittingly or unwittingly taught me and brought me to the thesis presented in this book.

Ecocultural Background of Papua New Guinea

PNG is a country rich in a wide diversity, over 850, Melanesian cultural groups. Each has its own language. Each has its own cultural values, practices, relationships, understandings, adaptations to other cultures, and systematic ways of interpreting, designing, explaining, measuring, enumerating, comparing, and using information (data) to make decisions. Tropical PNG is half the New Guinea mainland plus thousands of islands, large and small. The mainland and many islands are mountainous but there are also large upland and lowland valleys, fast flowing rivers from gorges to coastal plains, swamps, and coral seas. People have adapted to their environments, and their mathematics reflects their living in their places. Villagers are adept at living from gardens, the bush, and the river or sea to build their houses, supply their food, and engage in building reciprocal relationships. Pigs are an important form of wealth in many places. From the capital city to most other cities of the country, one has to fly or go by sea. Mobile phones are popular where services and power for recharging are available. Most villages do not have electricity, piped water, or road transport although these facilities are increasing. The country has rich mineral resources and many cash crops.

From the late nineteenth century, the Territory of Papua was under Australian authority while New Guinea became a Mandated Territory after World War 1. The Second World War was fought on their soils. In many places in the interior of the mainland, tribal law and fighting were the norm (Gammage, 1998). Land belonged to and was held onto by different clans of different language groups. Trade routes were well established. Many atrocities occurred before and after Australian administration infiltrated the regions. At all times, expatriates were reliant on Indigenous knowledge and labour, and maintained some form of western privilege such as salary or housing to survive. In the late 1960s, around the time that Aboriginal Australians were recognised by a referendum (1967), it was clear that Australia was sooner or later to give PNG its independence, having already established a parliament, a University, police, army, health and education services, and provided opportunities for leaders to learn about party politics and democracy. The Churches provided much of the education especially at the teacher training level and in remote areas, and this continues today. After a time of self-government (1973—the year we went to PNG), it became Independent on 16 September 1975.

Over the years, curriculum writers and mathematics school books attempted to give a PNG flavour to examples. Today, schooling under the reform begins with elementary schools (Elementary Prep, Elementary 1, and Elementary 2) built by communities with local teachers who are minimally trained and paid a small wage by government for a half day teaching. Schooling is generally in the vernacular *Tok Ples* or in cities in the lingua franca, *Tok Pisin*. Education transfers to English during Elementary 2 and continues when children start primary school (Grades 3–8). From 2014, the official government policy is to have English as the language of instruction from Prep. There are examination barriers to go to secondary school or to senior secondary school. The curriculum was adapted from overseas. Syllabi take account of diversity by being relatively general in outcome statements and using English words to explain, for example, position and direction. Unintentionally the syllabus is reinforcing English approaches through the language rather than a vernacular approach. Nevertheless, the syllabus says vernacular words and environmental shapes and places should be used. While this might occur, it is important to overcome the hegemony of school education mathematics and to give teachers an idea of how to bridge between the vernacular and the English-language approach to size, shape, position, and direction (see Chaps. 7 and 8).

Dubbo, NSW, Australia   Kay Owens

Chapter 1
Introduction: Visuospatial Reasoning in Context

> *Curiosity is an active habit—it needs the freedom to explore and move around and get your hands in lots of pots.*
>
> (Llewellyn, 1991)
>
> *If a man seriously desires to live the best life that is open to him, he must learn to be critical of the tribal customs and tribal beliefs that are generally accepted among his neighbours.*
>
> (Betrand Russell)

The Challenge

There is currently a renewed interest in visuospatial reasoning. Part of this interest has been generated by the use of diagrams in paper-and-pencil national testing but interest also comes from the application of mathematics to real life, and a renewed interest from dynamic geometry software. However, to appreciate the importance of visuospatial reasoning, we must first discuss the earlier research from the twentieth century. Research in spatial abilities and visual imagery during the last century was heavily influenced by psychological studies. This book draws from this literature but follows a less conventional line of argument by opening up new lines of enquiry which are demanded by other educational research literature concerned with the social aspects of learning and the cultural contexts of education. In this chapter, a preliminary definition of visuospatial reasoning is provided given that there are other closely associated terminologies. This chapter will provide an overview of the basic premises of the book. It begins with a short historical background that will be expanded in Chap. 2 on spatial abilities and visuospatial reasoning. How does visuospatial reasoning occur in the mind? How does it develop in children? These questions are further addressed in this chapter and Chap. 2.

However, an expanded understanding of visuospatial reasoning shows the importance of the contexts of education for developing visuospatial reasoning. Does the ecocultural perspective throw new light on visuospatial reasoning research? Will we reach a new richer perspective on visuospatial reasoning? An overview of ecocultural perspectives in education is presented in this chapter with an expansion in Chap. 3. Can visuospatial reasoning and an ecocultural perspective be drawn together through considering identity within an ecocultural context impacting on the mathematical learner? The argument is introduced in this chapter but developed throughout the book. Does visuospatial reasoning occur in different contexts when people undertake activities of a spatial nature? These questions are addressed in Chaps. 4 and 5 in particular.

Influential in this argument is the synthesis of the disparate literature on identity which can be rather simplistically classified as cultural identity and identity from a psychological perspective. Is this possible? This chapter begins this argument which is developed throughout the book but particularly in Chaps. 5, 6, 7, and 8. With the growing interest in dynamic geometry, there is a resurgence on how visuospatial reasoning occurs within that environment. However, will the emphasis on identity in developing a mathematical learner be appropriate for the classroom using electronic devices to pursue spatial concepts and geometry? Chapter 9 tackles this question for younger students. Chapter 10 provides an overview of the argument put forward throughout this book illustrating the importance and advantage of an ecocultural perspective on visuospatial reasoning for mathematics and mathematics education.

Historical Overview of Relevant Research on Spatial Abilities and Visual Imagery

In the early 1900, Binet and Simon were looking at intelligence in young children. They meant by this the ability to adapt and solve new problems. Their early items and scales focused on attention, memory, and spatial discrimination (Munn, 1961). Down through the years tests of intelligence continued to be developed. In many models, there was recognition of spatial ability versus verbal ability and other abilities but the critical argument lay around whether there was an overall general intelligence factor (e.g. Spearman's model) or not (e.g. Thurstone's model) (Watson, 1965). Importantly for the general argument of this book, these tests consistently used items of a spatial nature.

Piaget, one of the great influences on educational thinking, regarded "cognitive processes as also being expressed in thought and intelligence between which he (made) no sharp distinction" (Watson, 1965, p. 175). He reconciled the continuous scaling for intelligence tests as measuring "products and not the operations used by children" (Watson, 1965, p. 189). Furthermore he focused on perception and imagery when considering conceptual development (Piaget & Inhelder, 1956, 1971; Piaget,

Inhelder, & Szeminska, 1960). Hence Piaget from his earliest thinking about children's cognitive development recognised that visualisation and spatial constructs were embedded in cognitive processing, thought, and conceptualisation.

Many different spatial abilities have been defined as specific constructs, often determined through factor analysis models (Eliot, 1987). In mathematics education, there was recognition from the 1960s not only that such abilities may affect learning but also that these abilities could be, and indeed should be, nurtured. Meanwhile educational psychologists began to write about visual imagery or visualisation as important in concept development, often based on Piaget's earlier ideas. Many of these studies made use of reaction times when subjects responded to items anticipating, for example, mental rotation of objects. This simple approach was called into question when it was realised that analysis might be used as well as mental rotation during the measured reaction time (Shepard, 1975). Other psychologists developed more factors about imagery and spatial ability until Lohman, Pellegrino, Alderton, and Regian wrote:

> Although the number of potentially identifiable spatial factors is quite large (perhaps even unbounded), the number of distinct psychological processes required by spatial tasks appears to be much smaller. ... Visual stimuli must be held in sensory memory while encoding processes (or pattern matching productions) operate to identify all or parts of the stimuli ... We have six basic categories of processes: pattern matching, image construction, storage, retrieval, comparison, and transformation. (Lohman, Pellegrino, Alderton, & Regian, 1987, p. 273)

They concluded that:

> Spatial ability may not consist so much in the ability to transform an image as in the ability to create the type of abstract, relation-preserving structure on which these sorts of transformations may be most easily and successfully performed. (p. 274)

This synthesis of ideas brought a halt to the ever-expanding micro divisions that had been occurring and enabled educationists to start using these constructs in a meaningful manner. The mathematical learner was seen as making mental images that highlighted salient aspects but the learner went beyond the details of the image to imagine and reason visuospatially in the same way as an artist goes beyond the image (Goldenberg, Cuoco, & Mark, 1998).

Visuospatial Reasoning and School Test Performance

There was some recognition of the importance of visuospatial reasoning in testing regimes in the past. Kouba et al. (1988) noted that in the fourth National Assessment of Educational Progress (NAEP in USA), students made good use of visual clues in questions involving the equality of angles (75 % correct in grade 7) and sides (62 % correct in grade 3) but often did poorly on questions lacking visual clues such as questions on adjacent angles (14 % in grade 7 correct) and on points on a circle (26 % correct in grade 3; 35 % correct in grade 6). Sometimes, though, visual diagrams seemed to mislead students especially when concepts were not well established; thus, in some questions, students made errors because they seemed not to realise that the length of a diagonal interval was longer than the lengths of horizontal intervals on a grid or that a circle could be drawn containing any three given non-collinear points. In angle-matching tasks, students are more likely to perform better if they know analytical principles such as vertically opposite angles are equal and can decide quickly when an analytical principle is relevant and when visual rotation and decision-making are needed. Angle-matching tasks are more difficult if the diagrams are more complex, angles are dissected by lines, angle distractors are visually close in size, or angles are rotated, more distant, or external (when only internal angles of a polygon are anticipated) (Owens, 1998a).

Similar visual cues seem to have misled students in Basic Skills Tests (Australian Council for Educational Research, 1989–1991) in which the diagram of an object to be weighed was accompanied by larger diagrams of masses (Owens, 1997a). In other questions, for example, in determining the number of balls which would fit in a box, the visual cues were helpful to students. For visuospatial reasoning of grade 6 students, an interesting comparison can be made between two questions involving the joining of shapes to make a new shape. For one, 88 % of the answers were correct but for the other only 66 % were correct. One reason for the lower result was that students used an analytic approach noting the pointiness of the shape, and then chose a triangle without checking whether it was the correct triangle. Such an approach, in which analysis rather than a holistic method is used, can lead initially to more errors being made because the task involves more than one step and the student fails to continue to solve the problem. It also points to the attention of the graphic formation and its distraction for the problem (see also Diezmann & Lowrie, 2012). International comparisons such as Trends in International Mathematics and Science Study (TIMSS) and PISA (Mullis, Martin, Kennedy, & Foy, 2007; Sturrock & May, 2002) show that the low student performance in geometry at all levels is quite alarming (Lappan, 1999).

The report on the teaching and learning of geometry by the Royal Society and Joint Mathematical Council (2001) argues that "the most significant contribution to improvements in geometry teaching will be made by the development of good models of pedagogy, supported by carefully designed activities and resources" (p. 19). In fact, a primary cause of this poor performance in geometry may be the curriculum; both in what topics are treated and how they are treated but also in the poor preparation of teachers and their own lack of visuospatial reasoning in geometry. The Council noted that either work on three-dimensional objects is left out or aspects are learnt by rote. However, more research into effective geometry education is needed. While a minority of students may have only reached the basic level across various questions in NAEP testing, there was reasonable evidence of visuospatial reasoning being part of the curriculum and children having good attempts at the questions using visuospatial reasoning even if they were not correct (Brown & Clark, 2006). Furthermore, research continues to suggest that low socio-economic status can reduce the basic spatial ability and reasoning skills as a result of a lack of experiences outside of school related to conceptual and number learning required in measurement education (Casey, Dearing, Vasilyeva, Ganley, & Tine, 2011). Such results would indicate that an ecocultural perspective on visuospatial reasoning in geometry and measurement is needed to address inequity.

Visuospatial Reasoning in Mathematics Education Research

Mathematics educators since the 1970s began to develop theories that purposefully incorporated visual imagery (used in this book as synonymous with visualisation; see detailed discussion on definition later in this chapter). Initially visual imagery was often relegated to foundational levels of conceptualisation (e.g. Pirie & Kieren, 1991; van Hiele, 1986), although over time this did begin to change with Biggs and Collis, for example, revising their taxonomy recognising the importance of visualisation throughout the ongoing development of cognition (Campbell, Collis, & Watson, 1995). Pirie and Kieren discussed "folding back" to the earlier visual stages as concepts developed and then they developed both an action and verbal description for each level (Pirie & Kieren, 1991, 1994). Action is often a visuospatial memory or reasoning. Furthermore, at the higher levels of structure, and reification, visual imagery or visuospatial reasoning is significant (Presmeg, 2006). Others provided theoretical ideas that were relevant at all ages such as Goldenberg et al.'s (1998) coverage of visuospatial reasoning to include:

- Seeing processes that are both geometric and metaphorical
- Seeing quantity for verification and calculation
- Seeing pattern and structure for exploring conjectures and devising proofs

All of these are relevant to geometry and measurement. With these "habits of mind", more mathematical power is possible but it requires curricula "to treat visual imagery as a central ingredient in mathematical discovery, invention, and explanation" (Goldenberg et al., 1998, p. 39). This was indeed a feature of several US projects for middle and high school students and in Australia with the Count Me Into Space project (Owens, McPhail, & Reddacliff, 2003, see Chap. 2).

Nevertheless, the term "space" was causing difficulties in mathematics education and schools in general. The term was dropped from the UK and eventually the Australian mathematics curriculum although defined in geography. Furthermore, a new challenge arose for the theory and development of visuospatial reasoning with the rise of the twin movements of outcomes-based education and national testing. These took hold globally resulting in a focus on observable outcomes rather than internal thinking processes (Ellerton & Clements, 1994). Hence visualisation was relegated to external representations such as diagrams, computer-generated graphics, and concrete models (Zimmermann & Cunningham, 1991). A recent handbook (Clements, Bishop, Keitel, Kilpatrick, & Leung, 2013) developed the role of visuospatial reasoning only at higher levels through the use of dynamic geometry software and statistical packages but there was a brief reference to blind and deaf students' visuospatial learning. The role of visuospatial reasoning was otherwise not covered as was the case in an earlier review from Australasia (Perry, Anthony, & Diezmann, 2004). Although there was encouragement of the use of visualisation for early arithmetic, there was little mention for geometry in an important book on advice to teachers (Kilpatrick, Martin, & Schifter, 2003). However, there are textbooks emphasising visualisation across the curriculum subjects (Jones, 2012). Other handbooks from the International Group for the Psychology of Mathematics

Education (Gutiérrez & Boero, 2006) and National Council of Mathematics Teachers (Lester, 2007) provide substantial reviews on aspects of visuospatial reasoning in mathematics.

Despite the importance of reasoning about objects with representations (Battista, 2007a), outcomes-based education and paper-and-pencil testing have resulted in a lack of mention of visuospatial reasoning in curriculum. A recent critique of the Australian curriculum (Lowrie, Logan, & Scriven, 2012), although a similar critique probably applies to many other curricula, noted that,

> in fact, there is no reference to spatial or visual reasoning in the entire document. The lack of attention afforded spatial reasoning in the curriculum is compounded by the fact that no indirect mention of such processing is framed within the four proficiency strands. For example, there is no mention of "drawing a diagram", "imagining in your mind's eye", or any intent to promote reasoning which encourages students to manipulate or move objects within an internal, visual, space. Such processing is accepted as an essential aspect of mathematics reasoning. Without such reasoning, the depth of understanding within this mathematics strand is lost. Therefore, the "signposting" (for teachers) that spatial and visual reasoning is critical to this strand has been removed from both content and mathematical proficiencies. (Lowrie et al., 2012, p. 74)

Despite this retrograde step in curriculum documents and in the research literature, there has been a growing interest in visuospatial reasoning associated with dynamic geometry software (Falcade, Laborde, & Mariotti, 2007) with limited coverage in other areas. Interestingly beyond mathematics education research, there also continues to be an interest in visuospatial reasoning and its development through education. For example, because of an increase in representations of geographical information, the National Research Council Committee on Geography (NRCCG) (2006) commissioned a comprehensive report into this developing area of representation, knowledge, and conceptualisation. NRCCG and other researchers mentioned in this book would agree that the role of visuospatial reasoning from early childhood to adult career education has been underestimated. School education has failed to realise that there are a number of newly established ways of representing the four dimensions (three space dimensions and a time dimension) in a wide range of fields from physiotherapy to geography, and from geometry to number patterns requiring such reasoning (Shah & Miyake, 2005).

Some visuospatial reasoning is about external representations which have an ecocultural basis, often determined by the specific science such as medicine or surveying. However, the issue for this book is around what might provide a sound foundation in childhood education. The recent impact of ecocultural research has highlighted new ways in which we can perceive childhood mathematics education and new possibilities for mathematics itself (Ness & Farenga, 2007).

Defining Visuospatial Reasoning

Battista (2007a) noted that geometric reasoning requires "spatial reasoning which includes generating images, inspecting images to answer questions about them, transforming and operating on images, and maintaining images in the service of other mental operations" (p. 843). In other words, the mental operations on mental visual imagery are about space. The term "visuospatial reasoning" emphasises the reasoning associated with and dependent on visual imagery but also expressed and argued with spatial references. Spatial abilities, spatial skills, and both spatial and visual imagery are part of this reasoning. Although the importance of visuospatial reasoning was lost through the splintering of spatial abilities, its significance is now being recognised by educators such as the NRCCG (2006). Visuospatial reasoning is a necessary way of reasoning that develops during childhood. In my conception of visuospatial reasoning, there is no attempt to separate visuospatial reasoning that is about spatial visible things from visual thinking to understand other concepts in the way that Senechal (1991) proposed. Recreating the visible in terms of visualising from alternative perspectives, re-seeing, and reconstructing shapes is supplemented by representing abstract ideas such as multiplication of two-digit numbers or algebraic binomials, and then seeing change, process, quantity, or other structure. These are all part of visuospatial reasoning (Goldenberg et al., 1998). However, visuospatial representations overlap categories. For example, a graph or a number line representation of addition may represent arithmetic but at the same time they have their own inherent spatial values and meanings. The overlap in visuospatial reasoning that crosses the boundaries of mathematical categories and other knowledge categories in general will be more evident as we consider some of the ecocultural perspectives and emplaced understandings of space (see Chaps. 3, 4, and 5).

The NRCCG noted that "spatial thinking" results from early activities in physical spaces (a three-dimensional and time dimensional world) and went on to discuss the spatial thinking that occurs in intellectual spaces related to the physical spaces but involving concepts. Finally they refer to spatial reasoning about intellectual spaces where representations have not been generated from the space-time world but represent concepts, frameworks, models of processes, and relationships in a wide range of fields. For example, spatial representations may be a spatial form of geographic information, objects, or diagrams linked to knowledge areas other than mathematics. In addition, the Committee noted that spatialisation includes an attitude or long-term approach to considering things in a spatial manner.

This Committee considered reasoning as one aspect of spatial thinking along with imagery and spatial processes. I argue that visuospatial reasoning is more encompassing than the NRCCG committee's reasoning. The NRCCG (2006) defined spatial thinking mostly in terms of distance and measuring with a visual component, a symbolic component, and a reasoning component. However, the reasoning component is inadequately developed in their definition and this problem provides one reason why I consider visuospatial reasoning as more rather than less comprehensive than spatial thinking. The reason for this is to encompass the impact of all forms of sensory input and the variety of imagery associated with experiences in a physical world. Spatial imagery refers to kinaesthetic and motor imagery as opposed to visual imagery in careers such as aviation where physical response time is critical (Wickens & Prevett, 1995). In cognitive education fields, spatial imagery is also used to refer to static images about one-, two-, or three-dimensional spaces resulting from both visual and kinaesthetic experiences leaving the terms visual and spatial to refer to sensory input (see also Clements & Sarama 2007a, 2007b).

Some cultural studies have noted that spatial reasoning and imagery are linked to kinaesthetic experiences and resultant imagery. For example, tilting the head to see the stars (Worsley, 1997) or sensing the motion in a boat (Hutchins, 1983) embodies spatial information on position. Molnar and Slezakova (2012) refer to geometric spatial imagination as reproducing and anticipating static and dynamic images by using attributes of and relationships between shapes. Furthermore, spatial abilities and visual imagination play a part in visuospatial reasoning. Children's gestures assist with abstraction of space and geometric conceptions and in a metonymic way in communicating and imagining (Kim, Roth, & Thom, 2011). Spatial imagery can refer to the diagram or graph about which reasoning occurs in geometry in which experiment and change of features may be needed to reason about properties. Finally, the term spatial imagery has been used to identify its role within sociocultural place imagery (Tuan, 1977) but I argue that spatial and visual imagery do require a sociocultural perspective. In order to capture all these meanings the term visuospatial reasoning is used in this book.

There are many examples of visuospatial reasoning from everyday life and many definitions (Liben, 1988). The educational literature has blurred the meaning of visuospatial reasoning. Some of the literature talks about visual imagery and discusses it in terms of both internal and external representations (Gutiérrez, 1996; Owens & Outhred, 2006). On the other hand, definitions often focus on symbolic representations and diagrams. Thus spatial reasoning may be restricted to interpreting graphs or diagrams. In order to emphasise that reasoning is in the head, I use the term visuospatial reasoning to include the mental visual imagery research. At the same time, by turning to the fields of psychology and education the nature of visuospatial apprehension and cognitive processes is expounded (Mason, 2003).

Shah and Miyake (2005) consider "visuospatial thinking" from a psychological perspective including summaries of individual differences on psychometric assessments and the relationships between spatial and navigational abilities determined by these assessments on spatial activities like navigating. They also introduce the idea of comprehension of visuospatial representations or "spatial situation models" (p. iv) which provides a language perspective on visuospatial reasoning. In this book, I begin with a definition that is more encompassing; starting with the psychological literature but extended to have a stronger ecocultural perspective that encompasses culture, language, context, and ecology.

We begin with the following definition and description. Visuospatial reasoning incorporates a wide range of spatial abilities, spatial skills, visual and spatial imagery, representations and processes, and related concepts. Visuospatial reasoning is the mental process of forming images and concepts and mentally modifying and analysing these visual images. In the mathematical context, visuospatial reasoning is using visual images creatively in mathematical problem solving. Visuospatial imagery involves the relationship, position, and movement of parts of an image or sequence of images. The spatial component of imagery may result from bodily movements as well as visual perception. Frames of reference may be bodily rather than in terms of visual or cardinal (orthogonal north–south east–west) frames (Wickens & Prevett, 1995). However within this definition some extensions of meaning are needed, and I turn to these now.

Extending the Meaning of Visuospatial Reasoning

Mental often dynamic or patterned imagery that embodies relationships is part of visuospatial reasoning (Owens, 1993; Presmeg, 1986). Imagery signifies a schema of the relationships, one in which the cognitive processes, some innate, are not separated from the social context when reasoning (Walkerdine, 1988, p. 3). One example of visuospatial reasoning that illustrates that reasoning is a result of cognitive and sociocultural influences is presented in the following paragraphs.

Watson and Crick's development of the double helix as a mathematical model for DNA was seen by NRCCG (2006) as "the result of a brilliant exercise of imaginative visualisation that is constrained by empirical data, expressed by two-dimensional images, and guided by deep scientific knowledge and incisive spatial intuition" (p. 8). Imaginative visualisation, then, is influenced by conceptual knowledge not only of mathematics but also of the field in which the concepts are embedded, the result of being a part of a community of scientific practice. The imagining depends on visuospatial intuitions which are dependent on incidental learning and not just maturation (van Hiele, 1986). In other words, the spatial awareness, spatial abilities, spatial relationships, and spatial visualisation (see Chap. 2 for more details of these concepts) have been honed through experiences that may or may not have been explicated by the thinker. Watson and Crick's representation was initially two dimensional although later models were three dimensional (NRCCG, 2006, p. 2). The development of the representation indicates the sociocultural aspects of visuospatial reasoning as well as the fact that intuition and imagination which are influenced by culture and place are also aspects of visuospatial reasoning.

As a footnote to the story about Crick and Watson, I mention two other uses of the double helix representation for representing knowledge. The first comes from Lovat and Toomey (2009) who used it to describe the intertwining of quality teaching and values education. While the familiarity of the DNA representation was a context for their imagery, they provide examples of an ecocultural perspective in education with schools focused on values, community, and relationships. The second use of the double helix relates to the mathematics of the Yolngu, Northern Territory, Australia, as presented by Thornton and Watson-Verran (1996) in video format and by the Yolngu in a flattened string of "diamonds" represented in their own art forms (Fig. 1.1). The representations emphasise the importance that an activity occurred in the same place at the same time on each of the cycles of the double helix repeated yearly in both directions (past and future). Thus women wash the poisons out of cycad nuts in the same creek area at the same time every year. An Elder sits to negotiate with stories on the sand near the sting-ray-shaped lagoon, laying his spear into sand as his Elders have always done. Each act was visuospatial and rich with meaning, history, and relationships. Hence the double helix originated from visuospatial reasoning in very different contexts.

Visuospatial reasoning is also influenced by spatial imagery and context. This example is taken from the field of aviation (Liu & Wickens, 1992; Wickens & Prevett, 1995). In the cockpit, local guidance is received from the view ahead

Fig. 1.1 Drawings to illustrate Australian Aboriginal art about relationships

observed by the pilot and through instruments that provide two-dimensional (2D) representations of local space and position in a global system. Combined, these provide spatial awareness of the three-dimensional (3D) world. The pilot or navigator needs to mentally transform and align both sets of information rapidly before taking action. These actions are partially at the physical level and carried out automatically as one does after learning to walk or ride a bicycle. Much of this is determined by spatial imagery from nerve endings in the muscles and skin within the neuromuscular system but nevertheless linked to higher schema in the mind where adaptation needs to be made for a particular circumstance. Wickens and Prevett's study suggested that some information, for example, terrain details, was not recalled after the simulation because there was a focus on the flight path or the information was gleaned only for the working memory and not stored in longer term memory. The angle of the person at the time at which the 3D position information was supplied impacted on the speed and accuracy (clarity) of the scanning and generation of connections between the sets of information. These studies emphasise the bodily as well as visual awareness in visuospatial decision-making.

Visuospatial Reasoning and Geometry

In order to encapsulate the diversity of possible geometries that are particularly developing in modern contexts such as those related to medical imaging or tiling in computer graphics, a broad definition was given in the International Commission for Mathematical Instruction (ICMI) study (Mammana & Villani, 1998). Geometry is reasoning about the visual. It is clear how Euclidean geometry and analytical geometry (where graphs and algebra meet) are part of this definition. However, by this definition, geometry is about much more, and is more related to professional and cultural contexts. This definition allows for creativity and is less about learning definitions, symbols, axioms, and proofs. It is about all forms of convincing especially deductive reasoning including refutation by counter-examples, and making modifications because of good reasoning. Reasoning is about understanding, explaining, and convincing. It allows for inductive and visual reasoning and explanations, and for interactions with visualisations and constructions. Learners, the new geometers, interweave intuitive-visual reasoning and deductive reasoning.

Visuospatial reasoning provides a new way of looking at the situation in order to suggest a generalisation, and an explanation of why the generalisation holds. Furthermore students need to learn how to reason about visuals. It is also important to move beyond the initial, intuitive visualisation which might have visually dominant misleading aspects to a more advanced level of visualising by re-seeing the visual and reasoning from the newly attended aspects of the visual.

Bridging from the Psychological to the Ecocultural Perspective

Representations of space in terms of systems, measurements, projections, and other graphic features such as organisation of figure-ground features (i.e. the separation of a figure from the background) for readability, interpolation, and extrapolation (NRCCG, 2006, p. 3) are socioculturally determined. Furthermore the purpose for this visuospatial reasoning illustrates the importance of visuospatial reasoning in terms of sociocultural purpose, that of scientific advancement. Geographical advancements in the broad study of society linked to place are also served by mathematics. The Committee points out the importance of visuospatial reasoning in terms of a range of specialised skills that over time will become increasingly specialised such as those involved in interpreting MRIs (magnetic resonance imaging) and using the tools for capturing such images. There are many areas such as applied psychology, geography, physiotherapy, occupational therapy, medicine, architecture, design, computer science, semiotics, and animal cognition in which visuospatial reasoning is essential (Shah & Miyake, 2005). In schools, visuospatial reasoning impacts on a range of mathematical topics and some of these will be explored in Chaps. 8 and 9.

In many countries school curriculum refers to spatial awareness in primary schools as part of geometry or problem solving. It is recognised that visualisation of spatial attributes of objects and numerical representations assist the learning of concepts (Shah & Miyake, 2005). However, there is little recognition that these ways of representation and thinking have an out-of-school source (Pegg & Davey, 1998). Nevertheless, early childhood educators often refer to the context of education as influencing the thinking of the child (Brofenbrenner & Ceci, 1994). Children are immersed in not only a spatial realm but also one that has meaning to them in terms of their activity in the space and their relationships with others in the space. Education therefore has an ecological basis. This is further discussed in Chap. 3. However, the question remains to be answered in this book: How do ecological and out-of-school experiences impact on visuospatial reasoning? How does this knowledge broaden our understanding of visuospatial reasoning? And to what extent is it valuable for school education, especially in the early years, to build on this knowledge?

An Ecocultural Perspective on Learning

This book draws on work by myself and others to explore visuospatial reasoning in a cultural context and to explore how visuospatial reasoning is culturally constructed and culturally responsive. The term ecocultural is used to summarise the notion of responsiveness embedded in place and residing in culture and ecology. In other words, education besides recognising a school, system, and global perspectives as contexts may benefit from connecting to place and culture to understand and strengthen visuospatial reasoning.

This notion of ecocultural education implies a place-based, experiential education in which the learners construct knowledge, skills, and values from making meaning from direct experiences in their ecocultural places through the learning cycle of reflection, critical analysis, and synthesis. Conceptualisation of that place in terms of beliefs, values, taken-as-shared understandings, and language representations are embedded in meanings about space and geometry and of necessity are associated with that place. Relationships developed within and between places are also significant in establishing meanings. Thus a critical approach to education that considers place also provides for specificity relevant to the learner. It provides for an interdisciplinary curriculum which is ecological in terms of the contexts for learning both in the learning space and in the spaces in which the learner lives. Furthermore, a place-based education has a cultural context that is multigenerational and the boundaries between school and community can be crossed in a variety of constructive ways (Gruenewald & Smith, 2007). This is explored further in Chap. 3.

Throughout this book, mathematics of a variety of different cultures is recognised indicating a rich variety of ways of visuospatial reasoning. In that recognition, we confront issues of equity and social justice, of valuing difference, and of the cultural rights of First Nations before colonisation. Place-based education has particular relevance to First Nation or Indigenous communities because their worldviews are built on their relationships between their land, their place, and themselves.

Mathematics of cultural groups or identified social groups is referred to as ethnomathematics involving mathematical concepts and techniques used by a sociocultural group. These mathematical techniques may vary but could be said to involve some general principles of mathematics such as generalising patterns of relationships and variance (Johnston-Wilder & Mason, 2005). However, the extent and context of the relationships vary because of the ecocultural situation. For example, the relationship might be about a physical position or it might indicate a whole set of societal relationships as found with the way houses are positioned in a village such as those in the Trobriand Islands, PNG (Costigan, 1995), or with houses showing self-similarity in Africa (Eglash, 2007), or land identification showing connection to clan groups of the Yolngu, Australia (Thornton & Watson-Verran, 1996). The extraction of the relationship from the data or situated problem occurs in mathematical modelling (Rosa & Orey, 2012). While the extent to which the relationship should be divorced from the context is debatable, in terms of the ecocultural approach to mathematics, the debate is about how mathematics should be viewed, be learned, and be used. It is a debate about values in education (Adler, 2002; Atweh, Barton, & Borba, 2007; Barton, 2008; Bishop, 1988; Clarkson & Presmeg, 2008; Valero & Zevenbergen, 2004).

Matang (1998) argues there are two ways in which culture is connected to mathematics and mathematics education. One way is that the definition of mathematical knowledge is somewhat implicit, in that, mathematics is not a universal, formal domain of knowledge waiting to be discovered, but rather an assemblage of culturally constructed symbolic representations and procedures for manipulating these representations (Stigler & Baranes, 1988). Thus cognitively speaking, the advocates of this view argue that the incorporation of representations and procedures by children into their cognitive systems is a process that occurs in the context of socially constructed activities. In other words, the mathematical skills that children learn in school are the results of the combination of previously acquired knowledge and skills, and new cultural input, rather than logically constructed on the basis of abstract cognitive structures (Saxe & Esmonde, 2005). Accordingly, the notion of culture functions not as an independent variable that can promote or retard the development of mathematical abilities, but as an integral part of the mathematical knowledge.

The second way suggested by Matang is that the analysis of cultural influences on mathematical knowledge can demonstrate both the differences and invariance in mathematical knowledge across cultures (Bishop, 1988; Kimball, 1974; Nunes, 1992). Thus I contend that mathematics is defined as the ability to make inferences on the basis of these logical structures rather than the classifications, hierarchies, and procedures of mathematical content. Mathematics includes the organisation of the presentation of subject matter but also the relationship between teacher and learner. The conceptual conditions that provide humanity with its sense of identity also provide people with experiences that are meaningful in their worldview.

The key issue in this debate is not that of mathematics per se but of the ownership of the knowledge and how that might be shared in and out of school classrooms. In this respect, the people involved and their relationships are critical. For Indigenous communities, the role of the Elders and the roles related to specific relationships are central (Owens et al., 2012). It is also an argument about democracy and policy (O'Sullivan, 2008). Without specific policies that permit an Indigenous voice and a recognition of Indigenous knowledge and the associated rights, there will be little progress in establishing a continuity of education. These issues are illustrated through the examples of visuospatial reasoning presented in this book.

An ecocultural perspective assists teachers to be aware of the complexity of the ecocultural contexts in which students learn mathematics and to build successful school–community engagements through authentic open relationships built up over time under structures and plans that will sustain the relationships. Goals are thus mutually set within a social justice framework that provides equity and respect (Civil & Andrade, 2002; Gervasoni, 2005). Both the societal and individual constructions of mathematics are dynamic and so the mathematics may not remain static. A good example of the dynamic nature of mathematics is given by Muke's (2000) study of number among the Yu Wooi speaking community Mid-Whagi (Jiwaka Province, PNG) in which the language incorporates variants of the lingua franca Tok Pisin (Pidgin English) arising from practices of counting with larger and larger amounts of money for various new purposes. Similarly, Saxe and Esmonde (2005), in their genetic developmental work, provided evidence of how changes occurred in the Oksapmin's (Sandaun Province, PNG) counting systems and the use of the word *fu* (meaning the complete whole) over time developed to incorporate new meanings for the whole and for its usefulness in determining amounts and making agreements based on their counting systems.

Ecocultural Identity and Mathematical Identity

Ownership of mathematics like all learning requires personal connection and responsiveness that impacts on one's identity. Ownership and a sense of belonging are embedded in sociocultural experiences and those of being in and connecting to a place, family, and culture as well as to a school or learning community. Figure 1.2 (based on Owens, 2007/2008) provides a simplified view of the relationship between ecocultural identity and mathematical identity. The basis of the argument is that learners need to become self-regulated, confident learners with a sense of ownership of their mathematical problem solving in order to identify as mathematical thinkers. The social milieu and control of the social processes involved in learning play a significant role in this development. Furthermore, visuospatial reasoning is particularly influenced by ecocultural practices as I argue in this book. Hence the role visuospatial reasoning plays in learning geometry and developing a mathematical identity is critical and requires fostering.

Davis (1999) argues that identity is an enactive, dynamic, interactive ever-changing state of being. It is the doing of mathematics in an ecocultural context that is the identity. Ecocultural context surrounds and interacts with the learner whose values, beliefs, attitudes, and feelings control the cognitive processing involved in becoming a self-regulating learner together with and interacting with their ecocultural identity. The ecocultural context of learning influences the way a person thinks and feels about mathematical learning. For example, the form of questions (teachers and students), the expectations of the classroom, the available materials, and others in the learning space impact on the individual's cognitive and affective processing (Owens & Clements, 1998). The various cognitive aspects of the

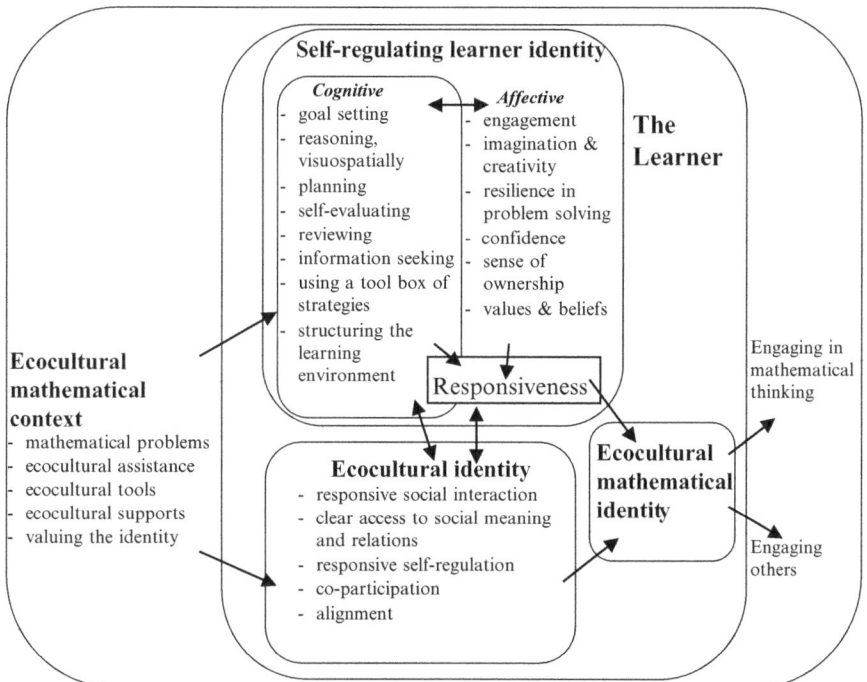

Fig. 1.2 Developing identity as a mathematical thinker

self-regulating learner (Jonassen, Peck, & Wilson, 1999) will incorporate a range of conceptual, heuristic, visual, and processing skills for working on individual problems (Owens, 1993; Owens & Clements, 1998) but over time self-regulation becomes an overarching aspect (Macmillan, 2009). Within these cognitive processes is reasoning and it is argued that visuospatial reasoning is a critical aspect of reasoning in terms of ecocultural influence. Responsiveness of the learner to the ecocultural situation and mathematical problem situation (referred to as situated learning by Lave, 1988) develops the ecocultural mathematical identity of the person.

The individual psychological aspects of the self-regulating learner interact with the individual ecocultural identity, which is also evolving (Davis, 1999). Both the psychologically and ecoculturally developing aspects contribute to the formation of identity as a mathematical thinker. The social context that encourages mathematical problem solving will still impact and surround this identity. Hence the identity is fluid depending on space (those people, power-relationships and environments around them) and time (dependent on experience). Identity may be relatively stable but it may develop and change at different velocities depending on the circumstances. An "aha" experience may change the identity quickly (Goldin, 2000), whereas thinking about the number and variety of problems that one has solved over a period of time may move a person more slowly towards an identity as a mathematical thinker. Problem solving within a community will also change one's identity when identity as a mathematical thinker is valued and tacit knowledge is explicated and communicated mathematically. Furthermore, in different contexts, such as a regimented drill-based classroom, identity as a mathematical thinker may not be as evident as in another type of classroom or social setting.

Learning through observation and listening, through imitating, and through practice develops psychological aspects of the self-regulating learner but ecocultural identity encompasses learning as belonging (community) and involves continuous aspects of social interaction with the community and the context of learning (Owens, 1999b; Wenger, 1998). This identity also incorporates learning as experience (meaning) by which language between members of the community provides the meaning for problem solving and learning. Learning as becoming is the resultant dynamic identity as a mathematical thinker.

Identity as a mathematical thinker influences the ecocultural context since the learner engages in mathematical activity and engages with others. "Being engaged to the fullest of one's identity is the source of creativity required for participation" in both the community of practice and "outside that community" (Wenger, cited in Kahan, 2004, pp. 30 and 33). Knowledge presented in a classroom is valued less than an "experience … that … often involves feeling like an integral part of a community" of practice (Kahan, 2004, p. 31). "A person's identity is engagement in the world" (Wenger, cited in Kahan, 2004, p. 36) which is unique and complex, dynamic, evolving, and enacted (Davis, 1999; Wenger, 1998). Is it possible to provide examples from PNG that establish an explanation of how an ecocultural perspective impacts on visuospatial reasoning in mathematics education? This idea has been pursued by Owens (2012a, 2014) and in Chaps. 7 and 8 of this book.

Moving Forward

This book is intended for mathematics educators rather than geographers or other vocations and the research from vocational fields is considered only when it elucidates our understanding of visuospatial reasoning in childhood space, geometry, and mathematics education. Does the suggested framework based on psychological and sociocultural perspective strengthen our understanding of education and identity and the role of visuospatial reasoning in learning? I argue both fields strengthen the education debate on visuospatial reasoning and so I present an ecocultural perspective that captures this rich understanding.

The focus is on prior-to-school, primary, and middle school education within a cultural context. For this reason, the informative research around dynamic geometry software or the software used to analyse numerical data sets is only used to strengthen arguments from research relevant to the younger age group.

Chapter 2 begins with the research of the 1980s and 1990s around perception, spatial abilities, visual imagery, intuition or incidental learning, noticing and attending, intention, and awareness. However, more recent studies of learning in context and its impact on visuospatial reasoning are used to extend our former understanding. Chapter 2 mostly covers a psychological perspective whereas Chap. 3 considers cultural psychology and psycholinguistic studies to present a new perspective on visuospatial reasoning that is not embedded in the psychological literature but rather in the ecocultural literature. Chapter 3 covers out-of-school and early school learning experiences involving visuospatial reasoning and transitions in visuospatial and geometric reasoning, designing and imagining in creative play, and in measuring. However, these ecological approaches will be developed in terms of place-based education and the geographies of learning spaces to present a critical pedagogy of place (Barnhardt, 2007; Ferrare & Apple, 2012; Gruenewald, 2008). This chapter establishes an ecocultural perspective of visuospatial reasoning and a social justice perspective on education.

These first three chapters provide a way forward in the argument to consider Indigenous perspectives as a fundamental ecocultural perspective for education. Cultural research on space as place (Tuan, 1977) and language are critical in understanding an ecocultural perspective in education. The particular focus in Chap. 4 is on visuospatial reasoning associated with representing spatial position. Language issues relevant to size and measurement are also discussed. It draws particularly on linguistic literature relevant to Pacific countries. Chapter 5, however, will focus on the visuospatial reasoning associated with Indigenous activities in PNG which is rich in having more than 850 different cultures. I synthesise research recently undertaken collaboratively with PNG colleagues and participant researchers to strength the ecocultural perspective of visuospatial reasoning. Chapter 6 extends this perspective by drawing on research within other Indigenous cultural groups around the world.

Nevertheless, the importance of this ecocultural perspective must be relevant to mathematics and mathematics education which is discussed in the rest of the book. Chapter 7 looks at the impact on understanding the nature of mathematics, mathematics education, and identity exemplified by ethnomathematics education projects in PNG. Chapter 8 considers the impact in practice for Indigenous and transcultural situations. The hybridity of visuospatial reasoning in two-way education is evident.

However, a test of this ecocultural perspective for visuospatial reasoning is by consideration of students in settings which are rich in digital technology. In Chap. 9, the authors discuss research from early childhood and primary or elementary school settings in terms of visuospatial reasoning and the role of the ecocultural technological context in development of identity. The concluding chapter, Chap. 10, draws together the research and discusses the implications of an ecocultural perspective of visuospatial reasoning on our understanding of mathematics and our implementation of mathematics education. The challenges for curriculum, teacher education, global citizenship, multicultural classrooms, and Indigenous education are addressed in terms of new perspectives on education such as place-based education.

Chapter 2
Visuospatial Reasoning in Twentieth Century Psychology-Based Studies

> *Ensuring that knowledge and skills are meaningful requires engaging the imagination in the process of learning.*
>
> (Egan, 1992)
>
> *I would say that all discovery requires imagination.*
>
> (Donald Coxeter, 1907–2003, cited in Hagen (2003))

The Challenge

From early in the twentieth century, there was interest by psychologists and educators about visual and spatial abilities along with other abilities perceived as valuable for learning. The scientific approach to research dominated the scene. Visual perception and spatial abilities were the main areas of interest for educational psychologists. Both constructivism and information processing theories were important drivers of research on visuospatial reasoning (or at least spatial abilities and visual imagery) in the twentieth century. Many mathematics educators emphasised that concepts are not passively received but are actively constructed as the learner uses existing schema to interpret information and draw inferences from this information (for example, Lohman, Pellegrino, Alderton, & Regian, 1987; Skemp, 1989; Steffe, 1991). In this learning, visuospatial reasoning plays a part when "the stored memories and information processing strategies of the brain interact with the sensory information received from the environment to actively select and attend to the information and to actively construct meaning" (Osborne & Wittrock, 1983, p. 4). The immediate context of the student was seen as relevant and it was accepted that memory was influenced by external prior experiences in a broader context. What was the legacy of the twentieth century from studies on visual imagery and spatial abilities? The influence of psychology on mathematics education was significant in this area of visuospatial reasoning but what impact could it have in the classroom? For some educators, Krutetskii's (1976) idea of visual and verbal reasoning was sidetracked into multiple intelligences or was there more to be learnt about visuospatial reasoning for mathematics education and in particular space, geometry, and measurement? In this chapter, I set out to research these questions, firstly through an extensive critical literature review and then via a number of empirical studies. Much of the work on visual imagery and spatial abilities was carried out in the 1970s and 1980s, so much of the foundation work for our understanding of visuospatial reasoning comes from that literature. A generative model of learning (Osborne & Wittrock, 1983) assisted to bridge the gap between information processing theories and constructivist learning theories. Other areas of research on visuospatial reasoning have been prompted by how children with disabilities learn visuospatial knowledge. Age-related studies are critiqued especially in terms of diversity of tasks in which visuospatial reasoning occurs and can be affected by the task. Then I explore in my studies how children are using visuospatial reasoning in school. I develop this research to show how students' attention and responsiveness are critical to their learning. However, it is salient at first to note the complexity of terminology generated by theorists and researchers in developmental psychology, factor analysis, and information processing studies on visual imagery, visualisation, and spatial abilities (Eliot, 1987).

Visuospatial Reasoning and Studies on Spatial Abilities

Terminology in these studies varied. For example, the word *visualisation* may refer to internal (mental) representations or external representations (Goldin, 1998), or to a specific spatial ability which was described and assessed by different kinds of testing items by different authors. It is worthwhile explaining this at the start of this chapter because it also gives the reader a greater appreciation of what is meant by visuospatial reasoning, a term that I say encompasses all these areas. The term *visual imagery* was usually used as an alternate to other forms of information processing or mental skills such as verbal processing. Spatial abilities were seen as a more stable intellectual quality than using visual imagery (Bishop, 1983) although training studies and age or maturation studies have shown spatial abilities can improve and change over time and with experience (Cox, 1978; Eliot, 1987; Lean, 1984). Problem-solving studies suggested some people preferred to process visuospatially while others preferred processing verbally (Krutetskii, 1976; Moses, 1977; Quinn, 1984; Suwarsono, 1982). This chapter teases out some of this complexity and then synthesises it drawing out important points for geometry education.

Visualisation or visual synthesis is contrasted with verbal reasoning in some intelligence tests but visualisation in other studies refers to one of the spatial skills—the mental rotation of a representation (visual image) of an object—in contrast to orientation in which the person considers the view of the object from another perspective (Eliot, 1987; McGee, 1979; Michael, Guilford, Fruchter, & Zimmerman 1957). Tartre (1990a) argued that the idea of limiting visualisation to mental rotation

Table 2.1 Visualisation and orientation

Tartre's categories	Descriptions and similar tests	Comments
Visualisation	"Mentally moving"	• Manipulation (Eliot & McFarlane-Smith, 1983) except alternative perspectives
Mental rotation		
• Rotating 2D shapes • Rotating 3D shapes		• More than rotation especially of 3D given it was often done by analysis
Transformation		
• 2D to 2D	• Form board tasks, integration of detail, tessellations, tangrams	• "Integration of detail" (Pellegrino & Hunt, 1991) and "spatial relations" (Johnson & Meade, 1985; Thurstone & Thurstone, 1941) except related to orientation—completing figures and fitting parts together
• 2D to 3D	• Surface development tasks	
• 3D to 3D	• 3D tessellations	
• 3D to 2D	• Unfolding tasks	
Orientation		
Multiple representations		See comment above
Re-seeing		
• Reorganisation of the whole	• Alternative perspectives	• Thurstone's spatial relations possibly
• Part of field	• Completing figures	• Pellegrino & Hunt's "adding detail", "deleting detail"
• Ambiguous figures	• Find part or fit part	• Lohman et al. (1987) have flexibility of closure (disembedding) as separate factor

• Hidden figures	• Figure-ground perception (Del Grande, 1990)
	• Also called "disembedding"
	• Recognition (Eliot & McFarlane-Smith, 1983)
		• Eliot & McFarlane's
		  – visual memory
		  – copying
		  – maze tests
		are not included in Tartre's examples

alone, especially of three-dimensional objects, is too limiting. She included all forms of transformation under visualisation and expanded orientation to include other forms of re-seeing shapes as shown in Table 2.1 which shows how various terms are used for similar spatial abilities based on examples of test items used by the various authors.

The term spatial relations is also used in different ways. Pellegrino and Hunt (1991) used it to refer to mental rotation tasks because, in terms of information processing, it is likely that, in fact, parts are rotated and checked in relation to other parts in sequence. Although "adding detail" and "deleting detail" were not classified by Pellegrino and Hunt (1991) with surface development and the integration of detail tasks, they do appear to be the same as Tartre's "part of field". Examples of items for assessing and investigating these spatial abilities can be found in my two tests: *Thinking About 2D Shapes* (Appendix B, see also Owens, 1992a, 1993) and *Thinking About 3D Shapes* (Owens, 2001a) discussed later in this chapter. These tests were for young children (5–10 years) and more like school experiences than most tests.

Visuospatial Reasoning from an Information Processing Perspective

While some information processing theorists' perspectives were incorporated into the discussion above on spatial abilities, they emphasised perceptual speed and the effects of speed and accuracy in spatial abilities. Poltrock and Brown (1984) suggested that individual differences were particularly due to the visual buffer (short-term memory of the image) and speed of processing. Measures of the processing for particular tasks depend on their complexity, speededness, and susceptibility to more than one solution strategy, so spatial abilities are reliant on creating structures which are abstract and relation-preserving and on which transformations can be easily and successfully performed (Lohman et al., 1987). Time is also significant for processing not only static spatial relations but also dynamic spatial relations which involve a time order and are generally studied by a series of computer images (Aust, 1989; Pellegrino & Hunt, 1991). New contextual areas requiring visuospatial reasoning include dynamic information presented, for example, in representing past and future weather patterns, and graphing data with traces.

Within the information processing theories, there are different emphases pertaining to visual imagery as a processing/storage medium. First, Paivio's dual-coding theory (Paivio, 1971, 1986) states that there is a non-verbal as well as a verbal symbolic modality for processing physical objects, scenes, environmental sounds and images, and general images. Kosslyn's surface representation theory (Kosslyn, 1981; Kosslyn & Pomerantz, 1977) suggests that during perception, units are abstracted, interpreted, and stored in long-term memory. In Pylyshyn's abstract transformational model (1979) the verbal and non-verbal information can also be transferred between modes by a set of propositions. A visual representation, sometimes accompanied by a verbal one, is generated by this proposition (Kieras, 1978). Imagery and propositions together with other memory structures interpret and are used for testing perceptions in short-term memory during learning (Gagné & White, 1978). Pictorial images then are not original photographic images but "quasi-pictorial representations that are supported by a medium that mimics a coordinate space" (Kosslyn, 1981, p. 46) explaining Bruner's (1964) notion of concrete, pictorial, and abstract representations. Support for visual images being processed, based on reaction time, in a way that is similar to manipulation of physical objects showed a linear relationship between the degree of rotation or number of transformations and the time taken to respond to the task (Cooper & Shepard, 1973; Shepard, 1971, 1975). However Shepard also noted that analysis rather than rotational methods could account for reaction time.

No matter how the storage of imagery occurs, the need to generate spatial representations is an initial stage in the processing of a spatial problem, according to the flowchart models of Egan (1979) and Carpenter and Just (1986). The emphasis is on part-whole relationships. For orientation, the model requires comparisons on dimensions one at a time while visualisation tasks require a search followed by a looping of transformation and checking. The particular task will affect the processing (Carpenter & Just, 1986; Paivio, 1971). Images can be generated by encoding a physical stimulus, retrieving a previously constructed representation, constructing a new representation from non-iconic (verbal) descriptions, or by some combination of these processes. Visuospatial reasoning is affected by the adequacy, efficiency, and accuracy of the encoding and the retaining of detail during transformations or comparisons. Some tasks do not require transformations but only assessment. Choice of frame of reference for encoding, consideration of size and proportion, and interpreting perceptual distortion are three aspects affecting processing and would be related to the ability of interpreting figural information which Bishop (1983) contrasted to the ability of visual processing.

Carpenter and Just (1986) based much of their work on detailed analyses involving retrospection and eye fixation, but a study by Poltrock and Agnoli (1986) further describes the importance of efficient imagery and what is entailed in it. They used structural equation modelling and found that a range of tests of spatial abilities required a number of visual imagery processes. The resultant model was used to relate the imagery-cognitive components as determined in laboratory tests to spatial-test performance by a linear regression analysis and then to a factor analysis of the spatial tests. Efficient image rotation and efficient image integration contributed to performance on all the spatial tests, but image generation time did not. This last factor was correlated with image memory performance. Adding detail and image scanning were two further imagery components suggested by Kosslyn (1983) and others (Brunn, Cave, & Wallach, 1983, cited in Poltrock & Agnoli, 1986; Poltrock & Brown, 1984). Visual memory and vividness of imagery did not correlate with spatial ability (Lohman et al., 1987) and Burden and Coulson (1981) also found that students used a variety of approaches to visual processing and that these processes could not be restricted to the processing methods suggested by Egan (1979) and discussed above.

Lohman et al. (1987) concluded that visualisation is the most general spatial-ability factor. The tests that load on this factor were quite diverse: tests of rotation, reflection, folding of complex figures, combining figures, multiple transformations, or no transformations. They listed another nine spatial factors: spatial orientation, flexibility of closure (embedded figures test), spatial relations, spatial scanning, perceptual speed, serial integration, closure speed, visual memory, and kinaesthetic memory. This list is not a complete list of spatial abilities and, indeed, Guilford's structure of the intellect was a model schematising a multifaceted intellect involving three dimensions—content, product, and operations—and it encompassed many cells with figural content that could be related to spatial abilities and visual processing (Magoon & Garrison, 1976). Lohman et al. (1987) stated that tasks which were complex tended to load only on the one factor called visualisation, but simple tasks, generally involving time, tended to involve more specific factors. While they summarised basic categories of processes as pattern matching, image construction, storage, retrieval, comparison, and transformation, Kosslyn (1981) listed other processes (rotate, scan, pan, zoom, and translate images, inspect and classify patterns). Among others, Carpenter and Just (1986) emphasised the use of analysis and checking in both orientation and visualisation procedures and this might explain the conflict between Tartre's classification and others. If this is the case, then visuospatial reasoning is not just a skill but it involves the understanding of concepts because analysis and checking are limited when images are not conceptualised; a point that is generally not mentioned in the literature but which is taken up in discussing types of visual imagery later in the chapter.

The question remains whether visuospatial reasoning is a spatial ability or a higher order ability encompassing spatial ability. Visuospatial reasoning can be used in non-spatial problem solving (Deregowski, 1980; Krutetskii, 1976; Owens, 2002c). The terms "imagistic processing" or "imagining" capture the creative use of mental visuospatial reasoning in solving problems (Goldin, 1987). The extent of visuospatial reasoning is reflected in the following statement:

> producing or using geometrical or graphical representations of mathematical concepts, principles or problems, whether hand drawn or computer drawn ... that is, the use of mathematical visualisation is intended to be a mental process but also to produce a drawing to assist in understanding or problem-solving. (Zimmermann & Cunningham, 1991, p. 1)

Visuospatial reasoning also incorporates "the ability to represent, transform, generate, communicate, document, and reflect on visual information" (Hershkowitz, 1990, p. 75) and to relate certain concepts to physical embodiment, pictorial or concrete through which each person would develop certain conceptualisations (Bauersfeld, 1991). Visuospatial reasoning then is a mental process that may come from, create, or manipulate physical representations (see also the discussions reported by Goldin, 1998). Visuospatial reasoning encompasses spatial abilities but goes well beyond these skills.

Studies of Learners with Disabilities

Another area that assists us to know about visuospatial reasoning is the studies with people with disabilities. Witelson and Swallow (1988) suggested that both hemispheres support spatial performance with maturation points at age 5 years and at puberty. Damage to the left hemisphere of the brain (often seen as dominant in language acquisition) reduces this performance. Landau (1988) noted that basic principles of spatial cognition of students who confronted their environment mainly by hand were the same systems as those of sighted children. By contrast, Stiles-Davis, Kritchevsky, and Bellugi (1988) showed right hemisphere-damaged infants display normal ability to identify class relations so long as these were not spatial relations. Furthermore, for spatial groups and relations, they were impaired compared to others whereas the left hemisphere-damaged children did not show the same difficulties. In the cases of a child with a disability that reduces spatial thinking, then appropriate language development seems to provide alternative pathways (Mandler, 1988). Visuoperception can be adequate for tasks like recognition of unfamiliar faces, perception of form, and closure for children with Williams Syndrome. However, visuospatial thinking is limited for these children as evident by their focussing on irrelevant features like height reduction to conserve quantities, by not showing connectivity of parts of perceived objects, and by not recognising transformed shapes indicating deficits in the visuospatial skills of drawing, spatial construction, line orientation, spatial transformations, and spatial memory (Bellugi, Sabo, & Vaid, 1988). This study in particular indicated a distinction between visuoperceptual skills and visuospatial reasoning.

Lillo-Martin and Tallal (1988, p. 437) also note that "while the well-known left- and right-hemisphere distinctions are upheld, some degree of plasticity, transferability, and compensatory change are indicated [by the studies reviewed by Stiles-Davis et al., (1988)]". In the area of attention, studies of subjects who were deaf and hearing who knew or did not know American Sign Language (ASL) provide further information. Deaf subjects showed compensatory mechanisms with occipital activity in both hemispheres while the hearing group with ASL (deaf parents who signed) had increased left temporal-parietal activity compared to the hearing group without ASL showing functional reallocation (Neville, 1988). In a further study (Poizner & Tallal, cited in Lillo-Martin & Tallal, 1988) there was no compensatory performance and Lillo-Martin and Tallal (1988) suggest this was due to the critical flicker frequency, lack of verbal labels, and the positioning of the visual stimuli on the eye. These last-mentioned researchers suggest that a spatial language still uses the left hemisphere although some brain reorganisation takes place. They conclude that "function rather than form dictates cerebral organization, at least for language and spatial cognition" (Lillo-Martin & Tallal, 1988, p. 438). While the acquisition of language and visuoperceptual functions are innate in certain parts of the brain, a limitation on that area may limit performance in early childhood but will lead to changes in brain organisation and limited plasticity. It may be that children without brain dysfunction or limitation may process differently and there is no implication for adult performance from these studies. In addition, timing might also affect performance on tasks related to motion and localisation in space (Anderson, 1978; Neville, 1988; Shepard, 1988). "Interactions of spatial processing with other, related areas, such as temporal processing, is an integral part of understanding spatial cognition" (Lillo-Martin & Tallal, 1988, p. 440). However, the studies suggest that context and social experiences in early childhood will dramatically affect development in the area of visuospatial reasoning.

Processes include spatial perception, object location, line orientation, spatial synthesis, spatial memory, spatial attention, spatial mental operations like rotation, and spatial construction (Kritchevsky, 1988). Spatial attention seems to be influenced by both sides of the brain and so does construction with one part particularly requiring more thought than the other to draw, for example, an image with adequate angles and detail. Objects can be located using both visual and verbal information. Importantly training that involves areas of the brain other than the perceptual, visual memory section assists in spatial construction needed for basic tasks designed to improve spatial attention, memory, and construction.

Healy and colleagues (Healy & Fernandes, 2011; Healy & Powell, 2013) have also studied learning of blind students. In a unit on symmetry they particularly noted

> There were differences between approaches to symmetry adopted by the two students. For example, the student who had never had access to the visual field tended to treat geometrical objects as dynamic trajectories and attempted to look for invariance relationships among the sets of points which defined the trajectories; the second student attempted to characterize the objects he was feeling in terms of objects he remembered from before he lost his sight. Nevertheless there were also similarities. Notably, both students tended to move their hands or corresponding fingers from each hand in a symmetrical manner over the materials they were exploring. (Healy & Powell, 2013, p. 78)

Reisman and Kauffman (1980) provided a range of visuospatial issues for consideration in this regard from work with disabilities. Visuoperceptual disorders underlie difficulties in spatial orientation, recognising position, discriminating figure from ground, and distinguishing near–far relationships together with sequential memory, visual spatial memory, or constancy of form difficulties. These difficulties impact on arithmetic skills and understandings as well as spatial-geometry understanding. Similarly Farnham-Diggory (1967) showed that alternative ways of reading using pictographs are possible although disability may slow progress. These studies on learners with alternative abilities indicate that visuospatial reasoning occurs using different pathways.

Age and Visuospatial Reasoning

While I argue later that strategies for visuospatial reasoning are found across ages, it is important to consider earlier studies and to build on them but at the same time show how modifications to assessing provide evidence to critique stage and age-related limitations. Piaget and Inhelder (1956, 1971) claimed that children who had not yet reached the concrete operational stage could not solve problems requiring mental rotation of images because this task required conservation skills. Visuospatial reasoning was linked to maturation and considered available only to those who had developed certain levels of thinking. However, Rosser, Lane, and Mazzeo (1988) who considered age as a predicting variable contributing to level of development actually found that young children could solve rotation problems which were not difficult (such young children may not be conserving). The children reproduced the simple models of two rods, which formed a T or an L, and a circle placed at the end of a rod or in the right angle. Most children aged 4 and 6 could reproduce a model present in front of them and when it was shown and then hidden while 8-year-olds could also memorise and represent an anticipated rotation (which was indicated by hiding and rotating a model), and represent another perspective by moving a model. Owens (1992a) developed an innovative paper-and-pencil test that used cardboard cut-outs in explaining the items and stickers for some responses. On an item inspired by Rosser et al. (1988), she found that these items were relatively easy (on a Rasch analysis) for children aged 7 and 9 years. The test incorporated items that linked to spatial abilities (Eliot & McFarlane-Smith, 1983) but more closely linked to typical classroom activities. It was developed in two equivalent versions, a copy of one is available in Appendix B.

Invariance of parts of a shape was more complex than that required by the Piagetian conservation of length task (the staggered lines test involving two equal horizontal sticks with non-vertical starting points). Kidder (1978) found that only a small percentage of conservers could choose the correct length of a side of a transformed triangle, and Thomas (1978) found that non-conservers (determined by the Piagetian task), irrespective of grade (1, 3, or 6), were less likely to be correct in assessing invariance of length of the side of a triangle under rotations, translations, and reflections than conservers in that grade. The older students considered the vertices as well as the sides of the triangle. This result suggests that conservation may not have been the most important determinant of the results of this study but some other factor such as the strategy used to make the decision or some features of the task.

van Hiele (1986) suggested that concepts in geometry such as equality of angles develop through the following stages and depend very much on experience. Students do not tend to reason about properties, although they may about parts, without first apprehending (attending and noticing) and reasoning visually. According to van Hiele, the stages are the following:

1. The student reasons about basic geometric concepts ... primarily by means of visual considerations of the concept as a whole without explicit regard to properties of its components. ...
2. The student reasons about geometric concepts by means of an informal analysis of component parts and attributes. Necessary properties of the concept are established...
3. The student logically orders the properties of concepts, forms abstract definitions, and can distinguish between the necessity and sufficiency of a set of properties in determining a concept. (Martlew & Connolly, 1996, p. 31)

Students with less developed approaches to concepts such as equality of angles may be operating in the earlier two stages. Several later studies suggested that development through these stages was concept specific (see summary in Owens & Outhred, 2006).

From a study of 2- to 5-year-old children's constructions and drawings of geometric shapes, Fuson and Murray (1978) reported that the verbal descriptions given by children were holistic and that, if an attribute was mentioned, it was in the context of describing a whole shape, for example, "the pointy one". The study showed that children could construct each of the shapes before they could draw it or analyse it suggesting that there were at least two prerequisites for drawing shapes:

1. The ability to discriminate the parts of the shape
2. The ability to operate on a mental image of a shape so that
   (a) The parts of the shape can be related in a sequential order.
   (b) The part(s) of the shape already drawn on the paper can be coordinated with the mental image of the whole shape that is projected onto the paper. (p. 80)

Support for an interaction between the visuospatial reasoning and the external actions (verbal and visual) as critical to our understanding of visuospatial reasoning comes from two interview and observational studies by Mansfield and Scott (1990) and Wheatley and Cobb (1990). Instead of determining the kinds of transformations that students could carry out by giving them test items in which students had to recognise transformed shapes, Mansfield and Scott's (1990) study observed 23 preschool to grade[1] 1 children selecting shapes to cover other shapes which were either marked with suitable divisions or not. (For example, a square could be covered by two right-angled isosceles triangles or two rectangles.) Although older children in this study tended to be able to solve more problems than younger students, this was mainly the result of their persistence rather than their more efficient or varied strategies. Covering shapes which did not have divisions was more difficult for children than covering those with divisions. Recognising shapes which would not lead to a solution and re-positioning pieces increased success. Rotating shapes and turning the pieces over were more advanced strategies. Children tended to use the same strategies in two interviews over time since persistence meant that a poor strategy could gain success eventually (Owens & students, 2007).

In Wheatley and Cobb's (1990) study, 24 children from first and second grade were given five pieces in the shapes of a right isosceles triangle, a parallelogram, and a square, and two smaller similar triangles which could be joined to form the other three shapes. The children were briefly shown a square with lines drawn to indicate that it could be covered by the three triangles. They were then asked to cover a blank square with the pieces. Wheatley and Cobb determined that the overt actions of the children represented images and conceptual structures. Students seemed to be using the following aspects of imagery and structures:

1. The divisions of the square could be thought of as being made up of two-dimensional space rather than just lines.
2. The size of shapes could be compared with imagined shapes.
3. Mental rotations could be used to anticipate how the space might be filled.
4. The whole is made up of parts in specific positions.

Wheatley and Cobb described the children's behaviour in terms of several levels: (a) imagining two-dimensional shapes as linear objects (matching shapes using their lengths); (b) covering the shapes globally (covering with overlaps or gaps

---

[1] This study and my own were undertaken in Australia where in fact grades are called Year 1, Year 2, etc. but grade is used here for consistency with other countries.

without aligning sides); (c) structuring an unfilled space as a shape (after positioning some pieces, children perceived the remaining space as a shape); (d) partially constructing images (mental images tended to involve only one aspect of the whole); and (e) constructing relational images (parts and properties were noticed as parts of the image). Such a description of students' visuospatial reasoning suggests a growing alignment between conceptual understanding and visuospatial development.

In attempting to provide a summary of children's uses of shapes, Clements, Wilson, and Sarama (2004) also suggested levels such as precomposer, piece assembler, picture maker, shape composer, and substitution composer but these appear to be a guide but not definitive levels in terms of students' behaviour or ways of thinking. Especially in terms of some consideration of the difficulties and nature of tasks is needed (Wilson, 2007). While these studies provide evidence of visuospatial reasoning, it is clear that trying to bring levels to these ways of reasoning is restrictive of students' diversity of thinking for any puzzle. However, experience, visuospatial reasoning, and decision-making are evident. One puzzle was to cover a bone shape by five regular hexagons. One student who had placed four isosceles trapezia on the shape but not as a hexagon was not immediately sure of covering it nor could he imagine where each trapezium would be placed. The visual and his train of thought may have prevented recall of other facts that he knew such as trapezia make a hexagon. His mental imagery was sophisticated already as he had begun a puzzle which required placing, imagining, and mentally counting trapezia in quite a difficult way compared to the tasks in my own studies (NSW Department of Education and Training Curriculum Support and Development, 2000; Owens, 1993).

Visuospatial Reasoning on Different Tasks

There is a further caution raised in comparing research using different tasks. Task features are significant factors in tests of spatial abilities. For example, students in all grades (up to 11) found it was very difficult to visualise the rotation of letters which had rotational symmetry (the S and N) and the horizontal reflection of the non-symmetric J. The half-turn clockwise also yielded greater differences between the grades than the two reflections or the counterclockwise rotation. Vurpillot (1976) explained that the use of a horizontal reference line in spatial perception tasks encourages subjective preference for distinguishing a "top" and a "bottom" of a shape while a vertical reference line encourages preference for homogeneity of perception favouring recognition of symmetry.

The need to consider variations in the type of transformation as well as the type of figure involved in the task was taken up by Schultz (1978). She varied the type of transformation, the mode (horizontal or diagonal), the lengths between positions before and after the transformation, the size of the configuration, and the type (meaningful, that is, the sailing-boat configuration, or not). The configurations were made of three coloured parts. She found the following: (a) lack of familiarity and unexpected sizes of shapes interfered with comprehension but not as much as type of transformation and features of the transformation itself; (b) "meaningful configurations apparently facilitated the operational comprehension of a task" (p. 205) and large shapes were preferred; (c) translations were far more "do-able" than reflections and rotations by 7-, 8-, and 9-year-old children; (d) rotations and diagonal reflections increased error rate or were found to be not "do-able"; (e) diagonal translations often resulted in re-orientation of the shape in the same direction; and (f) the distance of a displacement was a significant variable. However, the study did not give the significance of the differences in the percentages of different categories. Horizontal and vertical displacements in translation and rotation tasks were significantly easier than diagonal-displacement tasks for first graders but orientation of the figure made it even harder (see also Owens & Outhred, 1997). First graders' scores on subtests on the recognition of shapes and left–right orientation were relatively high but were low on subtests on perspective, figure-folding, and reasoning. After instruction the experimental group only improved significantly on the perspective subtest. Moyer (1978) found that explicit knowledge of the physical motion associated with a transformation did not necessarily help the child's ability to perform the transformation task.

Lehrer, Jenkins, and Osana (1998) considered children's reasoning for choosing two out of three shapes they considered alike. They suggested there were nine types of visuospatial reasoning with one kind of reasoning using properties and two kinds based on class of shape. The visuospatial reasoning was seen to vary with immediate context. For example, a skinny rectangle placed in an oblique orientation was considered similar to a skinny parallelogram with oblique small sides by a large proportion of children but when the parallelogram was enlarged, there was not the same degree of error in terms of definitions of shapes. Visuospatial reasoning was influenced by context within the page but also by children's schooling about what makes shapes the same. In other words, the school culture and the degree to which they had been enculturated into this Euclidean, definition-based system of shapes influenced their decision-making.

However, studies of children's intuitive behaviour yield other findings in terms of symmetry. Children from a very early age experience symmetry because it is an aspect of our bodies, of nature, and of many person-made constructions. Booth (1994) studied pre-school students' art and showed a natural tendency to paint symmetrically such as matching coloured lines on opposite sides of a central vertical line of symmetry and in patterns such as rows of coloured dots. Nevertheless, other ideas influence their paintings such as a desire to fill the whole page with paint. More formal, paper-and-pencil studies around 1990 showed children's difficulties with symmetry as illustrated by an analysis of grade 6 students' responses in New South Wales (NSW) on Basic Skills Tests. Two questions on symmetry involving mirror reflections were poorly answered by grade 6 students: 69 % were correct on a question involving a grid and a vertical reflection line, but only 20 % coloured in parts of a reflected face correctly. By comparison, over 80 % of students in grade 3 and grade 6 were correct on questions involving folding (Owens, 1997a). It seems that recognition of transformed shapes depends on the nurturing of natural symmetrical experiences.

The influence of visual skills, and diversity of means by which students can answer a simple angle-matching task should not be underestimated. Spatial skills such as disembedding or re-seeing were noted as helpful in using imagery and in solving spatial problems. Tasks themselves, especially the directions given to students, may encourage use of different kinds of reasoning; for example, novel tasks and tasks which relate to physical objects may encourage visuospatial reasoning (Paivio, 1971). Krutetskii (1976) pointed out that some students preferred visual methods, others analytical or verbal methods while other students preferred to use both methods. Lowrie (1992) found that students chose visual or verbal methods depending on the nature of the problem and how difficult they found it. Many studies (see, for example, Burden & Coulson, 1981; Lohman, 1979; McGee, 1979; Poltrock & Agnoli, 1986; Shepard, 1975) indicate that different people use different strategies for doing the same spatial tasks. For example, on tasks in which the subject has to decide if the object has been rotated, some subjects have rotated the visual image to the new orientation, others have considered the object from a different perspective, and others recognised features and used more *analytic* strategies. Studies by Egan (1979) and by Carpenter and Just (1986) have shown that part-whole analysis can be used in both "orientation" (other perspective) and "visualisation" (transformation) tasks. The skill of being able to disembed shapes and parts of shapes seems to be a different skill from those requiring mental manipulation of images (see Table 2.1; Eliot, 1987; Tartre, 1990a) but the tasks which seem to require this skill may still be completed by analytic procedures.

If this is indeed the case, then visualisation (used in the broad sense of all visual imagery) is a skill which can involve analysis and checking and hence concepts (Clements, 1983; Krutetskii, 1976). This point was not recognised in the earlier factor analysis literature on spatial abilities. Despite their differences both Pylyshyn (1981) and Kosslyn (1983) would agree that both verbal (analytic) and visual information can be processed, and that there is a means of mental storage which can be used either verbally or visually as needed in the working mind. Individuals vary in their preference for mode of mental representation whether by verbal, visual, or both mediums. Hence I incorporate these mental activities into visuospatial reasoning, avoiding conflict of terminology and emphasising these are using reasoning.

Personal Approaches to Visuospatial Reasoning

As Lohman et al. (1987) have suggested, visuospatial reasoning depends on a range of spatial abilities, visuospatial memory, and image integration and manipulation. In Poltrock and Agnoli's (1986) study, efficient image rotation, image integration, adding detail, and image scanning contributed to performance on spatial tests but image generation time did not. Numerous studies have assessed the impact of visual skills and choice of visual or analytical methods on problem solving. In the narrow area of spatial tasks, Barratt (1953) found that the choice to use imagery was important on tests with high loadings on a spatial-manipulation factor but less important on tests loading on a reasoning factor. Carpenter and Just (1986) found that those who solved tasks sequentially tended to have lower scores than those who rotated shapes holistically. However, Sheckels and Eliot (1983) found students who performed well on visual rotation tasks and processed visual materials analytically performed well on visual and combined visual/verbal mathematical problems.

One study of a personal characteristic, namely the preference for visual processing, that is, visuality, was carried out by Suwarsono (1982). His Mathematical Processing Instrument (MPI) consisted of a Mathematical Processing Test (30 verbal problems) and a questionnaire that asked subjects to choose between a visual and a verbal solution as similar to their own solution method. From the questionnaire a mathematical visuality score was obtained. He considered the effect of training in verbal and visual methods on performance and the use of visuality in mathematical problem solving. Suwarsono found that spatial ability and picture-completion ability were not related to mathematical visuality. This was also found by Lean and Clements (1981) with tertiary students in Papua New Guinea.

Suwarsono (1982) found that visuality did not assist or hinder mathematical problem solving. However, Lean and Clements (1981) found that students who used analytic–verbal processes tended to perform better than those preferring visual processes. As the MPI was designed for seventh-grade Australian students, it may have been too easy for the tertiary students of Lean and Clements' study (the mean test score in Lean and Clements' study was 11.1 out of 15 as opposed to 17.3 out of 30 for Suwarsono's sample). Furthermore, Tartre (1990b) found in a problem, in which the area of an irregular figure was to be estimated and calculated, that spatial-orientation ability (picture-completion test) was related to each of the following: the quality of the estimate, changing unproductive mind set, adding marks to show relationships, mentally moving or assessing size and shape of part of a figure, getting the correct answer without hints, and relating to previous knowledge structures. Barratt (1953) asked students to indicate the extent to which they used visual imagery. He claimed that those who used it extensively did well on tests with high loadings on a spatial-manipulation factor but no better than others on tests with high loadings on a reasoning factor. Thus there is no simple explanation for achievement but rather an indication of the complexity of visuospatial reasoning.

Students who have high spatial ability can still choose to use verbal methods of solving problems. In several studies, scoring on the test of verbal reasoning was the only variable explaining variance on post-training mathematical problem-solving performance except pre-training performance (other variables included pre-training mathematical visuality, spatial ability, and picture-completion ability) (Lean & Clements, 1981; Quinn, 1984; Suwarsono, 1982). The importance of verbal reasoning, at least on problems presented verbally, could be explained by better abstract thinking (as Lean and Clements have suggested) or by the nature and familiarity of the problem (as Paivio has suggested). Further support for the value of analytical thinking despite high visual processing ability comes from Sheckels and Eliot (1983) who found that, as only two visual variables—rotation and embedding—were related, the choice to use visual imagery (visuality) was unrelated to the ability to rotate visual material or to the preferred visual processing of material.

By contrast, Webb (1979) found that, besides mathematical achievement and verbal reasoning, only pictorial representation out of 13 variables accounted for a significant amount of variance. Moses (1977) also found that there were correlations for scores on the problem-solving inventory, measures of spatial ability, reasoning, and degree of visuality which were all significantly different from zero. However, she analysed students' written responses to the problem-solving tasks to determine degree of visuality but this procedure has doubtful validity, especially when it is considered that the problem-solving inventories were too difficult for most students. However Hegarty and Kozhevnikov (1999) have found that there are two types of visualisers: concrete imagery and abstract imagery affecting performance especially on items that did not require a high verbal skill. Why might this be the case? The key study described in this chapter helps provide an answer and explains the role verbal skills play together with visual imagery in problem solving.

A number of the above studies have used spatial-ability tests which could be high on reasoning factors rather than visual imagery. The type of task and level of difficulty make it problematic to conclude whether there is value in using visual approaches to solve problems. In order to overcome this uncertainty, training studies were used to assess the situation. This approach, together with exploratory qualitative studies of students involved in problem solving, has provided alternative methods of exploring visuospatial reasoning.

Training

Kyllonen, Lohman, and Snow (1984) found that short strategy training and performance feedback improved performance on a spatial-visualisation (3D rotation) task and a surface development transfer task but visualisation training was otherwise ineffectual. In general they found verbal–analytic training assisted more difficult paper-folding problems and for low visual–low verbal subjects a combination of enactive practice and feedback with visualisation strategy training helped. Higher aptitude students especially in verbal reasoning were already proficient in analytic strategies in the same way as Fennema (1984) found with the strategies of "encoding and classifying folds, rehearsing the sequence of folds, and deducing the solution using the rules provided by the analytic treatment" (p. 143). General spatial activities were as effective as short general training according to Baenninger and Newcombe's (1989) meta-analysis of correlational students. However, a three-week training programme did increase spatial visualisation for students in all grades 5–8 in Ben-Chaim, Lappan, and Houang's (1988) study. Lean (1984) comprehensively summarised studies on training in 3D visuospatial reasoning and concluded that general geometry courses are less likely to improve the skill of interpreting figural information (a term used by Bishop, 1983) than specific training courses. Furthermore, he noted that there is less conclusive evidence for being able to train visual processing. Lean (1984) warned that two major features could lead to misinterpretation of the value of training: (a) the training or testing may be indicative of skill in interpreting figural information or in some analytic skills rather than a visualisation skill (see also Deregowski, 1980), and (b) any improvement may merely be from practice rather than from a real improvement in visual skills as indicated by retention and transfer of skills to other tasks. (The latter argument was expounded by Piaget, Inhelder, & Szeminska (1960).) Cultural factors will also influence development of spatial skills (Bishop, 1983, 1988).

Nevertheless, kindergarten children showed an improvement on a perspective task after eight training sessions (Miller, 1977, cited in Lean, 1984), but in Cox's (1978) study with 20 individual training sessions, there was no transfer to a matrix task, prediction of a cross section, or the prediction of the water level in a tilted jar, and he concluded that the basic requirement for learning and achieving on the spatial tasks was not just operational thinking but spatial skills specific to the task. Retention scores (after 7 months) on the tasks which were similar to those in their training were also significantly different from the control group. Moses (1977) carried out a problem-solving training study in which grade 5 children improved their scores on spatial-ability tests as well as reasoning and problem-solving tasks as a result of the training (see also Lean & Clements, 1981; Quinn, 1984).

There have been a few articles outlining programmes developed to improve geometric and visual skills in younger children (Abe & Del Grande, 1983; Flores, 1995; Frostig & Horne, 1964; Kurina, 1992) but a carefully evaluated programme by Del Grande (1992) found that a course involving transformation of shapes did in fact improve the spatial visualisation (perception) of grade 2 students. The activities involved concrete shapes, geoboards, other common classroom aids, and pencil-and-paper activities. Similarly, Perham (1978) found that instruction in flips, slides, and turns (using activities involving tracing paper, geoboards, and free drawing as well as class and group discussion) assisted performance on tasks involving slides, flips, and reflections except those involving diagonal transformations, and some of those involving turns (see also Genkins, 1975).

Other training studies have involved older students. Although Lean (1984) concluded that general geometry studies tended not to show improvements in spatial abilities, a study by Bishop (1973) provided evidence that active participation in a geometry course did positively affect spatial abilities. A significant feature of this course was the use of manipulatives. Bishop's result lends support to the van Hiele's (1986) theory that recognition should precede analysis in geometry and that manipulatives and everyday experiences have an important part to play in this. Saunderson (1973) is another to make use of concrete activities at the post-secondary level in Papua New Guinea. His training programme involved both three-dimensional and two-dimensional activities and his tests also covered both areas. He used informal activities including three two-dimensional activities—tangrams, pentominoes, and enlarging tile shapes. Both the use of form board tests and the nature of his activities suggested that the improvement in spatial skills after training was linked to improvement in analytical skills. Rowe's (1982) training study considered the effects of different types of spatial programmes. The study involved grade 7 students, with one group undertaking training of spatial skills for transforming two-dimensional shapes, another group undertaking training on three-dimensional shapes, and a third group acting as a control. The group involved in the two-dimensional programme improved statistically significantly more than those involved in the three-dimensional programme but only on the test items involving two-dimensional shapes and easier spatial skills. Wearne (cited in Lean, 1984) found that the greatest improvement in scores for secondary students was associated with an increased number of analytic solution strategies. Caution is needed in applying studies and the van Hiele theory applicable for older students to younger students. The studies described later in this chapter address these concerns and provide a less structuralist approach to learning and using visuospatial reasoning.

Key Study on Children's Visuospatial Reasoning

In order to overcome this problem, I undertook a classroom study with children in grades 2 and 4 (Owens, 1993; Owens & Clements, 1998). The children came from three different schools in low socioeconomic areas of Sydney with most children having English as a second language. Within each class, based on their pretest scores, children were matched and randomly allocated to one of the teaching groups: geometry investigations working individually, geometry investigations in groups of three or four children, or number investigations. Children in the geometry groups participated in 10^2 one-hour investigative tasks requiring visuospatial reasoning over 5 weeks involving pattern blocks, tangrams, matchstick puzzles, and pentominoes while the control group undertook number problems. Children were also learning about shapes and angles by comparing them. The lessons are detailed below to indicate the kind of learning plans used to provide appropriate investigations for visuospatial reasoning:

1. Explore similarities and differences in the seven tangram pieces.[3]

    (a) Compare the pieces and decide what is similar about the pieces. What is different? What is the same about the square, parallelogram, and middle-sized triangle?
    (b) Notice what shapes you can make by joining two or three pieces together in different ways. Draw them.
    (c) Estimate how many small triangles are needed to make each of the other shapes, for example, the large triangle. Check it.
    (d) How many different ways can you make the large triangle with the smaller pieces? Draw them. When you wanted to make the shape, how did you move the pieces?
    (e) Extension: Make squares out of the pieces.

---

[2] All children participated in an introductory lesson, so the kind of interactive behaviour expected in investigations was established and children and I came to know each other. The class teacher taught the other half of the class and then we swapped.

[3] Tangram sets were made from cardboard with three sizes of right-angled isosceles triangles (two large, two small, and one medium), a parallelogram, and a square which combine to make a square. This is a well-known puzzle that can be used to make many shapes and pictures and the shapes have special relationships, e.g. the square, parallelogram, and medium triangle can all be made from two small triangles.

2. Explore the variety of pentomino shapes you can make with five squares.[4]

   (a) Take five square breadclips. Put them together so that the side of one joins exactly onto the side of another. When you make a shape, leave it. Take five more breadclips and make another shape. Keep making new shapes.
   (b) Check there are no two shapes which are the same although they are turned over or around another way.
   (c) How did you decide two shapes were the same?
   (d) How did you try to make new shapes?
   (e) What is the same about all the shapes in space?

3. Explore how squares have to be arranged to make more and more squares from the same number of matches.

   (a) Take 12 matches. Make one square. Now try to make two squares of the same size. Try to make three, then four squares. One of these number of squares can't be done. Which one?
   (b) Draw your answers.
   (c) Now take 24 matches and make one, then two, then three … up to nine squares of the same size. Which one can't be done?
   (d) Why did you decide to arrange the squares in a certain way?
   (e) Why does it help to join the squares?
   (f) When did you use a similar arrangement?
   (g) Extension: How did you know something won't work?

4. Explore ways of making each pattern block shape larger. How do you know the shape is the same but larger?[5]

   (a) Take one of each kind of pattern block. Next to it make the same shape but larger using a number of the same pattern block. Record or draw how you did it.
   (b) How do you know the shapes are the same?
   (c) Extensions: Is there another way of making the same shape but using different blocks?
       What can you say about the area of the bigger shape?
       Can you make the shape even larger? How many blocks do you think you will need?

5. Explore how to make angles using other angles of the tangram pieces.[6]

   (a) Which angles are the same, larger, and smaller? Which angles are the largest? Draw each in your book.

---

[4] Grade 2 started with four squares; square breadclips were used.

[5] Foam sets were used consisting of an equilateral triangle, an isosceles trapezium (equal to three triangles), a square, two sizes of rhombus, one of which is equal to two triangles, and a regular hexagon (equal to six triangles), a readily available set.

[6] Angles of shapes were marked by the thumb and forefinger to show size. The forefinger is rotated to line along the other arm of the angle.

Key Study on Children's Visuospatial Reasoning   37

   (b) Join the angles of the pieces together to make the angles.
   (c) How many of the smallest angle are needed to make each of the other angles?
   (d) Can you make them another way? Try making bigger angles.
   (e) Use an angle to draw angles in different ways on paper.
   (f) Extensions: Draw shapes which are different but have one of the angles the same.
   (g) Is there another way of making the same shape but larger?
       What do you notice about the shapes you used to make the large shape?

6. Explore how to make angles using other angles of the pattern blocks.
   (a) Compare the angles of the pattern blocks. Which are the same? Which are bigger than a right angle (angle on the square)?
   (b) Draw each angle in order of size.
   (c) How can we make each angle out of other angles?
   (d) How many of the smallest angle are needed for each of them? Write it down on your drawing.
   (e) Extension. Draw some shapes which have these angles but are different to look at.

7. See shapes in three different designs made with matches.
   (a) Two squares were joined at a vertex on the workcard.
      - Make the design.
      - Add two matches to make three squares.
      - Return to the first design, add four matches to make three squares.
      - Return to the first design, add four matches to make four squares.
   (b) A hexagon from equilateral triangles was on the workcard.
      - Make the design.
      - Remove three matches to get three equal shapes with four sides.
      - Return to the original design, remove four matches to leave two of this four-sided shape.
      - Return to the original design, remove four matches and leave two equal shapes with four sides but another kind.
      - Return to the original design, remove three matches and leave three triangles.
   (c) A square made from four squares was on the workcard. Make the design. Return to the original design each time.
      - Remove two matches to leave three squares.
      - Remove four matches to make two squares.
      - Remove two and leave two squares.
      - Move three matches to make three squares.
   (d) Extension: Try your own ideas.

8. Explore more about shapes by making their outlines (tangram pieces and pattern blocks).
   (a) Take one of each shape. Next to each shape, make the outline of the shape with matches. When you need different lengths, use the sticks.
   (b) Draw each shape without tracing.
   (c) Which shapes have sides of the same length?
   (d) What is the same and what is different about any two shapes?
   (e) What is the same and what is different about the triangles?
   (f) Extension: Join two shapes and make the outlines of the new shapes.

9. Explore lines of symmetry and other types of symmetry for the pentomino shapes.[7]
   (a) Guess where a shape can be folded in two so that the two sides lie on top of each other. Try it. Draw over the lines that you find make two symmetrical halves.
   (b) How can you explain the two halves match?
   (c) Are there any shapes which look symmetrical but don't fold so the two sides lie on top of each other? How can you move the piece so it lies on top of itself?
   (d) Extensions: Use pattern blocks to make designs with symmetry.
       Add a square to the pentominoes to make symmetrical shapes.

10. Explore why some pentomino shapes tessellate and why others do not[8]
    (a) Try to arrange the tiles of the same shape so there are no gaps. Will the same pattern go on in all directions?
    (b) Why do they fit together? Why don't they fit together?
    (c) Extension: Join two kinds of shapes so there are no gaps.
        Join one of each pentomino shape together to make rectangles.

A test (Owens, 1992, 1993; see Appendix B) was developed specifically for the study. Items that fitted well for an underlying trait on visuospatial reasoning based on a Rasch analysis were used for analysis. This test was deliberately designed to cover the range of areas discussed previously in reviewing the literature on visual imagery and spatial abilities but relevant and interesting to young school children. It was coloured and involved coloured stickers. It was introduced wtih carboard cutouts to match practice examples. The results of the test showed that grade 4 students reached a higher level of visuospatial reasoning than grade 2 as shown in Table 2.2.

An analysis of covariance with pretest scores and factors of gender, year level at school and different learning groups indicated a significant difference in scores for the groups in the delayed posttest ($F = 5.072$, $p = 0.026$). Furthermore, the confidence intervals of the means of the differences between delayed posttest scores of two-dimensional thinking and pretest scores showed that the mean gain scores of the students involved in spatial learning experiences were significantly greater than for students

---

[7] Each shape was printed on paper.
[8] Each shape was made from cardboard and a number given in each packet. Packets were swapped between children.

Table 2.2 Comparison of grade 2 and grade 4 students' scores

	PRE2D			PRE3D			POST2D			POST3D			Delayed Posttest 2D			Delayed Posttest 3D		
	N	MN	SD	N	MN	SD	N	MN	SD	N	MN	SD	N	MN	SD	N	MN	SD
Year 2	86	36.34	11.47	86	3.21	2.4	83	43.29	13.7	83	3.36	2.61	85	46.26	13.48	85	3.94	3.16
Year 4	99	45.56	10.38	99	4.11	2.5	96	52.77	9.68	96	4.25	2.92	93	54.89	10.27	92	5.40	2.96
t values	6.32**			2.51**			5.28**			2.13*			4.77**			3.17**		

Note. Maximum score for the 2D test was 75; maximum score for the 3D test was 11
N = the number in the group; MN = the mean score for the group; SD = the standard deviation for the group
* t values are significant, $p < 0.05$, two tailed test
** t values are significant, $p < 0.01$, two tailed test

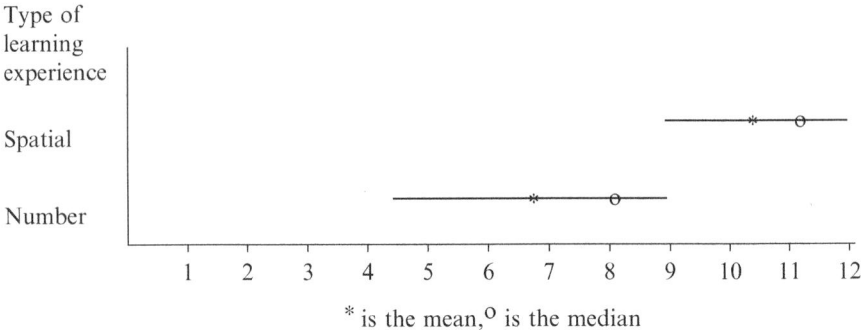

Fig. 2.1 Confidence intervals for means of the difference between scores on two-dimension delayed posttest and pretest for spatial versus number groups

participating in number learning experiences (Fig. 2.1). The learning experiences had a significant effect on children's visuospatial reasoning as assessed by this test.

Visuospatial Reasoning—Getting Inside Children's Heads

However, I also carried out a grounded theory study to explore how children were thinking during the investigations. Initially, I gave the problems to teacher education students and then to five individual children from pre-kindergarten to grade 5 in order to get spoken comments on visuospatial reasoning. Besides teaching in the three schools mentioned above, I also explored whether the findings were evident in a fourth school in another part of Sydney from a slightly higher low socioeconomic area and in a school in PNG. While some children worked individually on the visuospatial geometry lessons, others worked in small groups of three (or occasionally four). Groups in classrooms were videotaped but I also observed and videotaped[9] 12 groups of three children (from each year group, there was a group working individually although they could talk to each other and another working as a cooperative group sharing materials and findings). Following on from the problem-solving lesson each day, I used stimulated recall interviews in order to "get inside children's heads" and add to the observed behaviour and conversations. The use of materials meant that their reasoning was "out there on the table" (Richard Skemp in *Twice Five Plus the Wings of a Bird*) (Campbell-Jones, 1996; Skemp, 1989).

All incidents were replayed and analysed based on the children's descriptions and actions. A constant comparative method was used to make assessments of the nature of thinking. For example, if certain movements with materials were associated

---

[9] John Conroy, a retired mathematics educator from Macquarie University assisted with videotaping. All children were taught by myself. Lapel microphones were attached to children. To avoid class disruptions half the class working individually were taught followed by half the class working in groups on number or space problems.

with visuospatial reasoning explained in the stimulated recall of children being interviewed, then it was assumed that similar movements by another child were of this nature. The results indicated that there was frequent use of visuospatial reasoning of different kinds, some more than others.

Coding of over 1,800 incidents[10] (identified sections of actions or interactions with people or material) from all the videoclips indicated that visuospatial reasoning was involved in 540 cases and that three fifths involved holistic recognition and/or memory of visuospatial procedures but a half involved other types of visuospatial reasoning (Table 2.3). It should be noted that an incident could involve more than one kind of reasoning.

The study found imagery was important in reasoning, in creating new concepts, and more generally in directing the actions of children. The results supported the perspectives of Lakoff (1987) and Johnson (1987) who argued that imagination was a complex, embodied basis for making meaning about concepts and propositional judgements. Such a view suggests that visual imagery plays a pivotal role in conceptual development (Shepard, 1971; Tartre, 1990b).

Kaufmann (1979) has suggested that visuospatial reasoning occurs with parallel mental transformations enhancing problem solving more than sequential verbal processing. According to Kaufmann (1979) verbal processing is too bound to convention to allow for new ideas whereas visuospatial reasoning is

> more idiosyncratic, varied and flexible as to rules, and this fact makes it potentially more adaptable as a representational system for the transformational activity needed in solving tasks which possess a high degree of novelty. ... [This is not the] traditional Gestalt view of problem-solving as consisting of an immediate restructuring of the perceptual field. On the basis of our findings, we hold the view that the solution to a problem is obtained by building an analogous situation from other areas of visual experience. This process we regard as mediated by transformational activity effected through the visual symbolic system. (p. 79)

This kind of interpretation of problem solving provides support for the conclusion, which is suggested by the data in the present study, that the role of visuospatial reasoning is crucial in the problem-solving process. Dreyfus (1991) is another to argue that visuospatial reasoning plays a significant role in higher levels of thinking. According to Dreyfus,

> visual reasoning is not meant only to support the discovery of new results and of ways of proving them, but should be developed into a fully acceptable and accepted manner of reasoning. (p. 40)

This study illustrated the variation within visuospatial reasoning and how visuospatial reasoning develops and assists learning. While simpler names were used for in-school programmes based on this research, descriptions of different kinds of imagery were later confirmed as a useful tool for teaching and assessing (see later in this chapter).

---

[10] The videorecorded actions and interactions were described and spoken words recorded. An incident was a small self-contained segment of learning that could be described. After analysis, these tended to be a small cycle (context, context providing input, child or children's thinking, response affecting context), many of which formed a cycle within learning. (See Fig. 2.17 on responsiveness in problem solving towards end of this study.)

Table 2.3 Percentage of incidents involving different types of visual imagery

	Total number of incidents	Percent with visuospatial reasoning	Type of reasoning as % of incidents with visuospatial reasoning					
			Holistic	Dynamic	Action	Episodic	Pattern	Procedural
Tangram shapes	293	60	27	3	12	3	9	46
Pentominoes	218	47	35	8	11	1	14	31
Enlarging	274	76	27	6	19	1	26	21
Squares from matchsticks	260	30	33	9	4	3	34	17
Tangram angles	141	45	42	16	6	5	6	25
Pattern block angles	120	37	22	16	0	4	9	49
Outlines	135	81	45	2	6	5	2	40
Matchstick designs	85	52	36	0	11	0	18	36
Symmetry	198	45	34	4	1	3	9	49
Tessellations	109	64	16	0	28	0	41	16
Total	1,833	537	317	64	98	25	168	330

Holistic, Concrete, Pictorial Imagery

Students using concrete pictorial imagery (as named by Presmeg, 1986) tended to recognise the whole shape but some would make a shape but not hold an image in mind, and some would not recognise the configuration until it was completed. In the pictorial form, the image was often given a name that corresponded to a real-life object. For example, Michael,[11] in kindergarten, frequently named the pentomino shapes, "That's a cup" (for the C shape), while Sam in grade 2 named a configuration of tangram pieces as a sailing boat. This natural tendency helps to place the names of shapes into a wider ontological perspective. Indeed the pentomino activity especially helped children to realise that there were two-dimensional shapes without names or symmetry. This was a significant step in conceptualising the meaning of the word "shape". When Sam was making outlines of the trapezium and the parallelogram, he was pushing the pieces as if he were trying to get the pieces into place so the configuration matched his image. Holistic imagery generally did not enable students to recognise a lack of proportionality when they were making a trapezium or a parallelogram that was not similar to the given shape.

When students had made one large square with 24 matches and then had to make two or more squares with the matches, it was clear that often they made decisions on the basis of visual stimuli, with no counting or calculating being used. They seemed to use visuospatial intuition as a basis for predicting whether the required number of squares could be made with the matches that were left. Similarly, in the tangram activity several of the students, who had made the large triangle in two or three ways, responded very quickly to the question on its area by saying that four of the small triangles were needed to make a large triangle; it was only later that they began to reason verbally from their image. This visuospatial intuition is raised again in Chap. 5 where I discuss visuospatial reasoning in PNG.

James had a clear conception of the lengths of sides and this was strengthened by actually comparing sides. Later James and Victor made shape outlines for the tangram and pattern block pieces when Victor explained that James had not made a right-angled triangle, as James had thought, but that he had just made an equilateral triangle in another orientation. Victor himself had made the right-angled isosceles triangle with the long side horizontal and he checked it with the tangram piece which he put on top ("a lid", he called it). This discussion between James and Victor helped James to perceive the right-angled triangle in both orientations.

One developing visuospatial reasoning skill was the ability to recognise shapes in different orientations, including the more uncommon pentomino shapes and the right-angled triangle in unusual orientations (see Kathy, para. 4.03; James and Victor, previous paragraph). The problems themselves encouraged the use of this skill. For example, once students realised that two pentominoes were the same, they more readily avoided or recognised another pair of congruent shapes, either because the meaning of the problem was clearer or because they had developed that

---

[11] All names are pseudonyms.

visuospatial reasoning skill. Those students who had already imagined actions on shapes in their minds, especially the flipping of shapes, tended to manipulate materials more fruitfully.

Although James was part of a cooperative group, he began the second spatial activity somewhat competitively. He was thoroughly involved in making new shapes from four square breadclips and then in making pentominoes. He also enjoyed commenting and in other ways expressing his achievements and feelings of pleasure. ("Names" have been used for each pentomino shape and illustrated.)

Excerpt 1

1.01	James continues to count how many he has made, comparing his number with his friend's number.	
1.02	Using four squares, he makes a "Z", checks that it is all right, and then makes a "cross" avoiding repeating the Z.	
1.03	His friend points out "it is half her", so he changes it to a "T".	
1.04	He begins with five squares deliberately positioning the pieces to make a Z. Then he makes a "lineZ".	
1.05	He notes his friend's shape saying "yours has three columns. Mine has two; she copied me". (Each made the lineZ in different orientations.)	
1.06	Despite the teacher suggesting that they work together, he keeps making shapes quickly and happily, commenting on how well he is going. He uses a tactic of beginning a new shape with "three in a row". He counts his shapes and says "I'm beating her". He knows what he is making before he completes the shape, showing joy before he finishes making the shape. He places three in a row and claps as he makes a "C".	
1.07	He cannot recognise the "odd" shape in different orientations despite moving his body to assist orientation. He changes the shapes to make the easily recognised shapes "L3" and the "square-like shape", comparing the incomplete shapes with his short-term memory images of those he has made (that is, he is not physically glancing at his shapes).	
1.08	He changes his tactic from starting with three in a row to beginning with four in a row. He makes the "L4".	
1.09	He quickly grabs the last five breadclips so that he can make another shape.	
1.10	He wants to make a car but ends up with lineZ, globally deciding it is different and says "Oh, I can't make any more". His activity wanes when the teacher asks if they can find any shapes that are repeated in the group's work.	
1.11	He recognises the repeated lineZ and L4.	

Fig. 2.2 Children attending to parts of shapes. (**a**) Michael attends to the sides of the shape. (**b**) Children attend to the angle by marking and turning their finger away from their thumb

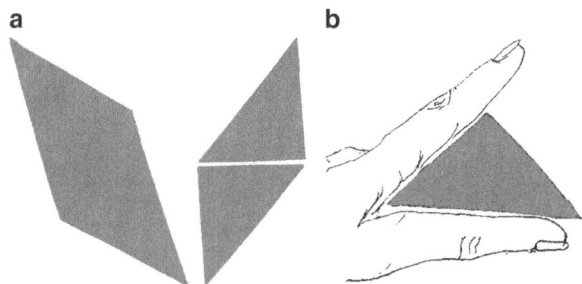

Similarly children marked angles with thumb and forefinger. James' affective responsiveness, visuospatial reasoning, and tactics are evident.

Michael in kindergarten was joining tangram shapes to make another shape. He had the two small triangles joined to make the parallelogram but he was concerned that they were not the same size when he put them next to each other although he could see they were the same when placed on top. He was pointing to the sides and saying, "no, they aren't, see" as he points to the two unequal sides near each other (Fig. 2.2a). In the next few moves he matched sides of the different shapes, disembedding the side from the shape and realising sides could have the same length although the overall shape and area were the same and a shape could have the sides with different lengths. Similarly children marked angles with thumb and forefinger (Fig. 2.2b).

Holistic concrete imagery assists students to learn about concepts such as a shape does not have to have a name or be symmetrical, or a side of a shape is not the size of a shape. In addition, holistic concrete images can play a part in visuospatial reasoning especially as a basis for size estimates and checks. The parts of an image may be recognised as parts of an everyday object or picture but they may be the geometric features such as lines and angles. However, it is soon enhanced by dynamic visuospatial reasoning and concepts as illustrated in the example below of Victor recognising equal angles on shapes that are turned. The important skill of disembedding parts from the shape and imaging concrete objects or pictures in two or more ways depends on past experiences, the current problem, and on which aspects of the objects or pictures are taken account of and which are ignored (Thomas, 1978).

Dynamic Visuospatial Reasoning

Dynamic visuospatial reasoning[12] is significant in problem solving. In the past, teachers have often regarded dynamic imagery descriptions, for example, "a rhombus is a pushed-over square", as inadequate and unhelpful. By contrast, this study indicates how students have made connections between images and associated concepts through dynamic imagery. A common example of a verbal description arising

---

[12] Presmeg (1986) referred to dynamic imagery as involving movement in remembering formulae such as moving letters in expanding a product of two binomials.

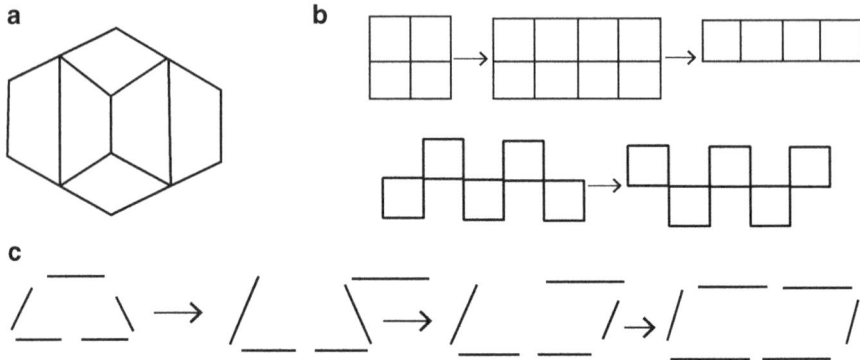

Fig. 2.3 Examples of dynamic visuospatial reasoning. (**a**) Sam's hexagon "like a square". (**b**) Michael's rectangles, and M to W. (**c**) Peter's trapezium to parallelogram

from visuospatial reasoning is the use of the phrase "a pushed-over square" for a rhombus. In this study, many students thought of shapes as being modifications of other shapes; for example, Sam described an irregular, symmetrical hexagon as "It's bigger. ... It looks a bit like a square" (Fig. 2.3a).

Michael in kindergarten has already established dynamic visuospatial reasoning. Before making something, he stops and thinks and later he says the shape that was in his mind.

Excerpt 2

2.01	Michael is asked to make shapes using four squares. He makes a square and when he is given four more squares, he adds them on, saying as he starts "It's a rectangle".
2.02	He then proceeds to use another four squares and says "I know. I could make a longer triangle, I mean rectangle". As he makes another shape he smiles and says "a rectangle. I made a skinny rectangle" (Fig. 2.3b).
2.03	Next he makes a shape and says "an icecream cone" and scoffs that others would call it a diamond. When I challenge with "but that is a square", he says "we made the square" (now modified as a rectangle). At first, he decides that the diamond is not a square but then concedes, commenting that names can be confusing.
2.04	He is given four more squares. "I know". He fiddles with them under the table and, having decided to make an M, asks for another square. He makes an M with five squares on the table. "It's a bit upside down for you". He modifies the M so that I can see the M on the other side of the table. "An M for you, a W for me" (Fig. 2.3b).

Michael's extension of the square to a rectangle (para. 2.01) and his use of symmetry to change the W into an M (para. 2.04) are examples of dynamic visuospatial reasoning. He found no difficulty in regarding both a thin and a fat rectangle as examples of a rectangle, and he appears to have decided on making the thin rectangle as a result of his visuospatial reasoning of the larger one becoming thin (para. 2.02). Nonetheless, a degree of analysis of a shape is needed with this form of imagery if one is to be fully successful in making and describing shapes.

Peter, in grade 4, also made use of dynamic imagery when he transformed a trapezium to a parallelogram by sliding one parallel side along and swinging the lateral match across to meet it (see Fig. 2.3c).

Dynamic visuospatial reasoning is an important step in extending prototypical images and concepts. For example, visuospatial reasoning provides for a diversity of triangles if one imagines moving the vertices of an equilateral triangle to other positions (see James and Victor's discussion about triangles above). Some properties of a square are invariant within a rhombus, but not all of them (Sam's efforts above). Similarly, the extension of a square to form a "rectangle" maintains some properties (for example, right angles and parallel sides) but deliberately changes others (see Michael's description and action above).

Emphasis has been given to the use of dynamic visual reasoning in many computer-based geometry experiences. In dynamic geometry software, for example, dynamic changes can be made to shapes, and students can see that the changes which occur in some parts of the shapes affect some properties while other properties remain constant. In addition, the basic notion of partial inclusion in visual reasoning permits connections to be made between shapes (Owens & Reddacliff, 2002; Tartre, 1990b). Furthermore, dynamic transformations can lead to property recognition; from this perspective computer-software microworlds which enable transformations to be carried out easily can be useful. However, one of the advantages of equipment like that used in the present study is that shapes can be changed physically. There is the disadvantage, though, that with static shapes there are no "in-between" positions. This disadvantage can be overcome to a certain extent with cutting, folding, uncovering, superimposing, and using a movable perimeter to produce "in-between" states. My favourite piece of equipment is a loop of thin elastic which each pair of children can use to make different triangles or different quadrilaterals. The use of a loop of string makes a thought-provoking comparison (see activities later in the chapter).

Visuospatial Action Reasoning

Presmeg (1986) classifies imagery, which has a strong emphasis on muscular activity, as "kinaesthetic imagery" but Wickens and Prevett (1995) suggested this is spatial imagery, rather than visual imagery. In fact, Kim, Roth, and Thom (2011, p. 207) have noted the following:

(a) Gestures support children's thinking and knowing
(b) Gestures co-emerge with peers' gestures in interactive situations
(c) Gestures cope with the abstractness of concepts
(d) Children's bodies exhibit geometrical knowledge

However, in the cases described now, the visual imagery holds spatial actions and whether or not there is spatial imagery, there is visuospatial reasoning.

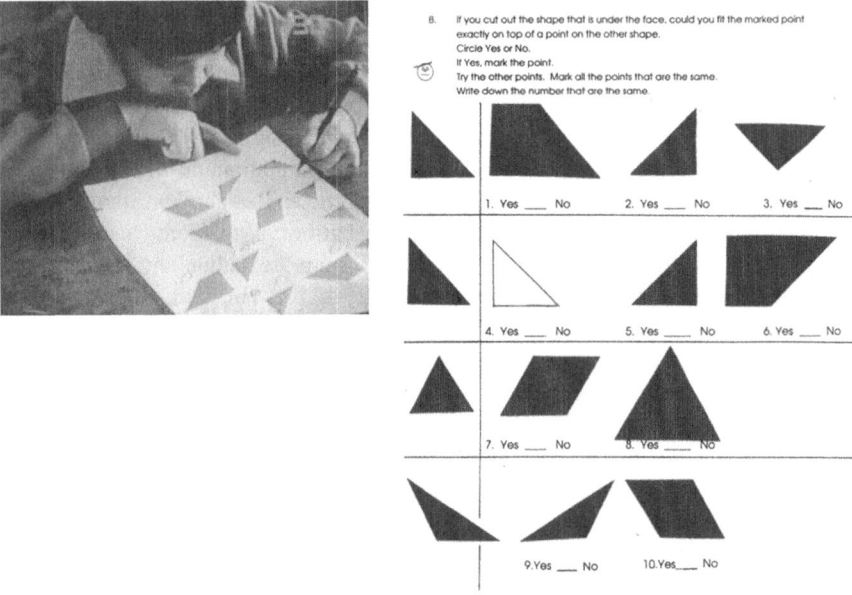

Fig. 2.4 Victor, grade 2, recognising equal angles on different shapes and on shapes in different orientations

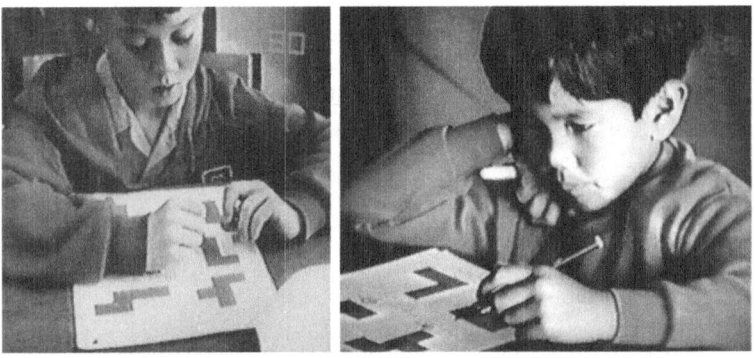

Fig. 2.5 Peter and Victor mentally folding the pentomino shapes to form an open cube

Action imagery is also important in children recognising parts of shapes in different orientations. When children were doing the test, it was clear they were thinking about the angles in different orientations as shown in Fig. 2.4 where Victor is concentrating on the angles of the shapes. Later when asked how he first did not recognise the right angle on the triangle in a different orientation but then self-corrected, he said, "because at first, I didn't recognise it as the same shape" and he turned the paper to indicate how he recognised it in his head.

When children were doing the test item that asked whether the pentomino shape on paper could be made into an open box, you could see them thinking and doing slight hand movements like Peter and Victor in Fig. 2.5.

In viewing the playback of her work with pentominoes, Sally who had just finished grade 4 commented, "In my mind, I pictured my hand moving the pieces around the shape". This clearly matched my observation of her efforts to obtain new pentomino shapes—she was reasonably systematic in moving the pieces around partly made shapes searching for new pentominoes. While solving the problems, she said that she was using "ideas in her mind" but she did not give the imagery names as she might have done if she had evoked whole shape images. In fact, this was commonly noticed with children making pentomino shapes and a comparison between adults and children raises some interesting points about experience on forming pentominoes (Owens, 1990).

In this study, I asked whether certain pentomino shapes prompt the making of other shapes? If so, was this due to the relative strength of a shape in visuospatial memory, the modelling of shapes with names or symmetry, the grid analysis of the shape, or the simplicity of the shape? The task was completed by 52 adults and 12 children. To investigate the differences for adults and children, each shape was given a value based on the order in which it was made (1–12 with a value of 13 if it was not made). The median scores for each group—adults and children—were calculated and correlation coefficients between shapes were calculated (see Fig. 2.6). The square-like shape was made early by the children but they often discarded it initially because "it really wasn't a square". Once accepted as a shape, it was frequently remade by children, often in different orientations, but they would recognise it as the same and change it. This shape correlated highly with the + shape as children were deciding what was acceptable as a shape, the T shape was made early, often first. Children tended to begin with three in a column (see left diagram in Fig. 2.6b) while adults often started with four in a column and hence the line (five in a column) and L with four were made early by the adults and often quite late by the children. The other common starting point was the three with two in one column and one in the next (see top of Fig. 2.6b). From both starting points shapes such as a T were made. Children often started the same way each time and when they seemed to have exhausted ideas, they would switch to this shape with either three or four in a column. The high correlations between the T and LineZ, W and LineZ, L4 and C, and W and C suggest that the movement of one square from the previous configuration to make a new one was common. From the L3 many shapes were made; the C gave some pleasure as a recognisable shape.

It was evident that certain tactics were used as illustrated in Fig. 2.6 as well as in the key study. For example, Jodie in grade 2 began with three squares and made shapes from this base until she could make no more. Then she tried another starting combination. Besides the figural similarities between shapes, the comparison of adult and children's data would suggest that imagery, short-term memory, strategy, and propositions (e.g. what constitutes a shape) influence order of appearance of different shapes. Thus from the observational data and the above analysis, it seems that experience influences children and adults' decision-making indicating an eco-cultural perspective to visuospatial reasoning is appropriate.

Students knew that pieces were to be joined or moved in a particular way even though they did not know the entire procedure to make a required shape. Once Kathy, in grade 4, was comfortable with the problem, she made deliberate moves to

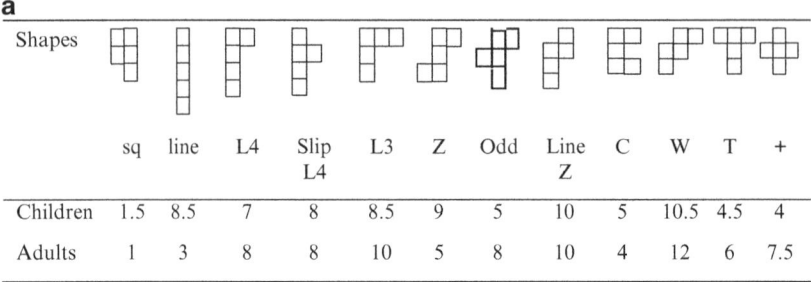

Fig. 2.6 Mean order of pentominoes being made and correlated data for adults and children. (**a**) Mean order of shapes being made by children and adults. (**b**) Correlations in order of appearance of shapes

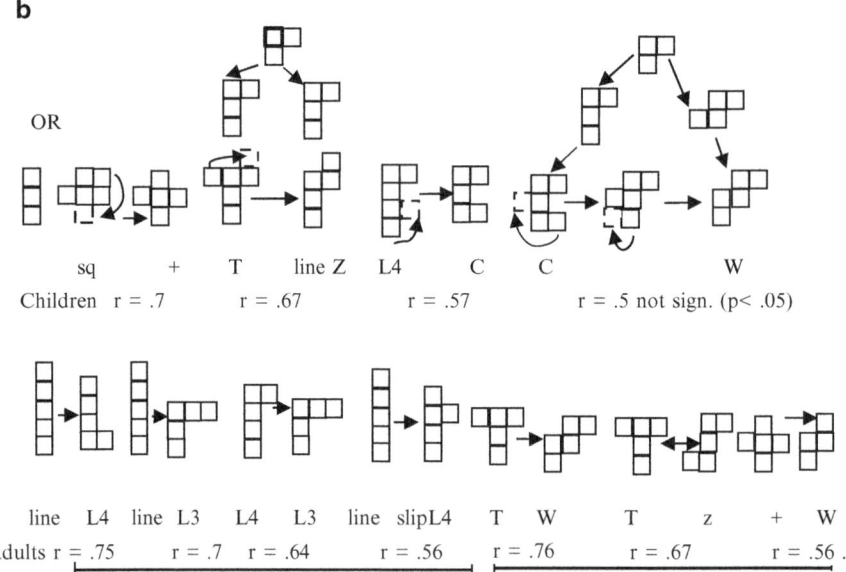

Fig. 2.7 Kathy's large triangle and rectangle from the tangram pieces

create various shapes. In viewing the part of the video which recorded her making the large triangle with smaller pieces of the tangram set (Fig. 2.7), she commented, "I was sort of moving them around in my brain. … Like I was just seeing the triangle in my brain moving and me putting the square there so I got it". In fact when

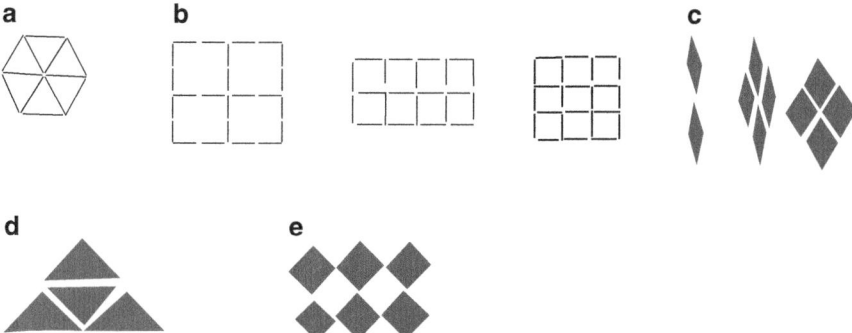

Fig. 2.8 Examples of pattern visuospatial reasoning. (**a**) Peter's hexagon. (**b**) Sally's 4, 8, and 9 squares with 24 matches. (**c**) Pattern repeated for both types of rhombus. (**d**) Right-angled triangle pattern that was also used for equilateral triangles by Sally. (**e**) Lena's pattern of squares

she was fitting in the last little triangle, I stopped her messing it all up with a "no, no" which gave her time to imagine it turned and placed correctly in place.

Action-based visuospatial reasoning occurred more frequently once students began to develop and implement tactics for solving problems. This reasoning was supported by concepts relating to the effects of operations or transformations on pieces and linked to dynamic or pattern imagery. Action imagery was a common means by which students solved the physical problems in this study. Action imagery is closely related to physical manipulations and to operational concepts. As action images were combined, procedural imagery was likely to develop. At this point, one of the paradoxes of learning occurred. While procedures are being developed, a high level of thinking takes place, but once they become automatised, and are reduced to algorithms, the level of thinking is reduced.

Pattern Visuospatial Reasoning

Pattern imagery[13] was evident when Sally carefully counted as she made a rectangular array of eight squares with the matches (Fig. 2.8b), and during the video playback stated "Like the picture was in my brain but it didn't work". In fact, she had interpreted the problem as meaning that the squares had to be in a square or a rectangle; she recalled "you couldn't have odd shapes like that—they had to be square or rectangle". Interestingly, she quickly succeeded in making both four and nine squares (Fig. 2.8b). In fact, students in PNG, who were less experienced

---

[13] Presmeg (1986) used the term "pattern imagery" and illustrated it with symbolic and numeral patterns.

with structured materials, frequently made the nine squares first and then tried to make the four squares. Perhaps the strength of the image of tessellated squares resulted from familiarity with squares (or diamonds) used in their pandanus weaving and string bag designs, providing support for an ecocultural perspective on visuospatial reasoning (see Chap. 5 for more details on PNG designs and coloured pictures).

Sally also used pattern visuospatial reasoning. Sally chats as she manipulates the tangram pieces. She gives a clear description of how she remembers about the triangle pattern (Fig. 2.8d). "Last year we made a Christmas tree out of triangles. That is how I know you put one up and one down. Four of these small triangles would make the larger one like this". This comment represents the use of imagery associated with a specific experience, and relates to what Gagné and White (1978) called episodic memory. Clearly there was a pattern involved as well as actions such as slides and rotations of triangles.

Similarly, Victor in grade 2 explained how he knew that three triangles made up the trapezium in the retention test, by referring to his making of the shape earlier with the pattern blocks (a similar pattern as the right-angled triangles). When Jodie, in grade 2, was asked why she had been able to make the triangle with pattern blocks so quickly, she said that she had remembered that there was a similar task before (in the pretest, an equilateral triangle was illustrated with appropriate lines for folding into a triangular pyramid). Jodie, Jonah, and other students called the pentomino cross "a box", relating it back to the net for an open box given in the pretest practice item. Pattern visuospatial reasoning was used by Peter when he was having difficulty making the hexagon outline with matches (Fig. 2.8a). He commented, "I know, I'll make it like the other day". And he proceeds to add one triangle next to the other as he had done with the pattern blocks and designs-with-matches problem.

In the tangram problem, students remembered patterns such as the configuration of the square and two triangles for making the large triangle, and they relied on this pattern when they rebuilt the large triangle for the class. Pattern visuospatial reasoning became important for students making the enlargement of the second rhombus. Generally they positioned the pieces to repeat the pattern of the previous enlargement rather than trying other possibilities. The following excerpt illustrates Sam's use of pattern visuospatial reasoning (Fig. 2.8c) especially when he explains to his friend but he also generalises about how pattern block shape enlargements can be made using pattern knowledge by using four similar pattern blocks to make an enlarged shape. It should be noted that Sam and his collaborative group have English as a second language; his language is the same as one of the other group members but he does not use this language in class.

Excerpt 3

3.01	Sam discusses whether they can make a square or not with trapezia and suggests that you can only make a square with squares. He makes a 3×3 square and claps.
3.02	He says "I'm going to make a triangle", and he collects triangles. He nearly has the triangle and knows how to complete it. When asked how he did it, he says "because it is a triangle and you make a triangle with little triangles and the corners are sharp so you can make it like that".
3.03	He turns to the blue rhombi (with 60° and 120° angles) and says, "You can make a square with these. Oh, no, you can make a diamond".
3.04	He takes two rhombi and touches points symmetrically, but misses seeing the enlarged rhombus and joins the sides.
3.05	He listens to the teacher talking to his friend and then concentrates on his own work and quickly puts pieces together to make the rhombus. He is happy, and the teacher praises him and asks if it is the first time he has made it and he says, "Yes".
3.06	He then describes to his friend how to do it "You put this here and this here" (touching the points of the rhombi). He goes on to describe how to make the triangle, "Up and down, up and down" to help his friend make her triangle. He is pleased with himself.
3.07	He now selects four trapezia and places three together but cannot get the fourth one in correctly. After a while he leaves it as a five-sided shape saying, "It looks like a trapezium", clearly knowing that it is not one yet.
3.08	He then moves both end pieces, leaves the symmetrical "butterfly", and then makes two joined hexagons… and then places triangles on the sides to make a long hexagon. The teacher asks him what is different about his hexagon and the yellow one. "It's bigger".
3.09	The teacher runs her fingers along the sides and asks about them. He says "It looks a bit like a square" [he is referring to the angles of a square]. When the teacher asks about his friend's parallelogram, made from joining six rhombi, he says "It is bigger and kind of like a square". Returning to the hexagon, he says, "It is unstraight". When the teacher asks about the sides of the parallelogram, he says "not pointy".
3.10	He places two trapezia together and asks his friend what it is called and is told "a hexagon".
3.11	The teacher asks him if he can make the brown shape (rhombus with angles of 15° and 135°). He says he has made it, pointing to the blue one but she says, "No, a skinny one". So he collects the narrow brown rhombi and quickly follows the same pattern to make it. "I'm the best in the world", he laughs.
3.12	He remakes the squares and then he remembers he still has to make the red one (the trapezium) but he thinks no one can, if he can't think how to make it. He sees that his friend is making the red one with rhombi and triangles, so he tries the same and makes a trapezium but not of the correct proportion. He realises and tries to adjust it without success.

Sam described or demonstrated several patterns and used pattern visuospatial reasoning to make the second rhombus (Fig. 2.8c), and this was supported by his notion that larger shapes were made by joining four similar smaller shapes. He also noted when a shape had some features like but not the same as another shape. For him, the angles of the shapes were more dominant than the lengths of sides, a feature noted in other students' work.

A group of children in PNG also seemed to make considerable use of pattern visuospatial reasoning. This was their first experience with pattern blocks (although the school had some dusty attribute blocks which contained a parallelogram, square, oblong, triangle, and circle).

Excerpt 4

4.01	Lena puts squares in a pattern, touching corners but not joining sides (Fig. 2.8e). Susan puts the brown rhombi with sides together and then touches the middle points and puts the third in place before disarraying quickly. The teacher says, "Close. You nearly had it. Do it again". They do and Nora places in the fourth rhombus.
4.02	Meanwhile Lena collects the blue rhombi, joins two with the 120° angles at the top, and then joins the third and completes the enlarged rhombus with the fourth. But then she spaces them apart and puts the rhombus the other way as if she didn't recognise the rhombus. … (Fig. 2.8c).
4.03	The teacher (not knowing what they have already made) asks if they have made "this diamond" (blue rhombus) yet. There is no reply but they don't make it and instead Susan collects up the narrow rhombi and puts four of them together but as an arrow. Susan looks up, satisfied and perhaps baffled. …

The initial pattern with squares was reminiscent of bilum (string bag) patterns (see Chap. 5 for examples). It is interesting to note that Susan did not really establish a firm pattern image of the rhombus after making it the first time, and that Lena did not recognise her first rhombus. Lena tried to have three obtuse angles on the shape (perhaps because of the usual positioning of the "diamond" with the obtuse angles to the sides). Nevertheless, she did finally succeed in making the patterns.

A specific type of pattern visuospatial reasoning is the image of a grid. Sally, like other students, used this form of imagery in the pentomino problem when she systematically searched for new shapes by imagining where the pieces would be on the grid in order to make a new shape (see Fig. 2.6). Recognition of the structure of the pattern is important for imagining the covering of a rectangle with tiles and understanding the ideas behind area especially linking the equal rows to calculating areas (Outhred, 1993). Simultaneously students were showing similar diversity of approaches with some evidence of becoming more efficient in the paper-and-pencil test used in my study (Fig. 2.9) (Owens & Outhred, 1997, 1998).

An analysis of the responses to the items on the test on covering areas revealed some interesting features about drawing and visuospatial reasoning. Figure 2.9 gives the items of the test referring to covering areas and a picture of a child completing the test. Children who drew on their worksheets were considered for analysis. Table 2.4 provides the percentages of responses indicating that covering with triangular tiles (Items 3, 4, and 5) was more difficult than with rectangular (including square) tiles (Items 2 and 7). The difficulty was particularly marked when the shape to be covered was the unfamiliar trapezium shapes (Items 4 and 5). On the first testing more than half the students thought the trapezia could not be made by tessellating the given triangles, and less than a third of them could give the correct

Visuospatial Reasoning—Getting Inside Children's Heads

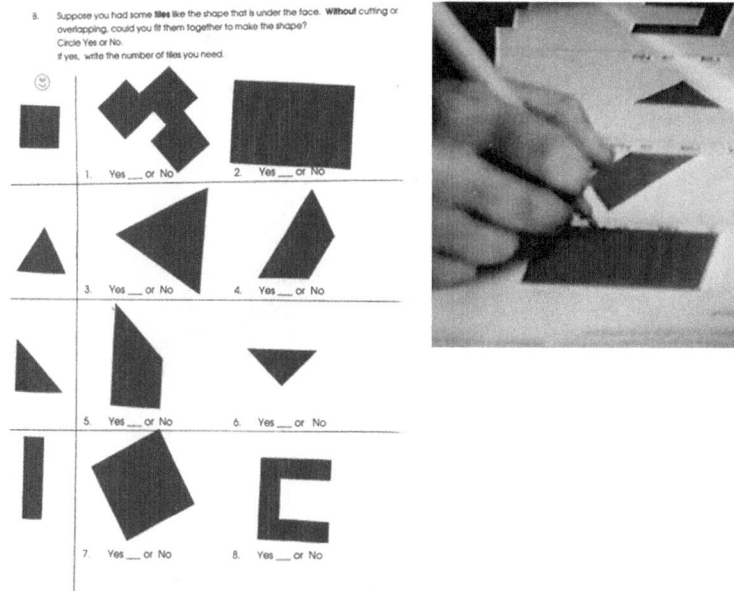

Fig. 2.9 Worksheet for tiling shapes (Form S) and grade 2 child completing the items

Table 2.4 Percentage of responses of children on pretest and on delayed posttest

Item	"No" response		"Yes" response, wrong number		"Yes" response, correct number	
	Pretest	Delayed posttest	Pretest	Delayed posttest	Pretest	Delayed posttest
2	30	21	42	39	29	40
3	45	35	27	32	28	32
4	52	36	19	16	28	48
5	57	53	15	17	27	30
7	33	26	28	25	39	48

Note. All these items (2–7) could be covered. Items 1 and 6 had no drawings so are not included in the table and 8 was a "no" answer

number of triangles. After teaching, students performed better. Although many students seemed to realise that the square (Item 7), the rectangle (Item 2), and the equilateral triangle (Item 3) could be made by tessellating the given tile, they were unable to visualise the tessellations to work out how many tiles would fit. For both the equilateral triangle (Item 3) and the square (Item 7), many students wrote 3 or 5 tiles as their answer. For the rectangle (Item 2), common answers were 8 and 9, but larger answers were also given which suggests that some students disregarded the size of the tile. Students' drawings frequently indicated difficulties with size estimates.

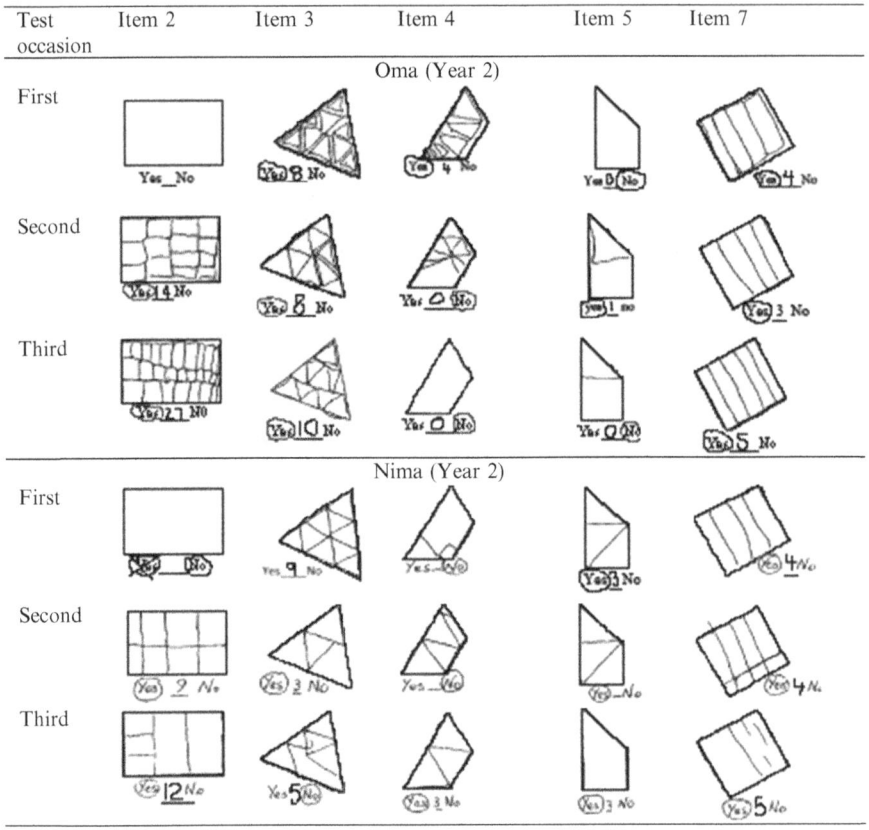

Fig. 2.10 Examples of children's responses to test items on covering areas with tiles

Figure 2.10 illustrates some of the responses from children in PNG. Students' responses were influenced by their cognisance of the following: (a) maintaining tile size, (b) covering without gaps or overlaps, (c) aligning tiles, (d) matching features of tiles such as angles, sides, and the triangular parts of the trapezia, and (e) relating the diagrams to various activities encountered during the learning experiences. Students needed to imagine or draw the relevant tessellation and to be aware of its structure. The triangle and trapezia tessellations were more difficult than the rectangular case as students had to consider the orientation of the tiling unit. Finally, students needed to be aware of the limitations of their own drawings. Students first considered covering a region with tiles by filling in from the sides and corners. Gradually they became more systematic, aligning tiles accurately and attending to features such as size and shape. Participating in activities or doing the test appeared to help students but not necessarily if they were remembering the visual image only. They might have been visualising as a picture rather than as a grid. In the interviews, several students who had been involved in the activities with concrete materials spontaneously remarked that they had made the isosceles trapezium from equilateral triangles in class and with their overall improvement from pretest to delayed posttest, it seems they had episodic imagery (Gagné & White, 1978).

The development of the grid structure for covering the rectangle with square tiles is particularly worthy of discussion. As an emerging strategy, a child might draw one to a few squares somewhere inside the rectangle, and then might align a square with an edge or corner of the rectangle. Children then tend to draw the squares one at a time in a row, row after row. There is a tendency for the rows to slant and narrow and for squares to get smaller. Children often recognise that there are too many squares in the drawing and may or may not discount the drawing as giving the possible answer. In some cases, children chose not to draw but could give the correct answer. Size and perpendicularity of lines improve as children continue to draw individual tiles until they attempt to draw a grid structure, partially or fully (Owens & Outhred, 1998). Occasionally, it is evident that the child's thinking is not fully reflected in the paper-and-pencil test score. This is evident in Oma's triangles on trapezium and her attempts for the rectangular tiles. It is also clear that she put more weight on her diagram when reasoning than what might have otherwise made sense.

By contrast, Nima discounted the small parts of square tiles on the rectangle in her second attempt. Her attempt to enlarge the triangles resulted in a less favourable structure than on her first attempt. Nevertheless, there is clear development in her attempts across all the diagrams based on various reasonings. For example, the tiles must not overlap (rectangle tiles) or my diagrams are not good enough to decide as she used a good sense of size and pattern in her mental imagery. However, she seemed to have difficulties in counting. Children were not required to draw and this part of the study used only examples where children attempted to draw but there was alignment with the simultaneous study by Outhred resulting in the findings presented in our joint papers and discussed above (Owens & Outhred, 1996, 1997).

Procedural Visuospatial Reasoning

The use of procedural visuospatial reasoning can be associated with parts being deliberately placed in relationship to each other. For example, in making the large triangle with the tangram pieces, some students deliberately turned the square so that the right angle of the square matched the right angle of the large triangle and having done that they then placed the small triangles. In comparison, less experienced students tended to place the square on the large triangle so the bases were together. Other less experienced students made many trials of the possible positions of pieces. This was partly due to the dominance of the features such as similarity and horizontal lines (see Fig. 2.11).

Students also considered overall size of area. Unlike Wheatley and Cobb's (1990) claim in their study with early covering, overall length was not dominating but rather overall size (area). Kathy, for example, stated during video playback of her covering the large triangle of the tangram with the smaller pieces, "I chose it (the square) because the others would not fit, they were too big [sic]". Kathy meant the other pieces would not cover enough and she went on to position the square so

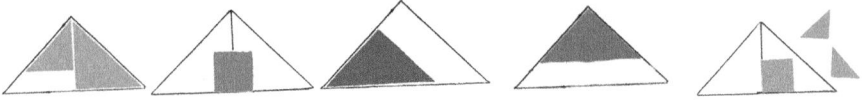

Fig. 2.11 Dominance of similarity and horizontal lines in initial trials before visuospatial reasoning improved

that the triangles also fitted. Visual analysis often began with students joining pieces together by matching angles and equal sides. This led to further analysis and subsequently to successful completion of the task. (See Fig. 2.7 which is the large triangle that Kathy made.)

Through their actions, children often had imagery of procedures in their minds, and this procedural image they used as they deliberately positioned pieces to remake shapes. Remaking the large triangle with different pieces (three only) was easy for most students. When a shape was made from a larger number of pieces, the students often remember part of the procedure. This was common with the square made from five pieces. Colin, in grade 2, remembered how to position two pieces but then had to re-image the rest; on the other hand, Tess and her group, in grade 4, remembered how all the pieces needed to be positioned. In fact, Tess' making of the square, with some help from her friends, indicated how action imagery developed into procedural imagery.

Excerpt 5

5.01	Natalie makes the large square (Fig. 2.12a) and she, Tess, and Damien decide to cover it. Tess puts the square into the corner and proceeds, saying "Wait, wait". (She completes the stages shown in Fig. 2.12b–d.) However, both Damien and Natalie also see the places for the triangles. She then shifts the square across (Fig. 2.12b) and suggests that they need another square, but they only have the medium triangle.
5.02	Damien removes the pieces and begins with the medium triangle in the corner, places the parallelogram next to it, but then the square won't fit, so he removes the square and parallelogram, and then the medium triangle.
5.03	Natalie suggests they put the parallelogram on the side, so Tess picks it up but returns to putting the square into the corner (Fig. 2.12e), and they remake the first configuration in another orientation.
5.04	Tess places the parallelogram against the medium triangle and then slides it across to the corner with the triangle (Fig. 2.12f–h). She flips the triangle as she moves it away. She continues to reposition the parallelogram on the large square.
5.05	Natalie says "perhaps if you turn them over". Tess places the parallelogram into the corner as she had before (Fig. 2.12i), seeing the various spaces that are left. Natalie gets bored and wants to use the book.
5.06	Tess picks up the square and fills in the top section carefully (Fig. 2.12j). She then collects the small triangles and restrains Damien from placing the medium triangle correctly, taking it from him. Although she is thinking she is unsure of herself, relying on action imagery rather than a completed pattern picture.
5.07	She flips the triangle into position (Fig. 2.12k) so that the point completes the right angle with the parallelogram. "Yeah, yeah", says Natalie. They all see where the small triangles will fit. Tess sits back, content.

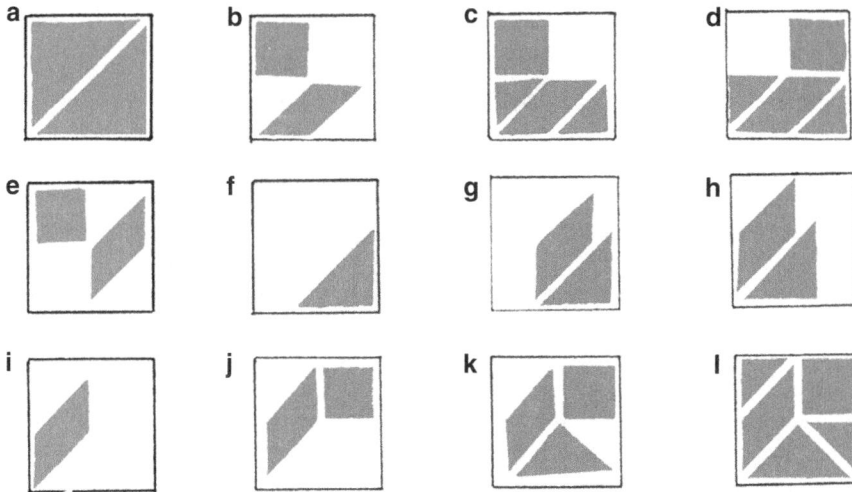

Fig. 2.12 Using action imagery to develop procedural imagery

Most of the positioning was done by Tess but Natalie, who made the initial square with the big triangles, also made some comments. Towards the end Damien began to manipulate the medium triangle, but Tess took it from him and positioned it herself. Certain procedures were more common than others. Initially, there was a tendency to match the right angles and to choose pieces to cover areas, but then came the realisation that two angles could be used to construct the right angle (para. 5.04, Fig. 2.12f). The juxtaposition of the two pieces and the observation of the ways that areas could be filled by the shapes assisted later problem-solving attempts. When the design was spoilt, the students had no trouble in reassembling the design, even though it was slightly rotated. The deliberate positioning of pieces into corners or against sides, and the checking of spaces that were left gave rise to the use of procedural imagery by all three students, whether they were watching or doing most of the manipulating. This kind of imagery is reminiscent of the procedures suggested by Carpenter and Just (1986) for recognising shapes in rotated positions.

Visuospatial Reasoning in Concept Development

Hershkowitz (1989) involved preservice and experienced primary teachers in her investigation into changes in the use of visual images which support concepts arising in the course of an activity. She investigated the use of concept images for a right-angled triangle, an isosceles triangle, an altitude of a triangle, a quadrilateral, and two shapes defined specifically for the investigation. She concluded that there are three levels of use of concept images:

1. The prototypical example is used as the frame of reference and visual judgement is applied to other instances. This seems to be the most common behaviour in the identification of right-angled triangles, where subjects failed to identify examples which contradict the vertical–horizontal prototype, and in the altitude task, where subjects failed to draw altitudes which contradict their concept image of an internal altitude.
2. The prototypical example is used as the frame of reference but subjects base their judgements on the prototype self-attributes and try to impose them on other concept examples. When this does not work, they do not accept the figure as a concept example.
3. The critical attributes are used as a frame of reference in the formation of geometrical concepts. In this case there is a chance that the individual will form concept images that are less (or not at all) visually biased (p. 74).

Hershkowitz pointed out the importance of ensuring that students see various examples of concepts. This procedure should prevent some students from imposing a visual bias on concept images.

In looking at the development of the angle concept in one of the collaborative groups in the key study described above, classroom interactions and use of manipulatives were predominant over the series of lessons (Owens, 1996b). There were a number of activities in which students were required to notice and begin to develop their concept of an angle. In particular, a case study of a grade 2 cooperative group provides a good example. Jodie, James, and Victor were asked to find small, middle-sized, and large angles on the tangram pieces. Immediately they checked the points of the pieces by overlaying them as the teacher had demonstrated in introducing the pretest where they were to mark angles on different shapes equal to the marked angle on a shape. Jodie called the 45°, the sharp angle, and associated it with "big". When the teacher called it the small angle and illustrated with the thumb and forefinger that they were only turned a small amount to lie along the arms of the angle (see Fig. 2.2b), she quickly readjusted her language pairing "sharp" with "small" (she was a bright child with English as a second language, so was used to learning new English words). The third member of their group, Victor, was absent from the previous lesson, so he was self-regulating and still doing the previous activity of making the large triangle with the other pieces of the tangram set as well as thinking about the angles. The discussion indicated how Victor, who seemed to know what was meant by the size of points (the word generally used by these children to refer to angle), temporarily considered that he should be comparing the size of the sides of the shapes. The interaction between students helped Victor to clarify what was meant by "the point of the same size". James, who had been able to match points in the previous activity, began this later session by choosing the wrong points, largely because he was choosing the small or middle-sized triangles. He established the meaning by listening to the teacher and to Jodie and by checking points with the drawings in their book. Thus the perceptual size of triangles dominated but the interaction with those around him and the visual representations assisted him to establish the meaning of angle size.

In their next activity, the children were able to order the angles in the pattern block set and draw them in order. They were encouraged to give them size values of the unit equal to the smallest angle (the 15° angle of the narrow rhombus of the set). In a later activity, the group made shape outlines and Victor explained that James had not made a right-angled triangle as James had thought but that he had just made an equilateral triangle in another orientation. Victor himself had made the right-angled isosceles triangle with the long side horizontal and he checked it with the tangram piece which he put on top ("a lid", he called it). The teacher asks what was meant by bigger angle. Jodie replied "more spread out" and picked up the tangram right-angled triangle and the pattern block equilateral triangle, put one on top of the other and said, "see it is bigger".

Interestingly, the children in different groups often noticed and recognised angles and they were perceptually more noticeable than length, or the starting and ending of a side, or straightness of a side. However, they had more difficulty to describe angles and without the use of the fingers would have struggled to show their understanding. Communication encouraged identification and representation of the angle-problem situations, and development of cognitive processes for solving the angle problems. Interactions with others and internal representations assist analysis which is an important aspect of concept development and problem solving (Krutetskii, 1976; Lean & Clements, 1981).

In a further study (Owens, 1998a), adults were required to identify equal angles in complex figures. Different conditions—visual, physical, aural, or spoken cueing—did not make a statistically significant difference on students' ability to solve the tasks. The reasoning provided by the adults suggested that prior experiences at school, often with negative feelings, and their view of themselves performing mathematically impacted on the their performances. Few of the adults felt comfortable about just perceiving the angles to be equal without "proof", so many drew on remembered knowledge about vertically opposite angles and angles in isosceles triangles. Other school-based knowledge such as angles related to transversals across parallel lines or exterior angles of polygons were not recalled. The adults who were given information in training audibly as well as visually had significantly less variation in scores on the test suggesting selective attention resulting from the additional auditory cueing played a part in their visuospatial reasoning.

These studies have shown that prior experiences and informal experiences can help students to establish visualisations. A unit on similarity for grade 5 prepared by Woodward, Gibbs, and Shoulders (1992) was to provide informal experiences for students to gain a good foundation for concepts about ratio and proportion. Such experiences included comparisons of angles and sides of similar figures. In fact, van Hiele (1986) maintained that the level of development in reasoning is more dependent on instruction or informal experiences (incidental learning) than on age.

Visuospatial Reasoning in Problem Solving

As the students in the key study described in this chapter continued trying to solve a problem, they seemed to use less pictorial/global imagery and more dynamic and action imagery, and finally more pattern and procedural imagery. Imagery, judging from the students' actions, involved recognition of different parts of pieces. Active involvement in the problems clearly increased students' use of imagery and their skills with images. Generally the improvement was associated with the following:

1. Students began to relate concepts to their visuospatial reasoning. Concepts associated with manipulations occurred mostly when action visuospatial reasoning was used. However, other conceptualisations related to size, angles, shapes, patterns, and symmetry were used. Concepts supported the imagery that guided tactics and manipulations, rather than vice versa. The meanings of verbalisations were not always clear suggesting that only limited conceptualisation had occurred.
2. Students, upon settling into the tactical stage of the problem-solving process, generally used reasoning other than concrete visuospatial reasoning.
3. Some problems encouraged students to manipulate visual images mentally while others encouraged the use of disembedding skills and yet others pattern or action visuospatial reasoning. Any one student may use a variety of types of imagery (see, for example, Sam and Tess in excerpts 3 and 5 above, respectively). Some children completed some activities more easily than others such as Sam enlarging pattern block shapes whereas he struggled with other activities.
4. Dynamic and action visuospatial reasoning developed into other forms such as pattern and procedural visuospatial reasoning. However, there is no hierarchy of types of visuospatial reasoning.
5. Pattern visuospatial reasoning especially provided the necessary connection between a visual image and an abstract conceptualisation, possibly because the processes of looking for, recognising, and describing patterns are basic forms of mathematical thinking.
6. Visuospatial learning experiences can assist in developing these mathematical thinking skills and structured materials, like those used in this study, can encourage recognition and use of patterns (Owens & Outhred, 1998).
7. The structured nature of the types of visuospatial reasoning described in this study not only reflected images of the physical embodiments which were used but also served as a way by which imagery was structured and used for reasoning.

Although imagery is necessarily individualistic, in the sense that an image "resides" in a particular person's mind, it makes sense to say that different people can have more or less common images and visuospatial reasoning in the same way that we say people have a shared understanding of a concept. Shared visuospatial reasoning particularly develops as a result of shared physical phenomena, problems to solve, body movements, and social interactions.

Visual representations in mathematics are not simply personal or disassociated images but they convey explicit knowledge structures that are constructed and negotiated in a context of visual representations that operate within shared rules, habits of seeing, and cultural practices (Voigt, 1994). Further, the different kinds of visuals that are generated depend on signs that are taken-as-shared but personally created (perhaps limited or enhanced by experience) although tools and hegemony (e.g. the common two arcs for a bird, the equilateral triangle on its base) may determine their nature. Visuals may be learned (e.g. the name of a type of triangle and a representative diagram), associated with a relevant experience (e.g. manipulation of string to form a triangle), or established through relational structural similarities (e.g. drawing a square as a rectangle with all sides equal).

Some visual representations require patterns that may lead to structures. Figural patterns often lead to a description or algebraic representation (Outhred, 1993; Owens & Outhred, 1998). Early generalisations are often additive but then multiplicative thinking occurs, at least in children from European backgrounds but in other cultures there may be a more multiplicative approach (see later reference to research on enlarging houses and counting groups in PNG, Chap. 5). Figurative patterns need a high Gestalt effect such as children picturing a bag of lollies for a multiplicative pattern rather than a series of dots although arrays can be physically created. In line with Dörfler's (2004) arguments, diagrams are valuable for visuospatial reasoning if they are structural and relational and the arrangement expresses the relationship. They need to possess internal meaning or rules for transforming the diagram. They have an external referential meaning, inside or outside mathematics. They need to be generic or visually general and transformed in a perceivable way. A visual template such as the circling of parts of each stage of a figurative pattern that is growing according to the pattern encourages pattern recognition and apprehension of the pattern (Rivera, 2011). The role of directing attention and then the self-developed selective attention are part of the visual reasoning process associated with diagrams.

Visuospatial Reasoning in Learning

Pirie and Kieren (1991) suggested the beginning of problem solving for new learning is "primitive knowing" and this links to intuition which is discussed later. Learners then make and hold an image to which they "fold back" (revisit) in order to go forward again with their learning through noticing properties of the imagery, and then "formalising", "observing", "structuring", and "inventising". Presmeg (2006) suggested that imagery was also a way of reifying conceptual understanding and could be considered part of formalising on which observations can be made, structured, and developed or used creatively. Perhaps Hegarty and Kozhevnikov's (1999) concrete and abstract visualisers are better understood as those who select one type of visualising more than the other. Visualising about everyday objects, concepts, and processes is different from mental objects, concepts, and processes

(Rivera, 2011). The latter can be associated with the higher levels of Pirie and Kieren's model when mentally formal definitions and structural complexities are recognised. The dynamic and pattern imagery associated with concepts, metaphors and metonymies reifies the concept ascertained from various, often concrete and practical, sources (Presmeg, 1997). This imagery develops through problem solving in which actions of a learner (e.g. to interpret or to construct by way of predicting, classifying, translating, or scaling), situations (e.g. abstract or contextualised), variables (e.g. the data type and whether concrete or abstract), and focus (i.e. the location of attention) are identifiable (cf. Leinhardt, Zaslavsky, & Stein, 1990; Owens, 1993; Rivera, 2011). Thus visualisation can vary in each circumstance.

Giaquinto (2011) analysed a number of diagrams to tease out visuospatial reasoning. He particularly noted that first there was perceptual recognition of a concept such as a square that depended particularly on symmetry in the horizontal and vertical plane but also in recognising the equality of angles. He noted that people have a tendency to attend to the vertical with the influence of gravity or external reference frames such as the page, table, or body position more than other positions and orientations (see earlier work, e.g. Vurpillot, 1976 and inexperienced responses in Fig. 2.11). If this experience was repeated it would become an acceptance of a square in a geometric sense. In my own study with adults (Owens, 1998a), perceiving equal angles was generally achieved but the context such as the complexity of the diagram or orientation of the angles or an exterior angle of a figure made it harder for equal angles to be perceived. In these cases, adults would recall school mathematical information to assist in identifying equal angles such as vertically opposite angles, angles of an isosceles triangle, or those associated with transversals of parallel lines, for some adults. Sides of triangles were more readily perceived as equal.

The adults used visualising and imagining to assist in decisions about angles. Similarly Giaquinto (2011) found visualising and imagining assisted to reinforce the dispositions or beliefs about the square. He also noted that dispositions could be given different degrees of support. For example, directly perceiving or remembering an experience that could be easily judged by memory, e.g. countable objects, may be easier than imagining a length compared to a remembered length. However, some explanation to support the comparison would be supportive of the disposition or belief. This might rely on a past experience or an intention to carry out a visual imagination. The intention focuses the attention and noticing of certain aspects such as the visual comparison of line lengths or the more holistic shape being translated or rotated or reflected. Squares can be both recognised when partially obscured drawing on visual memory or representations in the mind and imagined (see also Rivera, 2011). However, while physical proof and imagined proof require similar brain functioning, the imagined proof is often more convincing rather than perception of a physical representation because it is not likely to hold the imperfections of a physical representation. There can be difficulties such as imagining an object in an unusual orientation and not actually perceiving the imagined image correctly. It might be that another figure is in the mind distracting the visual imagination. There is also the possibility that the imagination is too complex or has too many steps or parts to be carried out easily and thus repeated in the mind consistently (see the story of the two girls re-enacting the string figure to recall the next step in Vandendriessche, 2007). It is also possible that a person imagines incorrectly because of a lack of conceptual knowledge.

Visual imagination can provide sufficient discovery and in some circumstances justification for a belief or proof in geometry. This visualising provides stronger reasoning than seeing because aspects of the diagram or object might not be generalised sufficiently and may require text to establish the justification. Giaquinto (2011) concluded that there is not a dichotomy between geometric and algebraic thinking but rather spatial thinking is used in conjunction with symbolic arguments. He used the four proofs of the sum of numbers as a culminating example to illustrate. While one algebraic induction proof was symbolic, the Gaussian proof of ordering the numbers in reverse order below and seeing that horizontally there are so many numbers and vertically the sum is $n+1$ requires spatial thinking as well as symbolic understanding. In the other proofs, the forming of a mat of dots by joining two triangular stairs of the numbers or by looking at the area of squares is basically spatial with the square one being more geometrical in nature. Nevertheless, spatial thinking is key in a proof requiring both vertical and horizontal reading approaches or when proofs involving arrays require both their horizontal and vertical size to be noted. This is also the case for rectangular area arrays (Owens & Outhred, 1996).

Although visuospatial reasoning is often usefully employed in problem-solving situations, such reasoning is not always recognised or regarded as legitimate. Often the person who evokes an image does not necessarily appreciate its richness, and the use of external representation to communicate its meaning to another person may not be successful (Dreyfus, 1991). Therefore visuospatial reasoning may have been undervalued in a number of twentieth century geometry studies.

Visuospatial Reasoning, Metaphor, and Metonomy

Johnson (1987) and Lakoff (1987) refer to the possibility of visuospatial reasoning in communication when they discuss the use of metaphors in thinking. Through metaphor, connections among existing image schemata are made and extended. Metaphorically (through new contexts) and metonymically (through partial representations) imagery develops mathematical thinking. While Vurpillot (1976) noted that young children in her study tended to categorise items on partial equivalence and indicated that this was a limitation in their conceptualisation, the excerpts and examples in this chapter suggest that metonymically part-whole connections of image schemata assist children to develop more abstract thinking. The act of problem solving, in itself, and interactions with others, tended to facilitate the formulation of alternative perceptions of concrete materials. Learning from problem solving is more than just associating conceptual knowledge to visuals. This view is suggested by Clements, Battista, and Sarama's (1998) careful analysis of young students' verbal and visual responses. The details of their report reflect those in my own study. Thus it is evident that imagery is likely to promote flexibility in thinking and creativity in problem solving. Nevertheless, inferences need to go beyond the imagery of the physical (Giaquinto, 2011).

In the key study described above, there were many examples in which ideas were creatively reconnected. Jonah, a grade 2 student, made two pentomino crosses in different orientations and said, "This is a box and this is a robot". Tess (excerpt 5) remade a right angle from two pieces despite the pieces having non-matching sides and Sally (discussed under pattern visuospatial reasoning) linked her image of her Christmas tree to the triangle enlargement. In each case, imagery did not appear to be primarily structured in terms of propositions; rather the proposition (such as Sally's description of the pattern) supported the image.

Metaphor and metonomy are often the genesis of connections. The connections enable features, for example, a pattern, associated with one configuration to be applied in a related situation. The so-called "concept images" (Fuson & Murray, 1978) can act metonymically for a concept, emphasising certain characteristics of the concept. Images need to be embedded in various visual and conceptual schemata if they are to provide a dynamic influence on a person's approaches to a problem-solving task. Without this, the concept image can limit conceptualising and creative thinking. Furthermore, the dynamic moving of images of shapes into related shapes can assist the development of conceptual relationships.

The continuous manipulation of materials meant that students were able to see where shapes could be added or taken away and this experience encouraged their visualising of results before trying the manipulations. The tangram tasks, in particular, involved children in a great deal of turning around and over pieces, and of matching angles in order to fit shapes together. The making of shapes, the comparing of angles, and the finding of shapes in designs improve students' visuospatial reasoning in that students were encouraged to disembed shapes and parts from more complex shapes and to imagine where other shapes could be (cf. Tartre's classification of re-seeing, Table 2.1). Students were using both their short-term and long-term visual memories in order to achieve greater problem-solving efficiency.

Based on Goldin's (1987) model of problem solving there are five interconnected language systems (the word "language" suggests "re-presentation" or processing of information). The categories are related here specifically to the processing of visuospatial problems which are likely to be met in early childhood:

1. Verbal/syntactic processing has input which can be verbal (as it is in word problems) but it can also be non-verbal in visuospatial problems. For example, students learn whether a diagram of a parallelogram is representative of all parallelograms or whether it is intended as a precise drawing such as a scale drawing of an area of land. The output can be imagistic processing or formal mathematical notation.
2. Formal notational processing usually refers to arithmetic or algebraic statements or to statements in geometric proofs such as AE ∥ BD. In the spatial area, another example would be the categorising of shapes and the schematising of these categories in a tree diagram or Venn diagram or drawing a triangle to represent all

triangles. Goldin (1987) points out the dangers of direct translation into this system without involving the imagistic configuration.
3. Imagistic processing involves the "feel" for the problem. In addition to visuospatial reasoning, this representation can include pattern recognition, and the matching of non-verbal sensory inputs to previously encoded information. The task content and context are incorporated into the processing.
4. Planning and executive control language include heuristic processes such as plans, strategies, tactics, and self-assessments. This category involves a recursive capability so that the processes can act as control not only on the other domains but also on itself. It incorporates the notion of metacognition described by Flavell (1987), Lester (1983), Mildren (1990), and others.
5. Each of these four processes is affected by and influences the affective system of representation. Affects include feelings, attitudes, beliefs, and values.

Context and Visuospatial Reasoning About 3D Shapes

Following on the analysis of how important noticing and imagining were in the development of children's angle concept over several sessions as described above (Owens, 1996b) and the impact of audible cueing in the adults' angle study (Owens, 2004a) described above, I began to explore how children make images and notice parts of three-dimensional shapes and how they might consider properties (Owens, 2004a). Here I report on the section of the study involving testing children in grade 3 in three different ecocultural areas: a girls' private school in an Australian city, a school in a lower socioeconomic, multicultural area in an Australian city, and a multicultural city school in PNG with children from a diversity of family backgrounds and with a greater range of ages. All classes had teachers who taught mathematics well. The study also involved individual interviews of six children immediately after they had completed each page of the test on 3D thinking (they came from the lower SES school in Australia). Prior to this study, test items were developed for different categories of visuospatial reasoning and those that had the best validity from a Rasch analysis were selected (Owens, 2001a). The test *Thinking About 3D Shapes* covered the following: recognising 3D shapes within shapes, joining two 3D shapes together to make a third shape (a rectangular prism) or an illustrated shape, tessellating a block to make a given shape, viewing objects from different positions, imagining folding, marking, and unfolding paper, and imagining folding a 2D shape to make a variety of 3D shapes or objects. The test was introduced by showing 3D shapes, ensuring the children knew the names, showing the children how to draw them on the board in isometric form, and illustrating the folding of paper, marking, and unfolding, and folding a rectangle and a net to form 3D shapes.

Some results are presented here. In the following tables, the lowest percentage and the highest percentage from the three classes are given. In this study, I made no attempt to link ecocultural background with results other than to note difference. The results showed that for recognising shapes in other shapes some degree of

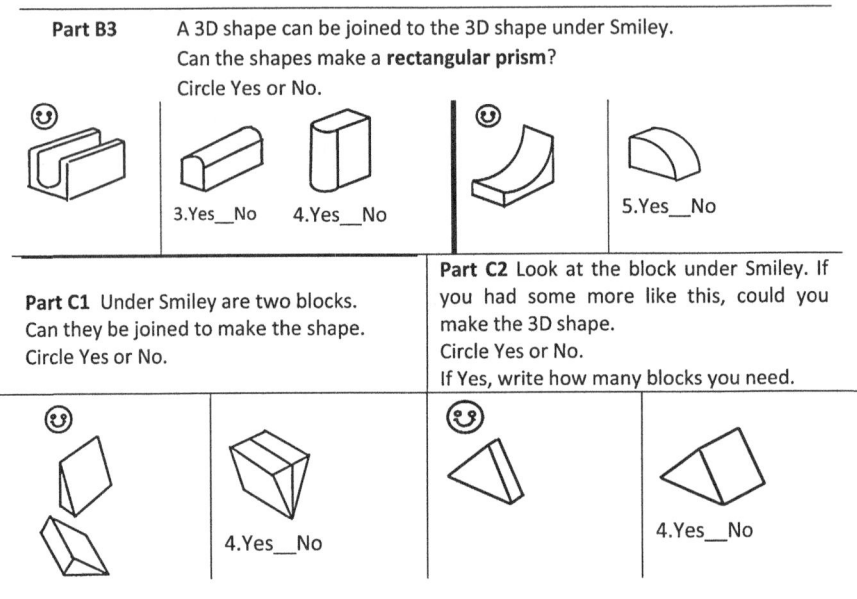

Fig. 2.13 Selection of test items: joining 3D shapes together to make other shapes

sophistication of reading diagrams was necessary because many children ignored the absence of a line in one diagram. For spatial relations or joining blocks to form a rectangular prism or illustrated block, children could explain how they were thinking (see Fig. 2.13 showing some of the items from the test, reduced in size). Ahmed at first could not see how these curved shapes would make a rectangular prism but then he referred to putting in the piece to make it smooth.

> Ahmed: (B3, 5) No because it is too long and can't make it, because it would be pointy at the top, so if turn over would be pointy at top?
> Interviewer: Yes, what makes you turn it over,
> Ahmed: So it is like this pointy.

It was interesting that the students considered the flat and in some cases rectangular surface that would be made by joining blocks together. Students did the items by analysis and mental rotation.

> Ian: If put that triangle [sic] same as that but facing other way, make a flat surface just like that (points to triangular face). (C1, 4)

Ian is aware that two blocks will join to give a flat surface together. Other students also expressed similar approaches. Visuospatial reasoning is important for tessellating blocks to fill a 3D shape (test, C2) in order to understand and calculate volume. Students generally found it easy to select whether blocks of the type illustrated will tessellate but are unable to keep the size stable in considering the number required, generally imagining many more blocks. Like the 2D tiling activities (see above, Owens & Outhred, 1997, 1998), students find it difficult to see how blocks

Context and Visuospatial Reasoning About 3D Shapes 69

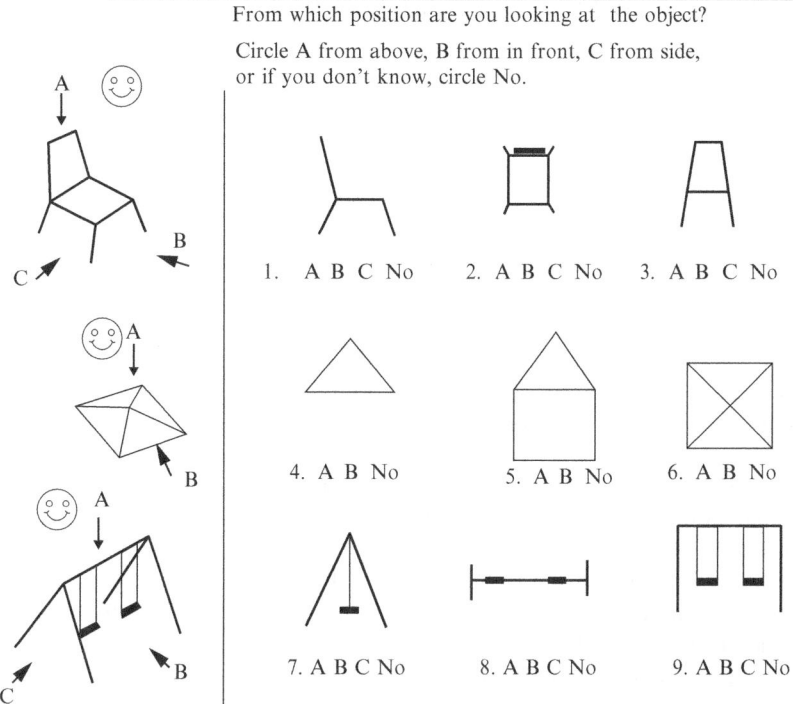

Fig. 2.14 Items for recognising shapes from other perspectives

join to give a distinctly different shape. For Item C2, 4, student difficulty was in seeing how the square face can be formed from the thin triangular prisms. Some students also find it difficult to consider size; and two interviewed students counted many rectangular prisms as making up the cube in another item whereas two rectangular blocks, one sitting vertically on and perpendicular to the other, had quite low percentages (36–53 %).

> Ahmed: (Points to C2, 4) If joined together, it will make that
> Interviewer: What was going on in your mind?
> Ahmed: A square
> Interviewer: You seemed to be counting? Can you explain what you were doing,
> Ahmed: It's like a book?
> Interviewer: Can you explain,
> Ahmed: (Counts) one two three four.

On the other hand, Joe says "but won't fit into square" still seeing the triangle fitted onto the square face of the larger triangular prism. He counts but is unable to draw to explain what he was thinking.

For Part C1 there was little difference between the schools. For Part C2, two schools fluctuated from item to item between the middle and lower ranking suggesting ecocultural context was influencing items differently.

Part D1 "Can you see it another way?" (Fig. 2.14) involved students in recognising shapes from different perspectives.

Table 2.5 Percentages of students who were correct on completing prisms and recognising shapes from different perspectives

B3	3	4	5	C1	4	C2	4 Yes	4 number=4
Lowest percentage	22	38	28		85		52	24
Highest percentage	87	73	73		95		73	53
D1	1	2	3	4	5	6	7	8	9
Lowest percentage	71	47	14	42	54	54	62	43	48
Highest percentage	87	63	47	67	90	67	93	67	80

Most items were around the middle of the difficulty range but some were quite difficult (confirmed by a Rasch analysis). Percentages in different schools are given in Table 2.5. The chair viewed from the front (D1, 3) was the most difficult one to recognise especially for children from the school with less experience with reading books and pictures but they did not have a difficulty recognising that D1, 5 was not a representative of the square pyramid (they were not the lowest scoring school on this item). Recognising the small parts in Item D1, 2 as legs, and the seats of the swing in D1, 7 and D1, 8 was a key difference between those getting these items correct and those that could not, reflecting the issue of the importance of lines in the earlier questions and Bishop's (1983) interpretation of figural information.

The last two parts of the test required visuospatial reasoning in mentally folding, punching a hole, and unfolding and then in making various 3D shapes (not necessarily closed) from a net or a rectangle. Lack of experience seemed to affect the results for the hole punching and opening questions since percentages for the two items for the different classes ranged from 29 to 75 %. Similarly for the items in Fig. 2.15, typical mistakes were not seeing the triangular prism as hollow prisms despite being told suggesting prior experience affected thinking in line with an ecocultural perspective.

Although most students drew the lines for the "table" (E2, 2) correctly, in other cases students drew lines horizontally. It seemed that the move from the isometric view to a flat paper was particularly difficult. Interestingly, only one interviewed student could imagine the rectangular paper being rolled to form a cylinder (E2, 3) reflecting the overall lower percentages for this question (25–67 %). Many students put lines on the rectangle, usually curved at the ends to illustrate rolling. The successful student, who knew there were no fold lines, was asked if he had done this in class but he could not remember and the current teacher confirmed this but they had used solid cylinders in building with blocks.

The items for folding open cubes were generally difficult. The order of difficulty was similar to that found on a previous test *Thinking About 2D Shapes* that had incorporated these items (Owens, 1992a). If students began to use a square that was not going to be the base, then they often struggled to imagine an alternative starting point for folding or to turn their shape to decide the base. The net requiring folding up and two sides to be turned (E3, 3) was the hardest (percentages for correctly

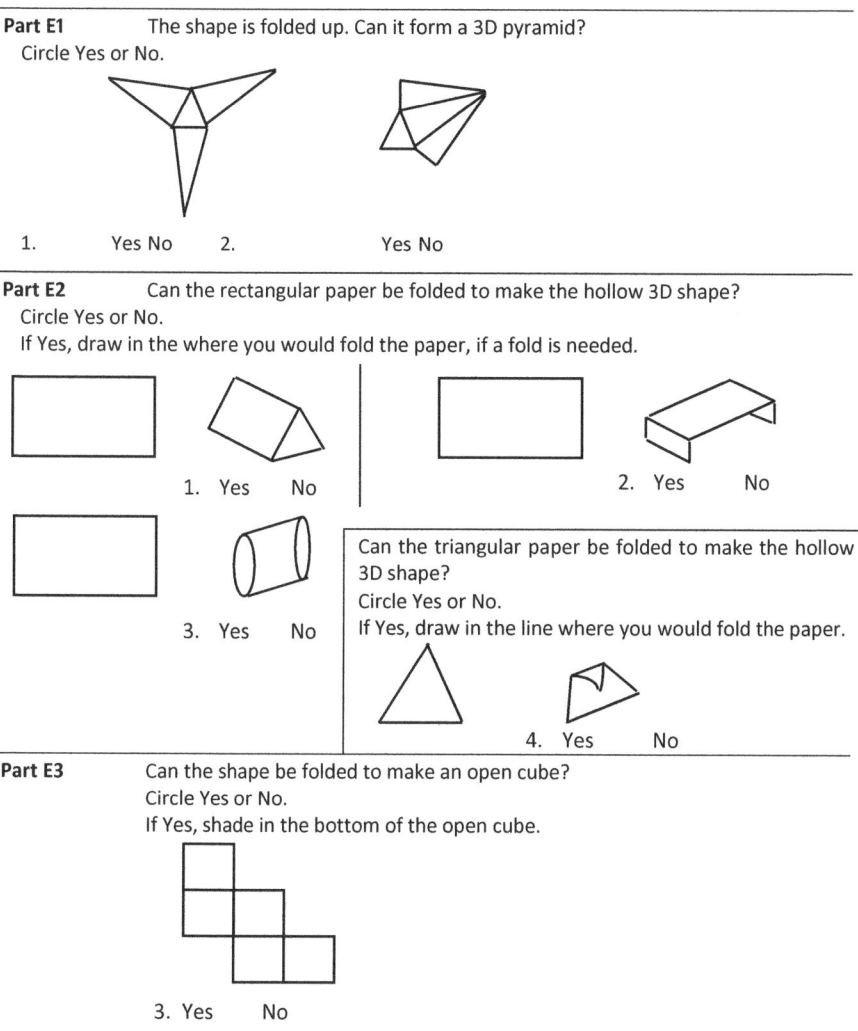

Fig. 2.15 Selected items from test: imagining folding 2D shape paper to make a 3D shape

selecting it ranged from 38 to 48 % with the correct base shaded being 33–43 %). These percentages were similar to the unusual net for the triangular pyramid, E1, 2. The shape (an L shape) that could not be folded to make an open cube was not consistent with the rest of the items but it provided a negative response to at least one of these items. Many students selected this item as a net but failed to imagine that some faces would be doubled with others left open.

The wide variety of items involved students in a range of visuospatial reasoning skills including fitting objects together, mental rotation of objects, viewing from another perspective, and mentally folding. The interviews indicated that students were mentally manipulating objects or parts of the figures that were perceptually accessible in the form of the diagram and they commonly analysed the shapes and considered parts. The test was not timed and so speed of image making and that of image scanning were not teased out as separate processes.

Re-seeing shapes and recognising shapes in different orientations, and noticing and imagining parts of shapes were skills being developed by this age group. Students appreciated parts of figures were hidden in some cases but could not imagine the piece that is hidden. It seems that details of figures are not always noted. This is evident when lengths, points of intersection, and size are not distinguished but shape is more important. Responses from students who were not strong (e.g. Ahmed) supported Tartre's (1990b) suggestion that low scoring problem solvers did not integrate analytical and spatial skills well.

Some recent studies have considered the impact of symmetry of objects (both familiar toys and abstract arrangements of blocks) from different perspectives. One recent European study (author unknown) suggested that symmetry might in fact assist students to concentrate on other features for determining perspective whereas asymmetry was less of a problem with a familiar toy than an abstract block arrangement. Children's reasoning for front/back perspectives was predominantly related to features of the object or the alignment of features within the object for both symmetric and asymmetric objects (animals and blocks) rather than an extrinsic alignment. However, for side views of symmetric animals, students in lower elementary school struggled to provide a reason or struggled to use a description whereas most students could describe characteristic differences for asymmetric animals. For side views of block arrangements, there were more descriptions of symmetric arrangements as well as asymmetric and front/back perspectives. It is suggested teachers should develop front/back perspectives using both symmetric and asymmetric objects but for side views it is worthwhile beginning with asymmetric objects. The left–right relation could also be stressed as what can be seen is not always successful. Importantly, the world around the child in terms of complex living creatures seems to provide more clues for reasoning intuitively.

School Learning Experiences and 3D Visuospatial Reasoning

The test was later used to show the effectiveness of a series of lessons from the Count Me Into Space project in NSW, Australia, on visuospatial reasoning of grade 2 students from five schools across three districts of the city of Sydney, Australia, matched with schools from the same districts (Owens, 2004b). While there was a significant difference between students who undertook the activities on orientation and motion (see below) including work on 2D to 3D shapes on the immediate posttest, there was no significant difference after 6 months. The confidence interval of the mean of scores for the groups overlapped. However, when the students were broken into three groups according to their pre-intervention scores, there was virtually no overlap for those students in the lowest group—intervention had confidence intervals for the mean of 38 ± 3.5 and non-intervention 31 ± 3.5 (see Fig. 2.16).

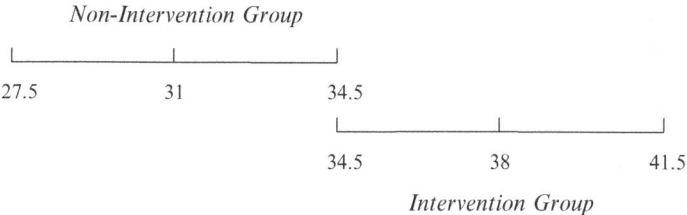

Fig. 2.16 Performance of low attainers from both groups on three-dimensional shape test

This confirms reports by teachers that the weaker students gained considerably from the classroom experiences. The group work, discussion, and hands-on experiences encouraged a sense of ownership of their work and helped these students to improve. The programme captured the essence of the research especially in developing imagery for (a) recognition of 2D symmetry and 2D and 3D shapes in different orientations, (b) modifying shapes that keep certain properties (dynamic changes), (c) perceiving parts of 3D shapes, and (d) imagining 2D nets of 3D shapes.

Given the differences in results from the three schools (grades 2–4) in different ecocultural contexts and the impact of school learning experiences in the studies discussed above, it seems important to pursue the influence of context on visuospatial reasoning.

Visuospatial Reasoning in Context

Kaufmann (1979) claimed that visual imagery did not necessarily lead to flexibility in problem solving, and that this might have been the result of limitations brought about by socially induced gender differences (see my discussion of many studies on gender and visuospatial reasoning, especially the meta-analysis by Linn and Hyde (1989) in my thesis (Owens, 1993)). Perceived rules of the classroom also impact on using visuospatial reasoning as well as children's interactions. From my key (1993) study as described above, this was evident from both the competitive approach of James in his group with making pentominoes (excerpt 2) and later Victor's discussion with him about the right-angled triangle with a horizontal hypotenuse. It was also evident in the classrooms when children moved the shapes too quickly to allow all the groups to think about the shapes. Tess was doing this initially in making a square from the tangram pieces (excerpt 5). Susan (in grade 2 in PNG, excerpt 4) tended to be a dominant figure within her group and her quick pulling apart of trial configurations may have prevented the students seeing shapes within shapes. Interestingly, none of the students in her group saw the nearly completed trapezium in enlarging pattern block shapes, and the complete trapezium was never made by the group despite several more attempts. (Most groups in all grades and schools were unable to make the trapezium and many were unable to complete it even when three pieces were correctly positioned.) In one class in PNG, a girl made one shape and sat with finger on her lips and hand up waiting for the teacher to give her praise and further instruction (until she realised she was to make many shapes if she could) (Voigt, 1985).

In my study (Owens, 1993), I found the classroom context had to be considered to fully appreciate the results. Classroom learning environments should provide not only receptive-language opportunities when students process another person's communication by listening, reading, interpreting diagrams, pictures, and actions but also expressive-language opportunities for speaking, writing, drawing, performing, and imagining (Del Campo & Clements, 1990). If this is the case, then students who manipulate and speak about their angle-matching tasks are more likely to perform better on angle-matching tasks in future. I decided to see if preservice teachers could put this into practice. (Later I will discuss a widely used programme Count Me Into Space which encouraged visuospatial reasoning.) The preservice teachers planned learning experiences after learning about the different types of visuospatial reasoning and the importance of substantive communication in the classroom. They used Wood's (2003) model involving strategy reporting and inquiry to prepare the learning experiences. The following extract indicates children, perhaps for the first time using visuospatial reasoning, to respond to teacher's questions. The following transcript from the pentomino lesson shows how she encouraged students to interact and give their opinions. (T stands for teacher.)

T: Is that the same shape or a different one
D: Same
T: How come it's the same?
S: It's been rotated
E: It's different
T: Why do you think its different E?
E: Because the square we're looking at is in the top row not the bottom row
T: Someone else
V: They're the same because if you rotate it's on the right side not the left side
T: What happens, yep someone else
J: If you flip it over and rotate it once.

In a later lesson the children were drawing examples of shapes, first in small groups and then discussing whether some given descriptors fitted the shape. In the process they tried to draw a 40°, 40°, and 100° triangle to visualise the obtuse-angled isosceles triangle.

T: So everyone got the isosceles triangle.
J: You know how you call it an acute isosceles triangle, doesn't it have to be acute?
T: What does everyone else think? Do you think you can have obtuse angled isosceles triangle?
R: No then it would be scalene. (Other students comment in the background.)
T: Then it would turn into a scalene
D: If both angles (pause) in the corners, it would go out like that (shows with hands)

Students continued to discuss other shapes on the paper deciding on whether they were irregular or not. In this extract and later in the lesson, students were initiating conversation. So conversations in the classroom can direct students' attention to features of shapes encouraging visuospatial reasoning to make decisions and develop concepts. I now consider attention in more detail. It is not just external attention.

Attention

Flavell (1977) commented that attentional processes become increasingly interwoven with other cognitive processes such as memory, learning, and intelligence. Attention is attracted by perceptually outstanding features such as nearness, isolation, size, special form, colour (Gell, 1998), number of items, and the inherent interest of the items (Bishop, 1973). As a result, people attend to certain features of a visual stimulus.

Selective attention is the result of focusing on both external and internal stimuli (Flavell, 1977). Selective attention can be affected by the visual ability of making ground-figure changes. For example, a student can change focus from a part of a shape to the whole shape. Less experienced students may focus on partial features to decide equivalence and may not be logical or recognise relevant orders such as size (Vurpillot, 1976). Selective attention can be improved by repetition and the recognition of a relationship which can be employed to solve a problem (Vurpillot, 1976). If students consciously or unconsciously assess information as incoherent, then they do not attend to the input (Egan, 1992; Lévi-Strauss, 1968; Mason, 2003). Such restrictions may reduce the effectiveness of selective attention in developing conceptual links but students' attention can be influenced by others through looking and listening to others as noted above with adults and children.

When students respond to problems that require visualisation skills such as those required in spatial problem-solving tasks with manipulatives or computer assistance, there can be an interference effect. Some researchers have contributed the difficulties to cognitive overload (e.g. English, 1994). English argues that the equipment can make excessive demands on the individuals' working memory and this cognitive overload interferes with the learning of desired concepts. Studies suggested that chunking material, practice, and reducing redundant and irrelevant material especially if it splits attention in the same perceptual mode can assist selective attention or learning (Sweller & Chandler, 1991). The nature of the material and its familiarity, difficulty, uncertainty, and modality of presentation also influence attention (Baddeley, 1992; Kahneman, 1973; Liu & Wickens, 1992). Disputes about selective attention were about the effect of early and late selection and about limited and unlimited capacity. However, Johnston and Heinz (1978) demonstrated that selection can be either early (based on physical characteristics) or late (based on semantic analysis) depending on the nature of the task, the instructions, and so on. Attention is assisted by ecocultural contexts that encourage observation, repetition, interest, and chunking material together from a holistic perspective.

Unlimited Capacity Model of Attention for Action

Then a more flexible view of attention was developed. Rather than identifying rigid upper limits, studies have demonstrated that our capacity to attend and use information is influenced. Allport (1987) argued that early selection is really about "the relative efficiency of *selective cueing* (which) is simply irrelevant to questions about the level of processing accorded to the 'unselected information'" (p. 409). Processing of both cued and non-cued information proceeds at least to categorical levels of analysis. Allport argued that unlimited capacity for perceptual attention for action explains results of experiments. He referred to

> crosstalk interference between parallel processes. ... Whenever the task-specified inputs are not the single most compatible among concurrently available inputs for the task-specified actions, (inputs need to) be actively decoupled from the control of particular actions. ... It is a radically different conception, however, from the earlier notion of a central, limited capacity, or even from that of multiple limited 'resources' (Allport, 1987, p. 411).

Selective attention has been described as like a spotlight on possible inputs and as a filter of sensory information. However, van der Heijden (1992), based on his experimental findings with short exposures, disagreed with both these metaphors for selective attention which imply limitation and loss of sensory information. Supporting Allport, he provided a model to avoid limitations and loss. The model involved the separation of location and identity for stimulus inputs and the importance of a feedback loop during processing from the location to the inputs. Thus attention can shift mentally to notice other information. Different sensory features of objects are coded automatically and spatially in parallel and are located in appropriate maps (Treisman, 1988; van der Heijden, 1992). Uncued information may take longer to locate but combinations of features specify objects through a master map of where features are located by neuronal activity selectively enhancing (not inhibiting or attenuating) processing (van der Heijden, 1992). Higher order centres involving past experiences and conceptualisations improve the locating. These centres involve expectations and intentions which influence selective attention. For example, if persons expect only to see an angle without a line dissecting it, they will not attend to angles that are dissected. Expectation influences the location (a) directly with verbal cueing, (b) via another module with attribute cueing, and (c) with a link from identity to the higher centre and then to location if symbolic cues are used. This theoretical position suggests a dependence on prior experience. The end result, though, is action (Allport, 1987).

Clements and Sarama (2007a) noted that mental maps are not like paper maps. They distinguish between the areas of the brain that note what an object is, "spatial visualisation" (its identity in van der Heijden and Allport's term), and the way upon which it is perceived "spatial orientation". While this may be a helpful distinction, it is not clear cut in that interaction with objects, their contexts, and people influences both skills. Part of visualisation and location involve recognition of objects. Furthermore language plays a role in such mental maps.

The above summary of van der Heijden and Allport's work provides a way of understanding selective attention in classroom settings as well as perception experiments. Various aspects of the classroom environment—words from the teacher or fellow students, the position of concrete materials, the expectation associated with a routine of classroom activities, and the task description—may influence selective attention. The student identifies, processes through higher centre schema, to give a location that leads to attending selectively to the inputs, with further loops as needed. Selective attention is influenced by expectation and intention as well as perceptual inputs and internal feedback through the higher centres. Expectations and intention are part of the inner visual system and alter internal and external feedback. Classroom and other social interactions form part of past experiences, and they frequently influence expectations and intentions. For example, the prior knowledge and feelings associated with the angle-matching tasks in the adult study on angles (this chapter) influenced students in the computer environment. Thus contexts and ecology of learning become important influences on learning.

Ecology and Visual Perception

One of the earliest theorists to discuss visual perception and ecology was Gibson (1979). He discussed the affordances that the ecology provided in perception. In particular, he noted the position of the head, the body, and the way in which the eyes were looking relative to the head in perceiving but he also noted the texture, curvature, and blocks to vision that the ecology produced that impacted on visual perception. Motion was integral to visuospatial perception and "ambient light" resulting from the environment impacted on visual perception. Thus like van der Heijden's processing model, further connection between context and perception results in visuospatial reasoning at a relatively basic physical level of the brain and nervous system. Ecology, however, impacts on the higher processes almost immediately as seen in studies with children crawling and viewing their surroundings (Cheng, Huttenlocher, & Newcombe, 2013).

Visuospatial reasoning can be involved not only in tasks with objects or drawings which are smaller than a person but also in tasks in which the person is part of his or her surroundings requiring spatial ability or visualisation in the larger spatial arena (Clements, 1983; Werner, 1964). Near spaces are first identified with recognition of where and what is there with developing discernment and discrimination (Newcombe & Huttenlocher, 2000). Thus we note that external contexts can feature in the development of visuospatial reasoning from an early age. Learmonth, Newcombe, Sheridan, and Jones (2008) showed when children were placed in a rectangular space, they were able to use geometric features such as the lengths of sides of the rectangular walls at around 18 months while they use the landmarks such as colour at around 5–6 years suggesting language was essential at this stage. Nevertheless, 18-month-old children can make decisions if the space they are using is small rather than large and so movement is available to them. A more definitive and comparative study suggested that children by 3 are able to use geometric features and by 4 they are able to use landmark information. In this study, walls of a rectangle with one wall being red were used but the child was not able to move outside the smaller rectangle so that nearness and distance were separated. As a result, they found that 3-year-olds were able to use previous experiences (four preliminary trials in which they were able to move to the outer walls) to successfully notice and use the features in a second set of four trials. This study suggests that experience assists students to make decisions. By comparing the various conditions of their experiment and three earlier studies, Learmonth et al. showed that effort was not an effect for the children not being able to make a correct decision in the larger room but the age of the children. The age at which students could make decisions based on features was between 3 and 4. "Spatial language may be one of these factors but not the necessary and sufficient condition for developmental change" (p. 424). The difference is not due to verbal versus visual strength. Instead an adaptive combination view suggests that both geometry and features affect decisions in which movement can assist attention to the spatial framework. Furthermore, young children aged 5–9 years have an ability to reason about nonlinear relationships.

Clements and Sarama (2007a, 2007b) noted that mental structures develop with what they call Euclidean or horizontal/vertical organisations associated with large and small objects. Clements and Sarama supported the point that interactions influence development but ecocultural perspectives best support a diverse range of findings that go beyond the more narrow studies of children influenced by western languages, perspectives, and built environments. Whatever the mental mapping, it seems that younger children (3.5 years) need to move through the space to show their visuospatial mapping. These are interesting results since later chapters (e.g. Chaps. 4, 6, and 7) establish the influence of Indigenous families moving with young children around their lands on the knowledge of space that these children bring to school (e.g. Pinxten & François, 2011; Pinxten, van Dooren, & Harvey, 1983).

Attention and Responsiveness

Attention for action (Allport, 1987) is an apt concept that links well with the model of problem solving that developed from my earlier studies (Owens, 1993; Owens & Clements, 1998). This is illustrated in Fig. 2.17 which suggests that responsiveness or action results from the complex interaction of cognitive processing. The term *responsiveness* implies a degree of understanding of the situation, involvement, and interest in the activity. The analysis of data indicated that cognitive processing embraced selectively attending, perceiving (e.g. listening, looking), visual imagining, conceptualising, intuitive thinking, and heuristic processing (such as establishing the meaning of the problem, developing tactics, self-monitoring, and checking). Responsiveness has an underlying affective aspect. With the changes in imagery, selective attention, and understanding, there is active progress in problem solving.

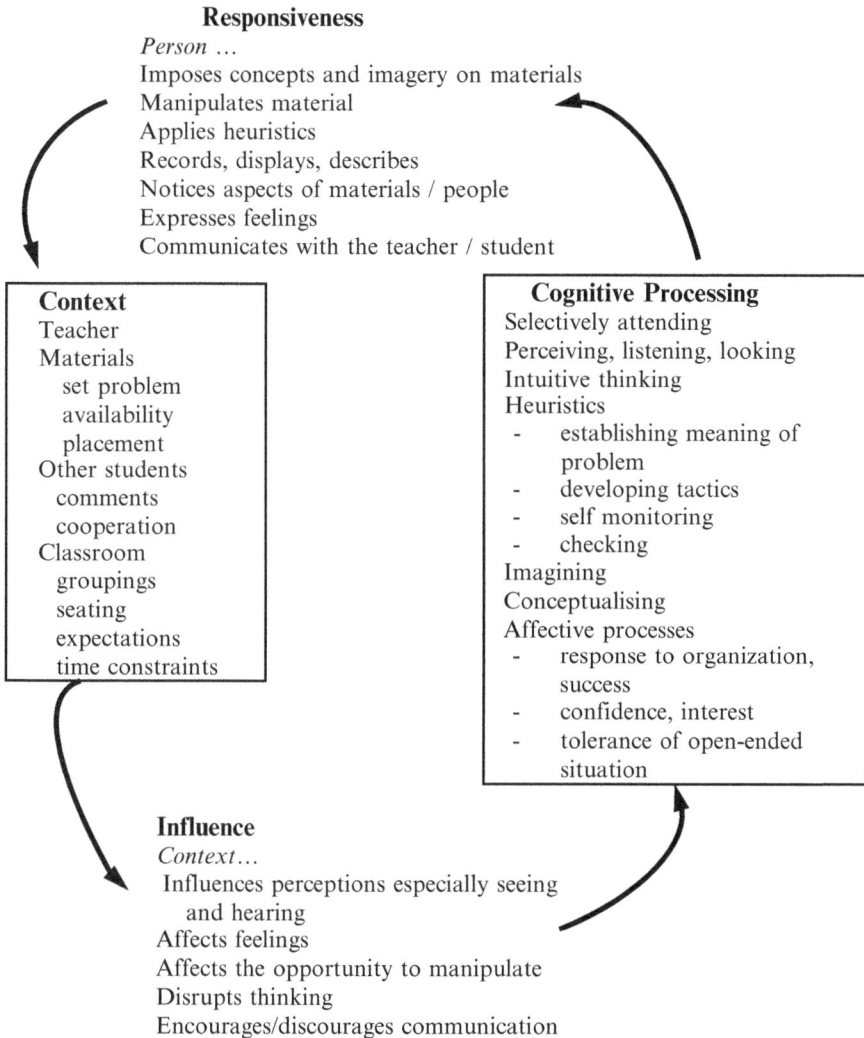

Fig. 2.17 Aspects of problem solving

Individual responsiveness also impacted on students' learning. There are several points to note about James' responsiveness in excerpt 1. First, a friendly competition existed between the students and this motivated them to participate and achieve (para. 1.01 and 1.06). Certain affective characteristics are evident in his behaviour—his responses to his successes (para. 1.01 and 1.06), his competitiveness (para. 1.01, 1.06, and 1.09), his desire to make shapes (para. 1.09), and his loss of interest at the end (para. 1.10). James' use of visuospatial reasoning influenced his responsiveness—not only his manipulation of materials (para. 1.03, 1.04, 1.07, and 1.10) but also his comment to his friend (para. 1.05) and his self-assessments (para. 1.06, 1.10, and 1.11) which tend to keep him on task. His visuospatial reasoning helps him to stay on task (para. 1.06 and 1.10). Third, he assessed or monitored his own progress on the task and this, too, influenced his responsiveness. He showed his monitoring by expressing how he was progressing (para. 1.01 and 1.06) and by changing his tactic in an appropriate way (para. 1.08 and 1.10). Finally, he expressed his understanding and knowledge (para. 1.03, 1.04, 1.05, and 1.11). The changes in his responses (para. 1.03 and 1.10/11) were precipitated by comments to him by his friend and by the teacher. Thus we see how his responsiveness was affected by (a) his understanding of the problem, (b) his use of visual imagery associated with comments by other students and the teacher, (c) his self-monitoring, and (d) his attitudes. At the same time, we can see how his visuospatial reasoning and tactics improved and influenced his responsiveness.

Materials or words spoken by others are important in students selectively attending and hence using concepts and images actively to solve problems (Owens, 2004a). While imagery has a role in generating intuitive responses, in inducing selection of and reflection on concepts, and in precipitating the direction of actions, the verbalisation of concepts often assists in interpreting perceptions and actions. In this way, conceptualising and verbalising are important in assisting meaning and later attention where analysis of the imagery is possible.

An example (already discussed above in the study of pentominoes from both adults and children) might clarify the role of selective attention and show why spatial concepts are constructed largely by idiosyncratic means. Students joined five squares to form different (pentomino) shapes. At first, some students made only symmetrical shapes or shapes with names. When they realised they were required to make more shapes, they realised that symmetry and a common name were not essential for defining "a shape". Their expectation influenced selective attention and initial schema location. Intentional and conceptual changes also occurred when they considered shapes in different orientations and the students developed their understanding of what constituted sameness and difference for that problem situation. (See Owens & Clements, 1998, for other examples.)

A subsequent study corroborated the findings this time in the context of adult students learning mathematics through interactive construction of concepts. An analysis of critical incidents revealed that interactions, affect, and responsiveness were important features of learning in a problem-solving classroom setting (Owens, Perry, Conroy, Geoghegan, & Howe, 1998). With further research, discussed later in this book, this model was modified to the diagram on identity in Fig. 1.2 taking even greater note of context.

Giaquinto also supported the argument that visuospatial reasoning requires an aspect shift (Allport, 1987; Owens & Clements, 1998). This particularly relates to disembedding and embedding as discussed earlier and motion as depicted often by arrows in diagrams although motion may be implied in other kinds of diagrams imaged with paper-based symbols and on computer screens (Giaquinto, 2011). Dynamic imagery entails motion in imagination (Laborde, Kynoigos, Hollebrands, & Strässer, 2006; Presmeg, 1986).

Shifts within an ecocultural perspective are evident in the thinking of Indigenous communities. For example, in Malalamai when asking about the relationship of houses to measurement and I pictured the square units tessellating depicted by squares with corners at the posts of the houses, the participant researcher Sorongke Sondo noticed half as much again as the size of the house which provided the additional floor space on which people sat, lay, and built rooms. In school mathematics, this perspective relates to ratio of areas rather than area units. When discussing the planting of crops, both Malalamai and Yupno people referred to the two equal lengths used for spacing plants at the points of equilateral triangles. However, comments were about the beautiful tessellating pattern of equilateral triangles represented by this planting. They had an overview of the shapes and the pattern but little way of connecting the geometry associated with the equal lengths to these shapes or patterns. The intention of the person was also influencing visuospatial reasoning. The villagers and myself attended to different aspects because our intentions in terms of cultural and school mathematics dominated our attentions and perspectives. Chapter 5 will provide other examples in which disposition, metonomy, motion, intention, and visuospatial reasoning impact on activity.

Visuospatial reasoning with number size and number lines is also cultural. While in western society most people recognise small numbers, it is less likely that one can immediately estimate larger collections as many Indigenous people do. In Chap. 5, I discuss the work of Paraide that shows that cultural context influences not only arithmetic knowledge but also the imagination. A similar result was found by Willis (2000) and by Treacy and Frid (2008) in Australia but not necessarily by others working with traditional representations on testing cards (Warren, Cole, & Devries, 2009). Furthermore, the cultural symbolism of a society impacts on visualising size of number (Giaquinto, 2011). For example, western societies are more likely to note the size of 0.45 than the binary 101101. This might not be the case in an oral society with a two-cycle system as found in PNG.

Furthermore, there is a tendency to have a left–right orientation of size for the number line in western societies. By summarising results from a number of studies, Giaquinto (2011) noted that participants' reaction times for deciding whether a number was greater or smaller than a given number varied when the smaller number buzzer was in the left hand compared to if it was in the right hand. The reverse was the case for Arabic monoliterates who read from right to left and reaction times were less strong for bilingual persons. It is interesting to note that societies with body-part tally systems such as the Oksapmin have strong visualisation of number (Saxe, 2012) but unlike the western number line, it might be considered that they have less of a sense of infinity since the last number tends to end at the point symmetrically opposite the first number such as on the little finger of the other hand from where the counting system starts (Owens, 2001c). While some PNG and Australian groups would want to complete counting at the end of the cycle, others considered counting people or reiterating the numbers since the notion of cyclic repetition was also important in the cultures and in those ways establishing an infinity in number.

However, image scanning, zooming-in, and extrapolating are tools available to be used on number lines when numbers out of current range are required (Giaquinto, 2011). While visual imagery, number sense, and the desire to illustrate concepts by drawing might be innate, the number line is based on cultural conventions. Non-written-symbolic cultures and young children will use a variety of representations of number, not necessarily a number line (Thomas, Mulligan, & Goldin, 2002). Thomas' study showed children's imagination with numbers written in a spiral but also school experiences such as a line of numbers and contextual experiences such as watching calculator screens changing with the constant addition of one. The whole recent movement on number learning (e.g. NSW Department of Education and Training, 1998), however, has emphasised the importance of figurative or visuospatial reasoning in the mind and much teaching and research is supporting this visuospatial aspect of learning arithmetic.

Developing a Theoretical Framework of Visuospatial Reasoning

Reviewing the earlier studies resulted in the development of a theoretical framework that could be used to inform teachers of young students' early visuospatial reasoning in geometry. The framework was also designed to build on ideas developed by The Count Me in Too project for arithmetic (NSW Department of Education and Training) through which teachers became familiar with such terms as emergent, perceptual, and figurative (imagery) stages. The success of emphasising both investigating and visualising together with describing and classifying for both part-whole and orientation and motion aspects of geometry is given in several papers (Owens, 2002a, 2002c, 2004b; Owens & Reddacliff, 2002). The framework is summarised in Table 2.6.

The actual activities (NSW Department of Education and Training Curriculum Support and Development, 2000) consisted of ten lessons where students make triangles, explore symmetry, build with blocks, and draw.

Developing a Theoretical Framework of Visuospatial Reasoning 83

Table 2.6 A framework for geometry based on visuospatial reasoning

	Investigating and visualising	Describing and classifying
Part-whole relationships		
	The student:	The student:
Emerging strategies	Attempts to put pieces together to see what is obtained	Matches shapes with everyday words, e.g. ball for a circle
Perceptual strategies	Recognises whole shapes used to build a shape or picture	Describes similarities and differences and processes of change as they use materials
Pictorial imagery strategies	Disembeds parts of shapes from the whole shape Matches parts of different shapes Completes a partially represented shape or simple design	Discusses shapes, their parts, and actions when the shape is not present
Pattern and dynamic imagery strategies	Develops and uses a pattern of shapes or relationship between parts of shapes Plans and dynamically modifies a shape to illustrate similarities between different representations of the same concept	Discusses patterns and movements associated with combinations of shapes and relationships between shapes
Efficient strategies	Assesses images and plan the effective use of properties of shapes and composite units to generate shapes	Describes effective use of properties of shapes to generate new shapes
Orientation and motion		
	The student:	The student:
Emerging strategies	Recognises shapes that match the child's fixed image(s)	Uses a shape word for a fixed image
Perceptual strategies	Recognises shapes in different orientations and proportions; checking by physical manipulation	Describes similarities and differences and processes of change as they use materials
Pictorial imagery strategies	Generates a series of static images of shapes in a variety of orientations and with different features	Discusses shapes, their parts, and simple actions when the 2D and 3D shapes are not present but recently seen
Pattern and dynamic imagery strategies	Predicts changes by mentally modifying shapes and their attributes using motion or pattern analysis Represent patterns and relationships of change by modelling or drawing	Describes a number of changes that will occur with one or more actions Discusses patterns and movements associated with combinations of shapes and relationships between shapes
Efficient strategies	Selects effective strategies to make changes needed to achieve a planned product	Describes effective use of properties of shapes to generate new shapes

Assessment Tasks

Teachers were also provided with assessment tasks for individual interview. These were also used to evaluate the programme. A number of carefully established cardboard cut-outs, drawings, sticks, and string are required and all tasks are presented so that students can show if they are using mental visuospatial reasoning before they are allowed to use materials in using perceptual strategies. Figure 2.18 provides some items from the test to illustrate how the task is presented with probe questions for extension or simplification. The first task is about recognising shapes (represented by cardboard cut-outs) in the environment.

The tasks did provide a range of strategies to be observed by different students. While students did not necessarily show the same type of strategy across all questions, there was a tendency for this to happen. Table 2.7 shows how one task could be used to decide what strategies were being used.

Task 4 (Fig. 2.18) shows how the skill of re-seeing parts is manifested in visuospatial reasoning while responses to the orientation and motion Task 2 (Fig. 2.18) indicated the development of orientation skills and noticing angles. Results for Task 6 (Fig. 2.18 and Table 2.7) on making triangles show how a carefully designed task can illustrate a full range of strategies. It was a particularly novel task for consultants and teachers.

Table 2.8 gives the results of assessment on the tasks (Owens, 2002b). These results indicate the effectiveness of the framework, series of activities, and teachers' professional development. The number of students who improved on each item and overall in the classes whose teachers undertook professional development and taught with the activities was significantly higher than those without the geometry lessons that emphasised visuospatial reasoning. This was the case whether professional development was through a consultant or a school facilitator. An attitude question also indicated that more students felt they were good at mathematics most of the time, more decided this because of self-assessment, and more recalled specific activities. Teachers confirmed that students enjoyed the lessons and remembered content well. Thus the framework implemented by teachers, the tasks, and assessments were valuable in increasing visuospatial reasoning but also in establishing self-regulation and positive attitudes leading to evidence of the development of a mathematical identity.

The tasks can be used for individual assessment or for the basis of activities for the class (see Owens, 2006a). The questions and probes can be used by the teacher to assist in students' learning and assessment during class experiences. The technology may be as simple as card cut-outs but computer-generated tasks could extend learning from previous activities with concrete materials. It is worth noting that Lehrer et al. (1998) had taken several different geometric and measurement tasks and had used probe questions and described different levels of assistance on each item, moving from more abstract to concrete to demonstrated responses. One of his questions related to transformation of a core square made up of four smaller different squares. Repeating transformations made a strip for a quilt.

Part-Whole Relationships
Task 3: Imagining shape completion

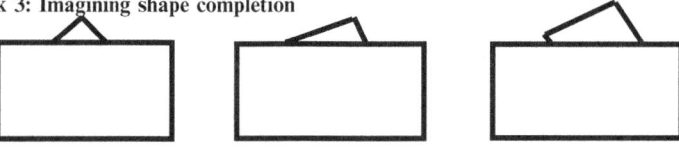

A square is gradually revealed. Each time, the student is asked what it might be and to trace where it might be. They are encouraged to give more than one answer.

Task 4: Reseeing shapes
Students use sticks of the same length to form 2 squares joined together along a side and then 2 triangles joined along a side. They are asked to draw the 2 triangles, while covered and are asked "If I take the middle stick away, what shape would I have?"

Orientation and Motion
Task 2: Angle recognition, visual memory, and rotation skills
Make the following diagrams on a circle using long and short sticks, point out the tab, let the student make the same diagram on their circle with tab mark aligned with yours. The first two are uncovered, the third is covered before the student starts, and the fourth is shown to the student, covered, and turned before the student starts.

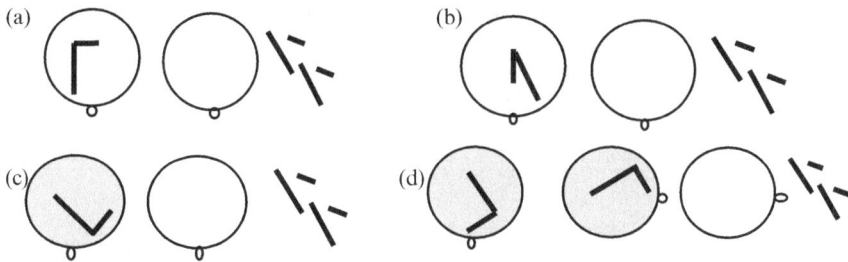

Task 6: Dynamic imagery
Use 40 cm string, joined to form a loop; a firm stick.
Place the loop of string on the table and hold two points firm, about 12cm apart.
Provide the student with the stick.
"Use this stick to pull the string tight and make a triangle."
How would you describe the triangle you have made?
Make other triangles?
How would they change?
　Probe: If the student cannot explain, let them use the stick to demonstrate and tell about the triangles they are making.
Point to one of the sides of the triangle.
Tell me what you would have to do to make this side shorter.
Point to the other side.
As the first side is made shorter, what will happen to this side?

Fig. 2.18 Tasks for assessment in Count Me Into Space

Table 2.7 Examples of different visuospatial reasoning strategies for a task

Visuospatial reasoning strategies	Indicators of investigating and visualising	Indicators of describing and classifying
Part-whole relations: task 3—imagining shape completion by tracing possible hidden shapes		
Emerging	Traces an edge	Says any shape name
Perceptual	Attempts to trace hidden shape or traces visible triangle	Says triangle
Pictorial	Traces for a triangle or square or rectangle or two of the same kind	Says triangle or square or "diamond" or rectangle
Pattern and dynamic	Traces possible shapes of varying sizes	Explains how the shapes change by lengthening or shortening the sides
Efficient	Indicates tracings and various changes	Readily explains how different shapes could be underneath
Orientation and motion: task 6—dynamic imagery using a stick to move a loop of string		
Emerging	Moves stick but does not make or recognise a triangle	
Perceptual	Makes a triangle	
Pictorial	Makes two or more triangles, e.g. right angle, isosceles	Knows names and properties of different types of triangles
Pattern and dynamic	Automatically slides stick to make different triangles Makes both acute and obtuse-angled triangles	Comments on changes to triangles and gives names of different types of triangles
Efficient	Shows an arc of points to shorten side	Explains why continuous range of triangles can be made in general

During implementation of the tasks, a grade 2 girl, who mostly showed emerging strategies and seemed to have most difficulty with describing and classifying, was able to show perceptual or pictorial imagery strategies in other tasks such as Task 6. These were novel questions for her and may have been less associated with her general struggle with learning shape labels in English. The assessment provided the basis to plan suitable activities for her. For example, she needed experiences in sorting and grouping many different kinds of triangles, squares, and rectangles (two kinds at a time); talking about the reasons for grouping, e.g. four sides or four corners; seeing shapes within shapes in matchstick type puzzles; doing more jigsaws; and making geometric shape like squares with tangram pieces.

On the other hand, a boy in the same class generally showed pattern and dynamic imagery strategies and an ability to see shapes within shapes assisted by good general language. However, his recognition of diversity when referring to a shape like "a triangle" still needed extension. He needed activities like matching parts of different shapes in order to notice similarities and differences, and to develop properties. He also needed more language to describe the parts and types of shapes. Interestingly, in Task 6, he showed some hesitation in positioning the stick to mark the vertex of the triangle to shorten its side, trying to indicate that it would be further

Table 2.8 Student improvement on assessment tasks

Task	Number (%) who improved with school-based facilitator		Number (%) who improved with consultant	Number (%) who improved without programme	χ^2 value comparing consultant and non-intervention group
	Group 1	Group 2			
Part-whole relationships	$N=135$	$N=193$	$N=140$	$N=75$	
Task 1	89 (66)	129 (67)	87 (62)	31 (41)	8.54*
Task 2	63 (47)	130 (67)	85 (61)	25 (33)	14.65**
Task 3	74 (55)	113 (59)	72 (51)	18 (24)	15.10**
Task 4A	95 (70)	131 (68)	74 (53)	22 (29)	10.94**
Task 4B	73 (54)	122 (64)	84 (60)	27 (36)	11.26**
Three or more tasks	79 (64)	141 (73)	77 (55)	20 (27)	15.83**
All tasks	17 (14)	40 (21)	19 (14)	0 (0)	**,†
Orientation and motion	$N=136$	$N=160$	$N=73$	$N=34$	
Task 1A	57 (43)	73 (46)	33 (42)	9 (26)	4.48*
Task 1B	63 (49)	98 (61)	Not included	Not included	
Task 2	41 (31)	69 (43)	43 (59)	13 (38)	3.97*
Task 3	73 (54)	94 (59)	42 (58)	9 (26)	8.97*
Task 4	72 (53)	81 (51)	44 (60)	12 (35)	5.80*
Task 5	66 (49)	80 (50)	38 (52)	8 (24)	7.70*
Three or more tasks	70 (53)	103 (66)	37 (51)	9 (26)	5.55*
All tasks	16 (12)	19 (12)	8 (11)	0 (0)	**,†

*Difference assessed by chi-square analysis is significant at $p<0.05$ level
**Difference assessed by chi-square analysis is significant at $p<0.01$
†No chi-squared value calculated because $n=0$ in one cell of the table

away than the line of the string (tending towards efficient strategies). He was ready to use properties to establish that squares are rectangles and that the same names apply when the shapes are in turned positions (a problem that can be exacerbated by the use of words like diamonds), and to use words like rhombus, trapezium, quadrilateral, or four-sided shape.

Moving Forward

This chapter has synthesised psychological literature around spatial abilities and around visual imagery especially from the last century. These were heavily influenced by psychological studies and experimental designs. Visual imagery research was particularly common from the information processing theories of psychology but several theorists have linked it to perceptual and contextual aspects of learning.

In my research, I attempted to draw all these psychological literacies together to discuss children's thinking when problem solving. I drew on qualitative research in order to get inside children's heads to see how they were visuospatially reasoning. Visuospatial reasoning relies on four skills (Wessels & Van Niekerk, 1998) that I elaborate as follows:

- Visual skills especially seeing and re-seeing aspects of the environment, objects, and shapes
- Verbal skills that support comparisons and decisions with words, and encourage interactions about the visuospatial reasoning
- Tactile skills such as cutting, joining, and folding that support or provide affordances in the visuospatial reasoning
- Mental skills especially mentally manipulating spatial images

Encouraging these skills together strengthens measurement and geometry education. These skills come together through pattern and dynamic imagery used in visuospatial reasoning supporting the learning of processes and concepts in measurement and geometry and expressed in conjecture, explanation, argument, and proof.

Visuospatial reasoning emphasises reasoning associated with and dependent on visual and spatial imagery but also expressed, developed, and argued spatially. Visuospatial reasoning is the important part of reasoning with visual and spatial imagery or imagination. It is a mental process linked to physically seeing and doing in a spatial world that has spatial relations. Geometry is about spatial relations. We reason not just in verbal written proofs often associated with high school geometry such as congruent triangles and trigonometry or circle theorems but with perceiving and interpreting diagrams. In primary school that reasoning relates to shapes, both two-dimensional and three-dimensional, their interrelationship, and lines; and to transformations and symmetries. It also relates to interpreting drawings. A drawing may be used as a metonymical representation of a class of shapes thus "knowing what a triangle is, is more than being able to label an equilateral triangle sitting on its base as a triangle" (D. McPhail, Count Me Into Space videos). Initially we know that students cannot always verbalise why a shape is, for example, a triangle—they seem to have a global understanding much as they do that a chair is a chair. On the other hand, a young student may just focus on the pointiness without seeing the whole or noticing other important properties. Students may also have a fixed image that needs to be developed by experiences. For example, one young boy making a triangle with a loop of elastic thinks that a right-angled triangle must be placed with horizontal and vertical sides. Students will realise that a variety of examples of a shape can be categorised as one particular shape. Students will begin to associate more and more properties or parts as necessary for that shape. They will also begin to decide what is not necessary for a shape to belong to a particular category. None of this is restricted to the school mathematics shapes. These comments could be noted in other ecocultural environments.

It is often thought that children need to develop words first but they in fact develop a visual image of a shape before they have the language to talk about it.

When children talk about their images, their explanations help them to clarify what is in their images and to develop their concepts. Children often say about triangles that they have properties like having "three sides" by rote (note how often children leave out that they are straight and intersect) but children need to be able to perceive these sides separate from the whole shape and to reason visually often by running their finger down each side as they count. A good example of physically representing visuospatial reasoning through dynamic imagery is that of pulling a vertex of a triangle formed on a computer screen or a piece of thin elastic. There are an unlimited number of triangles. Prior to reasoning in that way, children might only recognise a couple of images of triangles or think they are the shapes with "pointy bits and not corners". Without extending children's imagery of triangles they may have a prototypical first image and procept (Gray & Tall, 2007) or beginning conceptual understanding.

Visuospatial reasoning occurs when a child seeing part of a hidden shape says, "it can't be a triangle because it has two corners" (pointing to the right angles of the partially revealed shape) (Count Me Into Space video) or when the same child in seeing one "corner" and a triangular section of the shape can show that "it could be a larger triangle or an even larger triangle underneath or even a rectangle or a larger rectangle or a square underneath". Every time the child told us what shape it might be, she traced with her finger where the shape might be. In the research on this hidden shape task, one child from grade 2 said "it could be any shape". When asked what he meant, he called it by an imaginary name and traced out a zigzag line at the end of the imagined extended sides. (Being an English-as-a-second-language learner, this child had learnt to "play" with words and this strengthened his visuospatial reasoning.) Children can mentally slide, rotate, and turn over shapes or reflect them. By talking and pointing, students indicate that they notice parts and visualise their relationships. These are skills required in visuospatial reasoning.

Students learn to attend to the more important aspects of images, overcome initial static perceptions in favour of pattern and dynamic ones, and acquire appropriate mathematical conventions in developing and conceptualising visuospatial reasoning (Hegarty & Kozhevnikov, 1999; Owens & Outhred, 2006). Episodic and illustrative visuospatial reasoning is important in transforming visuospatial images to new situations as shown by the above examples such as Sally's tangram and pentomino problem solving. Diagrams need to be manipulative whether mentally and/or virtually and then visuospatial reasoning can be applied through the use of structures and propositions to new situations (Dörfler, 2004) as illustrated in the examples in this chapter. However, it depends on the valuing of the visuospatial representations and reasoning whether these remain significant in memory and purpose (Rivera, 2011). Gestures in cultural practices are mathematical representations in use and constitute the interface between embodied and cultural aspects of knowing and learning geometry (Kim et al., 2011). Significant are the manifestations of visuospatial reasoning, especially through actions, when two communities of practice merge whether they be western and Indigenous or community and school as the chapters that follow develop (Civil & Andrade, 2002; Gutstein, 2006; Téllez, Moschkovich, & Civil, 2011).

Visuospatial reasoning and a move away from stereotypical images and practices is important for visuospatial reasoning to be evident in all areas of mathematical problem solving. The following example illustrates this well as it links to limitations in both geometry and number as a result of teaching practices that fail to encourage students' use of visuospatial reasoning. M. Clements (whose work has been discussed earlier in this chapter) reported on a study by Zhang with grade 5 children (2012, p. 14). The teacher used a textbook that used area-model representations of fractions (circle and rectangle) following a Standards curriculum. At the end of the teaching, the children, and the teacher to a lesser extent, could only represent fractions and not use fraction concepts to solve a simple problem, "find a third of the way around an equilateral triangle". He noted that "These students thought about simple fractions in terms of parts of a circle, and many of them knew of nothing else" because of the overuse of one kind of "visual algorithm". A similar limitation has been found with base 10 block representations of fractions. However, to correct this through a verbal, especially symbolic representation, would be worse and curricula that encourage multiple visual representations should not be crowded so they result in visuospatial reasoning in only one context or medium. Rather there needs to be a visuospatial reasoning approach in which problems that require some visualising are set but then students are encouraged to act through heuristics such as to draw, compare with other representations of a third, compare with other fractions of this representation, and represent with another model. This argument also applies in geometry and measurement education.

The complexity of visuospatial reasoning and the way it relates visual imagery, spatial abilities, and other forms of thinking is important. Nevertheless, the case is established that the context, both within the classroom and in the community and indeed the school with its curriculum and teachers and government policies, is impacting on visuospatial reasoning. In fact, attempts to recognise visuospatial reasoning in the geometry area of mathematics, at least in Australia, have met with structuralist theories of development, rigid thinking of two categories of 2D and 3D separately, poor teacher content knowledge or pedagogical content knowledge, paper-and-pencil testing, and the view that visuospatial reasoning cannot be assessed by such testing without even realising its role in the test (Lowrie, Logan, & Scriven, 2012).

Some of the authors cited in this chapter made reference to the importance of context, in terms of perception in small and large spaces, in terms of development and reasoning, and in terms of classroom routines and expectations. My own research in classrooms indicated a strong influence of teachers, peers, and materials on children's ways of thinking and learning but also the role of expectations in learning. Learners rely on "deep, personal, and situated structures" (Goldenberg & Mason, 2008, p. 183) to provide a possible variety and range of examples of a concept but at the same time their attention needs to be drawn to the generality whether intuitively or by interaction with an external source.

However, what are the possible impacts of family and community's shared knowledge, values about aspects of education, and ways of teaching on visuospatial reasoning? In the next chapter, we will establish a case for considering visuospatial reasoning from an ecocultural perspective. Examples of culture and ecology and theories of education related to an ecocultural perspective will be developed.

Chapter 3
Changing Perspective: Sociocultural Elaboration

> *A paradigm shift occurs when a question is asked inside the current paradigm that can only be answered from outside it.*
>
> (Goldberg 1997)
>
> *... what young children learn and remember are things that arise as a "natural" and often incidental consequence of their activities ... Setting out deliberately to commit a body of information to memory is quite a different affair from such examples of natural or spontaneous remembering, where what is subsequently recalled is something one literally handled, attended to or in some way had to take cognisance of in the course of doing a practical activity.*
>
> (Dave Hewitt, 2001)

The Challenge

In the previous chapter, I illustrated that earlier research on cognitive psychology was limited in explaining how we learn about and use visuospatial reasoning. While it helps to establish how the brain may be processing information of this nature and how it is attending to aspects of visuospatial information, there remain questions about how the world outside our minds is indeed interacting with our mind. An exploration of intuitive, tacit knowledge shows the importance of situated learning. Much of this learning occurs in activity. Questions arise about bodily involvement in visuospatial reasoning and how practice—participation in activity—influences this reasoning.

There are also questions about how different ecologies, environments, societies, and cultures impact on visuospatial reasoning. There has been an argument for a sociocultural perspective in education for some time and these aspects need to be considered for visuospatial reasoning. I wanted to understand the impact of place and culture on reasoning. Since place has a spatial aspect, it is expected that visuospatial reasoning should be understood in terms of place and culture. First we need to appreciate how a critical approach might impact on our thinking especially about space and place. To do this we draw particularly on elaborations of a critical pedagogy of place, that is, how a questioning approach to the relevance of curricula and classroom practice might influence our view on learning and subsequent social justice issues (Gruenewald, 2008; Gruenewald & Smith, 2007; Somerville, 2007). Furthermore, people with a sociocultural perspective also espouse quality education as the child belonging, being, and becoming as expressed in Australian early childhood education (Department of Education formerly DEEWR, 2009) and no doubt elsewhere (Radford, 2006). How is identity perceived in these terms? What does this mean in terms of visuospatial reasoning?

Much of the work in this area results from anthropological, cultural, and semiotic research in cognitive psychology as an explanation of meaning and learning. Has it remained too focused on the psychological aspects rather than the sociocultural perspectives that might link, for example, intentions and self-regulative approaches to learning? (Macmillan, 1998). Part of our thinking in this area is also challenged by the genetic developmental and anthropological psychologists. These psychologists investigate by asking specific questions to different cohorts in order to look at the sense of meaning both for different age groups and also within the diversity of the cultural group at any one time. These approaches were used by Wassmann and Dasen (1994a, 1994b) and Saxe (Saxe 1985; Saxe & Esmonde, 2005) in exploring classification and number in remote communities of PNG. Saxe's (2012) theory of change over time but with recognition at any one time of diversity within a cultural group informs an ecocultural perspective of visuospatial reasoning.

However, there are questions about how there is change over time in our visuospatial reasoning in terms of both age and experience of fairly common activities such as schooling and what might change in terms of new types of experiences. The question can be addressed from a sociocultural perspective particularly by considering cultures where rapid change is now occurring. However, there is a need to explore how this ecocultural perspective can be integrated theoretically with the psychological research.

Local Context

One issue that van Hiele (1986) grappled with in his book was that of intuition. He felt that intuition was often incidental learning that occurs in everyday activities. A parent who uses the term "cylinder" to talk with a child when a particularly noisy tanker goes past encourages the child to note the shape and often the word. The parent is likely to draw a circle in the air by hand and describe it as round reinforcing the nature of the shape. The noise attracts the child's attention and the significant person, the parent, provides words and representations of the shape. Thus a reinterpretation of van Hiele's work illustrates that the ecocultural context of learning is

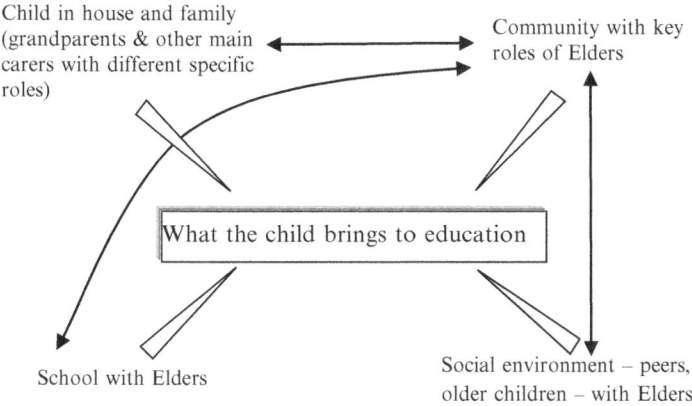

Fig. 3.1 The role of Elders in the education of the child. *Source*: Owens et al. (2012)

critical for geometric understanding. However, explicating significant knowledge is important as illustrated by the tanker example (Aikenhead, 2010).

Similar to other researchers' changing perspective (for example, Bishop, Barton, Gerdes, & Saxe), experiences with Indigenous people, their worldview, and their metaknowledge have resulted in my change in perspective about visuospatial reasoning from that of an internal way of thinking within an individual to that of a sociocultural phenomenon. A similar account of how people develop their local identity as one born and bred in a place, but not Indigenous, required Garbutt (2011) to be confronted by what it personally meant including resolving the problem of Indigenous ownership. Spatial links to an area as well as specific ways of thinking, kinship relations within the area, and a history in the area were all relevant. However, from my experiences in PNG, the term "local" referred to Papua New Guineans, generally born and bred in the place. Local was used to avoid the term "native", then regarded as derogative. Hence, I see spatial knowledge that is local as primarily Indigenous.

However, I was to be further confronted. In a forum with local Indigenous community Elders and others in Dubbo (where I now live), the group moved away from the Brofenbrenner's (Brofenbrenner & Ceci, 1994) ecological perspective. They gave prominence to the role of Elders in all spheres of life that supported a child. The child was not in the centre surrounded by the systems of school and family as Brofenbrenner had illustrated but the child in the family was supported by the family with the Elders' influence, the school with the Elders' influence, and the peers with the Elders' influence (Owens et al., 2012, see Fig. 3.1).

The child's knowledge was not independent of the sociocultural context. The child's learning was not just represented in the child's mind but the child in the family. The child did not just internalise knowledge shared by others but the knowledge was part of relationships within the family and with the Elders. Local takes on a meaning that incorporates the family and Elders' knowledge and their worldview and interpretation of space in terms of place. Thus space designations are part of relationships with people. The close connection between culture and place and people is identified. Local, however, can extend in boundaries and in time. Hence, contemporary culture and not just historical culture are part of the sociocultural context. The historical perspective is not forgotten in Brofenbrenner's model or in Indigenous or community's local knowledge, whatever the community may be. The historical perspective is embedded in memory and language and frequently embodied in routines, tools, play, practice, and activity.

Furthermore, border crossing or generative changes in spatial concepts and visuospatial reasoning occur. These changes result from the reasoning or even dogmatic or hegemonic representations of more powerful individuals such as teachers and curriculum writers (Aikenhead, 2010). Indigenous cultures have developed meaning and reasoning in mathematics and these ways of thinking are not necessarily found in western or schooled societies. However, individuals cross cultural boundaries in everyday activities and in school requiring education to be aware of the social injustices and the loss of world knowledge that occurs with dominance and lack of continuity of learning for individuals.

An Example of an Ecocultural Practice

By taking an ecocultural perspective, it is possible to make sense of what seems like discrepancies in PNG counting systems from the same language group. From my investigation into Lean's (Lean, 1992) thesis on counting systems (Owens, 2000b) and my own data, such as the sets of Alekano (Gahuku-Asaro) counting words given by different speakers (Eastern Highlands Province, PNG), there are not universal sets but a diversity of ways of conceiving certain numbers. Table 3.1 provides some numbers with associated embodiments. The discrepancies do not reduce the fact that there is mathematics but a more important ecocultural and individual identity is at play. Both men could understand the other and I as an outsider could understand and accept both based on my broad knowledge of PNG counting systems.

"*Logosigi* squared means 2 plus 2" but "*logosigi* to power of 4 means 2 plus 2 plus 2 plus 2". Since reduplication is common, many words are phrased as squared, e.g. kau^2 for *kaukau*, sweet potato. This is part of PNG humour even if it might be a visual representation that is confusing for school mathematics. It is likely that the variation in Table 3.1 is influenced by a neighbouring language (Bena Bena). Nevertheless across Gahuku-Asaro area, in different villages, slightly different counting words were given. Other words of interest are those words for large numbers. In Gavehumito village (2004), the teacher noted "100 *asasi ligizani luga luga* (stick or 10 hand finished finished); 200 *go' hamo* (*bilum* one); 1,000 *mulisi* (*hip*=heap)".[1]

More common were the various ways of combining twos and ones for 6 and 7 especially in languages with 1 and 2 as the frame words for the counting system or there is more than one way to say 4. In Wiradjuri, NSW, Australia, 4 can be expressed *bula bula*, or *bungu*, or *magu* (Grant & Rudder, 2010). However, there are more complex systems. For example, Paraide (2010) explains how the complex ways of measuring and counting in Tinatatuna, the Tolai language (East New Britain, PNG),

---

[1] *Bilum* and *hip* are words in Tok Pisin, the lingua franca of PNG.

Table 3.1 Alternative numbers in Gahuku

	First man	Second man		First man	Second man
1	Hamo	Hamako	8	Nigizani hamo asu o'oko makotoka logsive hamo ol omalago	Logosi[4] means logosi is repeated 4 times "means 2 plus 2 plus 2 plus 2 plus 2"
2	Logosita	Logosi	9	Nigizani hamo a su o'ko logosive logosive oli'o malago	Luguhagi luguhagi luguhagi
3	Logidigi hamoki	Luguha (logosigi moka)	10	Nigizani logosi asu igo (nagahuni hamo)	Golaha
4	Logosivi logosive	Logosigi[2] "Logosigi squared meaning 2 plus 2"	11	Nigizani logosi asu oko ligisaloka hamo oli'o malago (2 hands finished and one on the leg)	Golohaki hamakoki
5	Logosigi logosi hamo or nigizani hamo asu igo (one hand finished)	Logosigi luguhagi	15	Nagahuni makoki logosi logosi hamo or nigizani logosi asu 'olo ligisa hamo asuigo (2 hands finished–one leg finished)	Golohaki luguhagi logosigi
6	Luguha luguha	Luguha logosi	20	Nagahuni logosi or nigizani logosi aso'oko nigisa logosi asu igo (2 hands finish, 2 legs finish)	
7	Luguha luguha hamoki	Luguha logosigi makoki (segininaga)			

builds on both 10s and 12s (3 groups of 4). For example, 12 coconuts are called a *pakaruati* (one lot) and then 10 or 20 of these lots are named. These forms of counting are associated with practical activities and with the visual representations of the sets. For Hagen and related languages, there is both a (4, 8 cycle) system and a 10 system. Gestures are associated with both systems and different morpheme combinations are used for numbers depending on what is being counted, the ceremony, or purpose. From these various examples, even simple mathematical activities like counting are a way of expressing identity and the individual identity within the rich identity of the cultural group.

Values in Education

There is often an emphasis on the cultural conflicts that occur in schooling where colonial or western perspectives dominate. Furthermore, acculturation of mathematics allows school mathematics and those in position of power within the school system to dominate. However, I contend that there are important strengths of cultures that must be highlighted before acculturation or multicultural stances are taken. I am not sure that any of the choices of strategy generally espoused for intercultural relationships, that is, integration, assimilation, separation, or marginalisation, satisfactorily achieve the ecocultural perspective that is contended in this book. Like the Elders in my local Wiradjuri community and others whom I have met from other Indigenous communities (in Australia, Sweden, New Zealand, PNG), relationships between people are dominant. Many Pacific people are relational beings legitimised by

> Sacred relationships built on the values of tapu (sacred bonds), alofa (love and compassion), tautua (reciprocal service), fa'aloalo (respect and deference), fa'amaualalo (humility), and aiga (family); to them, culture is the core of their very existence, both individually and collectively. Nevertheless, being cognisant of the negative consequences of colonisation and forced acculturation among Pacific cultures is critical for working towards balanced intercultural relationships that can lead to positive outcomes for people of the Pacific. ... [Educational developments should align] with Pacific cultural values of shared responsibility, reciprocity, and interdependence. (Vakalahi, 2011, p. 87)

Thus I suggest that a strong recognition, extension, and valuing of cultural mathematics in a school setting are important for effective and efficient learning of other mathematics through an effective transition approach.

Signification, Meaning, and Becoming

One of the challenges for cognitive psychologists has been that representations of objects in the mind have not remained constant over time or place or people or indeed within a person. Objectification is indeed a probabilistic determinant or decision of best practice of people's responses to a stimulus. In other words, the debate between subjectivity and objectivity around meaning and interpersonal communication is around the certainty of words or diagrams signifying objects without dispute. Objective knowledge was challenged by radical constructivists and social constructivists (Davis, Maher, & Noddings, 1990; von Glasersfeld, 1991) through either an emphasis on personal, radical, individualistic construction of meaning or the social impact of others on meaning-making. One approach to finding a way forward was to talk of taken-as-shared meaning (Lerman, 2001). In that situation, cultural and social influences were recognised in discussions on mathematical concepts that might have been taken as a universal mental possibility for all persons even though some might favour visual or verbal or other ways of solving problems and knowing.

"Concepts of meaning indeed are based on presuppositions concerning the relationship between the cognizing subject and the object of knowledge" (Radford, 2006, p. 40). Referencing systems are related to the relationship between the person and the object and its position in space. Generalisations of patterns found in experiments (Peirce, 1998), no matter if embedded in everyday activity, are often associated with numbers but are established in terms of the sociocultural contexts and what they might mean for future communities. Intentions then closely monitor the meaning that is established for the person individually and the person within the cultural group. Thus, for example, the mathematics involved in building a village house in PNG requires consideration of many construction principles but also reciprocity, recognising relationships in terms of assistance and gifts. In a collaborative group task in a classroom, the value placed on collaboration, purpose, relevance, sharing, and appropriate communication will influence the resulting mathematising from the task.

However, by building on and modifying the constructs of Peirce (1998) and Radford (2006), I suggest nature (the environment) might indeed provide possibilities and restrictions on the experimentation, abstraction, and interpretation within mathematics. An ecocultural perspective is more comprehensive than either a psychological or sociocultural perspective. Furthermore, it is possible to extend these generalities and concepts by relating meanings and by making logical possibilities. Thus intention in learning is controlled by more than sensing and perceiving as the psychologists suggested in generative learning theories (Osborne & Wittrock, 1983) though these information processing theories provided a good theoretical background for earlier work on visuospatial reasoning. Furthermore, according to Radford (2006, p. 42), "ideas and mathematical objects ... are conceptual forms of historically, socially, and culturally embodied reflective, mediated activity". In that way, language plays a part in the establishment of meaning and representation and reasoning. Habits and patterns of language and activity provide a mediated realm for signifying objects and their position in space. This point is taken up further in Chap. 4.

Furthermore, it is possible to have an intuitive sense of relationships based on our intentions, according to Radford (2006) in expanding on Husserl's notion of praxis. Thus reflective practice increases the likelihood of learning and making meaning. For example, geometric generalisations are ideals that develop from the more specific examples that are generated through reflection from one to another. Such generalisations occur through visuospatial reasoning about the examples and how they can be modified to provide a different example but with certain properties maintained and represented in both words and diagrams. For example, a triangle made from a loop of thin elastic can be modified by pulling points or moving their position in a dynamic way as occurs if the loop is held by three fingers. A similar occurrence occurs in dynamic geometry software. The person may generate some triangles by chance but the construct is best appreciated when the person compares the examples and deliberately tries to make specific examples. Thus through praxis, conceptual meanings are generated. Through intention, we attend to certain aspects of the phenomenon or generalisation in order to create meaning and conceptual understanding. The intending and referring that occur in practice are creating meaning not only for the individual but also for others. It is the historicity of culture that provides the meaning for problem solving and learning and inter-subjectivities. "A set of morphological instruments, syntactic and lexical systems, literary genres, figures of speech, forms of representation of events, etc. that are part of our cultural inheritance" anchor the referring and signifying of the ideal of the object through interaction and negotiation (Radford 2006, p. 53). This set of representations provides our way of referring to ideas embedded in culture. In other words, "mathematical objects are conceptual forms of historically, socially, and culturally embodied, reflective, mediated activity" (Radford 2006, p. 59).

However, alternative perspectives, different ways of attending to the object or concept, and different people can reach different but probably related ideals. It may be that the spatial textures of the object or environment are what draw the attention or affect the intention of the person making meaning. This spatial aspect is encompassed in an ecocultural perspective.

Taking this critical perspective further, Soja (2009) suggests the meanings of human spatiality and related concepts such as place, location, locality, environment, and geography are best understood in terms of thirdspace, a metaphor to keep an open perspective on geographical imagination to incorporate multiple postmodern views as a space for race, gender, and class without privileging. Following Lefebvre (1991), Soja suggests social production of social space as a spatio-analysis. He envisaged multilevels of the right to difference (from body to nation). Firstspace—perceivable spatial dimensions and direct connections such as measuring; secondspace as spatial representations, cognitive processes, and symbolic meaning in the dualistic geographical imagination but now thirdspace as a critical discourse that problematises spatiality and avoids reductionism in interpretation. Thus an ecocultural perspective on space perceives and works with the overlapping perspectives of the various descriptions of a place. There are different ways (western and traditional) of describing a route that is also part of a mythological story of how the place was formed.[2] These multiple spaces intersect to provide new mathematical knowledge such as noticing unusual flat stones, the movement of water, the surrounding vegetation and the provision of food from the plants, distances embodied in walking the track, and the ownership and sharing of the land.

This open perspective on spatial knowledge provides for radical, creative mental spaces and relationships. Mental spaces consist of perceived spaces of practices and social relationships; conceived spaces that are representations; and lived representational spaces that embody clandestine meanings. The thirdspace is a social space that transcends the first two dimensions (physical and mental)—all three are "real and imagined, concrete and abstract, material and metaphorical" (Soja, 2009, p. 52). Soja presents "being" as having spatial, social, and historical aspects that are balanced in the three spaces in spatial imagination. This trialectic ontological assertion of space, time, and being-in-the-world impacts on epistemology and empirical analysis. It provides for a recognition of othering and an emancipatory change which is never fixed but permits lifespaces and life views in epistemologically derived situations such as education. Spatiality as a term prevents the dualism of physical and mental and becomes the perceived, conceived, and lived spatiality as a practical sense of the spatiality of social life. It is incorporated into our sense of visuospatial reasoning.

---

[2] An example are the song-lines found in Australian Aboriginal and PNG stories such as the half-man story, their river and valley told by men of Kaveve village, Eastern Highlands Province, or the songs of tracks associated with particular Foi hunters, Southern Highland Province.

Thirdspace provides resistance to hegemonic power. A new spatial awareness emerges as a product of spatial imagination that combines the physical and mental in a lived space, or location of culture (Bhabha, 1995). The ecocultural perspective as a thirdspace provides a critique of the dominant culture's liberal permission of cultural diversity. Thus the Yolngu women[3] wash cycad nuts in the same place at the same time of the year every generation, so the place has more meaning and more mathematical generalisations than just a western description of location. Lossau (2009) links thirdspace to borderlines and border crossings (Jegede & Aikenhead, 1999) as a strength rather than a limitation or negative compromise in learning and she suggests it provides an outward and expanding position, even as a hybrid. Thirdspace implies the myriad of circulating crossing of boundaries and movements from centre to periphery. However, it is important to keep multiple perspectives rather than reduce them to a common form. Lossau explains that the spatial metaphor permits ordering of ideals, connections, and relationships in the knowledge. It permits difference indicated by distance between ideals but spatial imagination can only be imagined by a hidden reification. However, I am concerned that difference is, like any representation, reducing difference to dimensions and positioning in all its implied meanings. To that I would bring into the metaphor the notion of topology rather than dimensions as a way of understanding difference in the thirdspace. Topology emphasises relationships rather than metered distances on various dimensions and thus reducing the ordering of different perspectives.

Furthermore the ecologies of the classroom itself, especially during cooperative group work, may also be important in terms of the positioning of students including the various mathematical and social identities of the student (Esmonde, 2009). The relationships between people in the classroom have variation, not only as a result of gender and language and prior experiences but also in terms of the contextual ecologies of the students. As Macmillan (1998) showed in early childhood settings, discussions were assisted by mathematical language and other aspects of game or role positioning within the classroom but language proficiency also positioned students in the social interactions. Teachers extended the students' play and thinking, thus permitting "identity formation by encouraging tolerant and accepting co-participation, responsible self-regulation and clear and negotiable access to resources, activity and meanings" (Macmillan, 1998, p. 122). The mathematical activities and discourses then lead to individual motivations of curiosity, challenge, choice, and imagination resulting in interpersonal motivations and socio-regulative interactions.

A Genetic Approach to Ecocultural Perspectives

A genetic approach considers how form and function of signs (language, objects, and symbols) change over time. In some ways, in a simplistic understanding of a genetic approach, I think it is a sad perspective in that it loses difference and

---

[3] Northern Territory, Australia.

multiplicity of visuospatial reasoning. It indicates a hegemony of language and practice and an acculturation or enculturation often into school mathematics rather than an enhancement of school mathematics. It is often marked by a loss of language and the role of perceived power (see Saxe's (2012) example of the teacher's authority in giving the changed counting words for Oksapmin). However, it does indicate that cultural identity is fluid just as change is inevitable. Language becomes an important part of change, or at least it is a visible sign of change. The use of language and signs does not necessarily precede the development of thought. Similarly, practice may or may not determine thought and skills. However, both language and practice facilitate communication and may facilitate thinking. However, we explore Saxe's genetic development model in more detail as it illustrates the impact of ecocultural change on valuing and reasoning.

Saxe (2012) does not think the amplification metaphor, that suggests signs and systems assist in the extension of thought, is appropriate for explaining change because the circumstances and the systems may be quite varied. It may not be the result of the sophistication of a system or representation or its function. Each system in practice develops its own complexities. To keep culture and cognition separate and to suggest that one influences the other fails to draw adequate analysis from the sociohistorical cultural and cognitive developmental processes. Saxe maintains that both culture and cognition participate in constituting the other. Both are processes. One thing to note is that there is variation in culture. Descriptions of cultural practice frequently ignore the diversity within the society but Saxe's explanation is similar to de Abreu's (2002) explanation about the common valuing of a practice in a particular ecological circumstance. Culture has representations, practices, and artefacts but diversity is also a property of culture and boundaries. As Saxe puts it, these properties of culture are osmotic in many ways resulting in variance and change. Cognition is a process and is not internal as opposed to external as in representations. Thus the representation is part of the cognitive process and it has cultural forms and cognitive functions for both individual and group activity. Form–function relations as processes in motion are constituted and shifted in the processes of microgenesis, sociogenesis, and ontogenesis.

Saxe has written clearly, engaging the reader in the first chapter with his personal journey into a little known realm of the world, the Oksapmin of the highland reaches of the west Sepik area of PNG. He follows this with his theoretical journey referring to various theories that approach psychology from an anthropological or cognitive (constructivist) perspective to present his genetic model that incorporates microgenetic development at the personal level and how this relates to the sociogenetic level as individuals in the society influence each other and over time to an ontogenetic level that leads to changes also in the individual and at the social level. This criss-cross influence is illustrated diagrammatically in Fig. 3.2.

Furthermore, the interplay between positional identities and the elaboration of mathematical goals may be explored. At the microgenetic level, change is not a mere transmission from individual to individual as in one person suggesting a new use for a particular idea but change for the individual is the result of development in society of the construction and use of ideas and their representations and

A Genetic Approach to Ecocultural Perspectives

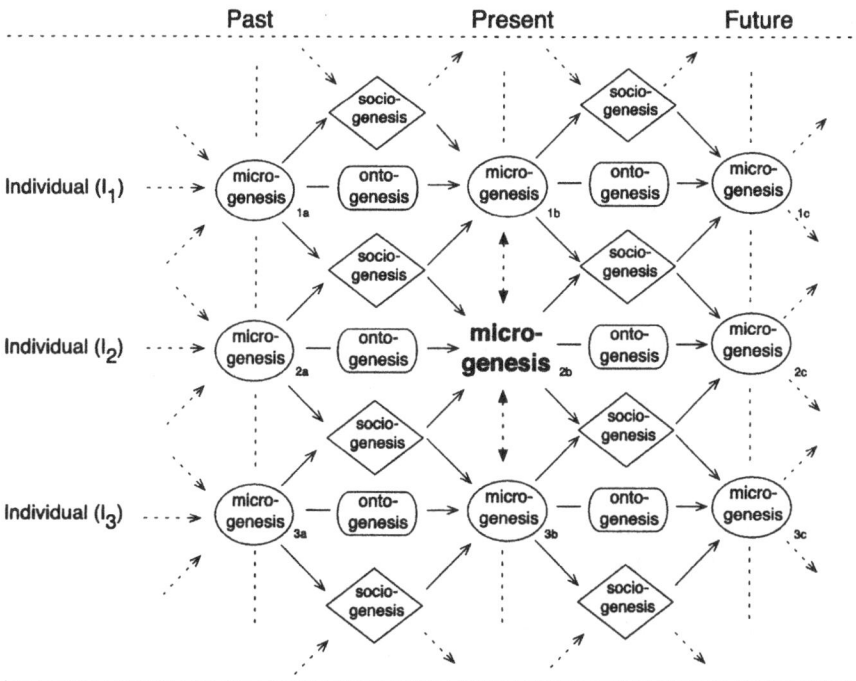

Fig. 3.2 The interplay of genetic aspects in cultural contexts. *Source*: Saxe (2012, p. 33)

modifications. This is the sociogenetic level. These of course limit the individual possibilities and provide an understanding of the form–function relations (ideas, objects, and signifiers and their uses) over time. Ontogenesis is the individual development over time. A blind person who uses a cane to navigate makes adaptive changes to spatial thinking through the physical manipulations and responses of the stick on objects during practice and in new ecologies. Practice in various situations permits adaptation to new situations but there may need to be sound or other guidance for the adaptations to be effective and efficient.

Saxe used an interesting research procedure. Initially in the field he observed everyday activities but also asked people to specifically explain, for example, how they counted which was quite visually demonstrated. They showed their system of tallying objects against parts of their body in a fixed order starting at the thumb (which is not a common starting point for body tally systems in PNG) of one hand to the small finger to the wrist, lower arm, inner elbow, upper arm, shoulder, ear, eye, nose, and then across and down the other side but going from the thumb to the small finger (not a mirror image which is also unusual among the body-part tally systems) (Lean, 1992; Owens, 2001c). Sometimes they stopped at 20, the elbow of the other arm, because that matched with 20 shillings equalling a pound (during the former Australian administration) and then adjusted that to $2 and finally to K2 (two kina note—all amounts below this are coins). Others went back up the arm to the elbow to 30. The visuospatial connection between shillings and 10 toea (the coins looked the same in size and colour[4]) were also used as a standard value for a pile of food in the market. In this study, there was already variation among the people. Saxe then selected specific groups of people to support the view that certain experiences were likely to be associated with certain approaches but to also show that it was a relatively common approach or valorisation within a specific age group.

Having come from an empirical tradition of psychology, Saxe describes his methodologies for a range of studies which he grouped into two areas. The first related to the influence of the outside money economy found in trade stores and used by others who had worked outside the community and to a lesser degree by those who sold garden produce and paid school fees only. The second area related to the two-way influence of societal and school ways. On his visits, 20 years apart, he selected certain divergent groups to interview that illustrated different degrees of contact. This also related to different visits across time providing a check on the genetic explanation of change. However, he did not lose the serendipitous opportunities and the opportunity to review videotapes. Furthermore, he has shared his visual research on the web (Saxe, nd).

I select here to evaluate the findings from his first study on the money economy. The number of interviewees was large, nearly 80 in four categories requiring walking a considerable distance between trade stores and villages. The addition and subtraction problems with or without 10 toea (10t) coins (still called shillings by some) were carefully counterbalanced. Each oral word problem was presented in terms of the trade-store buying and each number indicated on the body-part equivalent. The visuospatial reasoning using body parts and the trade-store experience are evident. Saxe's graphs present interesting differences in results but a good proportion of all groups with coins present correctly answering most questions, with more coaching of older people and of schooled young adults for subtraction (quite a few had been systematic but gave the wrong addition answer). Saxe noted a common understanding of the story problem without actually designating whether the problem was a difference or sum when coins were not present. The visuospatial relevance of coins and the oral communications were evident. It seems that the body-part tally system is indeed a built-in ruler-type representation for numbers that is easily used for problems to which in school we would apply arithmetic concepts of addition or subtraction. This method was used mostly by older adults and to a lesser extent by young adults and plantation returnees but not often by trade-store owners whose practice with costs allowed them, I contend, to visualise or recall in ready-reckoner form the ratio of values for different amounts of goods or for commonly added amounts.

A more intriguing method of responding to problems was the use of double-enumeration or matching of parts obtaining a result by one-to-one correspondences of body parts. Some people used words rather than body parts to match providing a more efficient response. For these people, it was clear that the body parts were a

---

[4] The coin is sometimes called a shilling and it is not unusual for a younger man to give the kina and toea values in pounds and shillings to the Elders, without any apparent effort in calculating.

number system. This method was used by a significant number of people across groups except the older adults. The body-tallies led to an interesting compensation method where both arms were used with the shoulder (10) being a critical part in the designation and transfer of parts (e.g. 16 might be 10 on one arm and 6 on the other). This method was used by a few of the trade-store owners and by very few others (mainly plantation returnees). In addition, by providing scores for advanced strategies that used body parts as a system to provide accuracy, Saxe was able to show a significant difference between groups. An analysis of variance, supported graphically, showed greater use by trade-store owners, with less use by plantation returnees and young adults with no advanced strategies or use of the body parts in a numerical system for older adults with problems without coins being present. The question remains whether this was lack of practice or indeed genetic development from social experiences. For either explanation, the visuospatial representation and reasoning are evident.

Saxe indicated that hegemony and position of people such as trade-store owners, teachers, missionaries, patrol officers, and returning employees can have a stronger impact on the development of practice than other people. Thus Tok Pisin methods of referring to currency were universal unlike the diversity of ways of developments in naming and valuing currency. The example of K5 is particularly indicative of people "playing" with words to develop a phrase to represent the K5. Unlike the restricted variety indicating ontogenesis for K2 or K20 resulting from the established 20 parts to the whole, K5 was referred to as a combination of K2, K1, and 50t, often with doubles such as two K2 and two 50t. While he might suggest how this illustrates both personal developments that become new forms of knowledge over time (ontogenesis) influenced by sociocultural developments, there are also important values in the play of words when children are learning about money and values. Similarly the various numeric equivalents for *fu*, "a complete group of plenty", are associated at different times and by different people with differing amounts in a consistent way. However, *fu* is also associated with doubling. There is no explanation about how this developed although doubling occurs to turn K1–K2, K5–K10, and K10–K20, each of which is marked by coins or notes. Doubling is notified by "married". The grouping of 20 is common practice among various language groups of PNG and for many travelled participants 20 became the *fu* rather than the original 27 in the body tally system. Visuospatial representations and words were linked through the double arms. With activity and interactions between people, there are changes both of signification and meaning but in each case the meanings between variant ways of naming the currency are in themselves important in reasoning.

Part of the reasoning or "play" with words for different amounts subsumes the visuospatial reasoning. For example, the use of the name "hole flat" for the K1 coin (which has a hole) or the word "leaf" for the green two kina note assists people to know these denominations and use these to make K5. It would be instructive for teachers to make use of these various forms of describing K5 to encourage learning rather than to just replace with English or Tok Pisin (very similar counting systems based on base 10).

In a similar way, when children are first learning to make and remember additions, use of the body-part tally system as a number line (equal spacing is not important because it is signifying counting numbers) in conjunction with English can provide a good visuospatial reasoning platform. The children may learn the order of the words because they are also body parts. They can see them. They can count to a number like six and count on eight by matching corresponding parts of the other hand and/or orally counting up another eight. It is worthwhile noting that it passes the 10 (shoulder) which takes on special significance now that there is a regularly used base 10 counting system in place. I have regularly noticed across different language groups in PNG young adults (whose schooling was in English and not their own language) using their traditional counting systems to make words (string of words or morphemes) for larger numbers by using different combinations of morphemes or words that represent smaller numbers (see discussion earlier in this chapter). In this current period of change, various systems are being used. So this "play" with counting words seems to be strengthening the arithmetic. If one person in Gahuku-Asaro says 18 is three groups of 5 and 3 and another says it is 2 groups of 5 and then 5 and then 3, both are correct and providing different arithmetic expressions. Both have visuospatial representations behind their reasoning (Gahuku-Asaro has groupings of five in the counting system). In forming larger numbers once the pair of hands has been used, people will "borrow" the hands of others standing around in the group with a nod at each person keeping track of their pairs (two hands) of fives. Thus we find the ecologies of people are influencing their construction of visuospatial reasoning behind their number combinations and descriptions and physical representations.

Saxe's theory does allow for the differences within a society and the changes that occur within a society. This was an area that Montiel and Managal (2011) were concerned about in looking at cultural identity. The degree to which there is continuity in change can be significant for the integrity of a person's identity of self-worth by which they make effective decisions. However, in terms of mathematical identity as outlined in the first chapter, self-regulation and the ongoing sociocultural interaction embedded in sociogenesis will impact on mathematical identity in terms of practice, values, satisfaction, and effectiveness. If practices are valued by the person and society in their ecocultural context, then mathematical identity will develop and the loss of cultural identity and ecocultural mathematics will be avoided.

Visuospatial Reasoning in a Navigating Team's Sociocultural Knowledge

In PNG, we had occasion to travel on sailing single-outrigger canoes along the coast or across large rivers as well as canoes in inland rivers and lakes. At other times we travelled on dinghies with outboard motors and larger coastal boats. But it was the use of sail, wave, and rowing around particularly difficult points that amazed us most. These canoes are not easy to steer, let alone balance and manage waves. The recognition of wave, current, and place even in the half light of the early morning and the automatic responses of the sailors and rowers as a group illustrated visuospatial reasoning that was shared among the men and embodied in their strong paddling and steering actions.

The breaking down of boundaries of inside/outside (social/mental) in cognition was a critical premise of Hutchins (1995) in his account of the cognition of a team of navigators on a US navy ship. His study of the team and their cognition showed that the calculations of position and the various knowledge and roles were partially represented by the breakdown in the tasks to be done in the social order of the ship's crew and relationships as well as in the navigation tools that were used. He provided several examples of the breakdown in the flow of information that illustrated well this inside/outside view of cognition. One was of the need of those having access to the chart, assisting the position bearer to find a required landmark. Access to experience and the chart permitted one crew member to give an approximate bearing by which the attention of the bearer was narrowed sufficiently for him to locate the required landmark. In other incidences, the reasoning prompted by a possible mishearing of a position or the inability to know what precise point was required for a bearing was overcome by an interaction of people through a written and oral exchange of information and through a physical gesture as well as a description. The reasoning was completed by a team working on the spatial problem together. In another incident, it was obvious from plots on a chart of bearing lines that three close beam landmarks were not as good as using a beam and another landmark closer to the line of travel. The chart facilitated one crew member pointing out this situation to another. To complete discussion of the team's visuospatial reasoning, he described how the team created their own shortcut or modularisation for speeding up calculations, albeit with an error initially under the pressure of a large moving ship that had lost power. In this critical incident, the team developed and shared a new mathematical approach which was not, unlike many other activities and tools, part of the legacy of navigation techniques. Nevertheless, it would be easy for another person or team to develop this method again.

One of the influences on Hutchins in interpreting the results of his study was his study of Micronesian navigators. He outlines how different cultures establish their representations of position, direction, and distance and I will return to this in Chap. 6. Firstly, in western society, familiarity with small maps representing large space presumes a position will be fixed by orthogonal axes. It is accepted that the Mercator representation of the earth is visually incorrect for representing distances and areas the further one leaves the equator for higher latitudes. Hence radio-beacon, the shortest distance between two places on the spherical globe along great circles, rhum directions which is by line-of-site, or straight-line reckoning are used, often simultaneously, to decide position. Over a relatively small area, these are virtually the same. Navigation charts become a computational tool on which lines and circles of distance are represented and by which computation and decisions can be made about time to reach other destinations at certain expected speeds. These charts and the visuospatial thinking of navigators who use them come from a history of western mathematics. Tools also include previously recorded information, tables for allowing for magnetic north effects on the compass, and routines for sharing tasks.

By contrast, the navigators of the Pacific, specifically those of Micronesia, use alternative but equally effective methods to navigate to unseen places without an extensive array of equipment found on large ships and yachts. There are in both navigation systems certain constraints. There are representations and computations in both systems. This is discussed in Chap. 6.

Visuospatial Reasoning in Measurement

Area measurement is a particularly problematic issue for children if visuospatial reasoning from ecocultural contexts is not incorporated into the learning. Rahaman (2012) showed that students' greater reliance on formal strategies to undertake problems related to area in other contexts reduced the use of their own strategies. She suggested the importance of incorporating contextual and visual experiences into the curriculum. One issue is that people can have two different visual images of a rectangle, one as a border and one as a filled-in rectangle associated with a two-dimensional space having area, an image required for measurement of area. Doig, Cheeseman, and Lindsay (1995) also found these alternatives occurred with children. They reported that more children placed given tiles around the rectangle as if a door frame but if they were given a drawn square, then more attempted to fill in the rectangle. Martin and Schwartz (2005) showed that students tend to realise the grid arrangement and its link to multiplication of areas with square units on small rectangles but not for larger areas when students resorted to addition rather than multiplication. Furthermore, the similarity of a rectangular unit also led to alternate physical and mental processes. It could be argued that ideas are not so stable in different environments requiring learning in different contexts and adaptable processes. In Chap. 5, I will discuss the ecocultural issues associated with visuospatial reasoning used by PNG people in comparing areas.

Visuospatial reasoning is particularly important in estimating and draws on ecocultural experiences. How important is estimation? A study by Adams and Harrell (2003) asked 17 people in different occupations:

1. For what kinds of tasks do you frequently engage in estimating?
2. Why do you choose to estimate instead of using a tool to obtain a measurement?
3. Why do you choose to use a tool to obtain a measurement instead of estimating? (p. 229)

The reasons for estimation are mainly to save time but they also do it to verify the validity of the measuring tools and methods. It might also be that precise measures are not possible or relevant or a quotation for a customer is needed or just because it is enjoyable. However, at other times, it is not appropriate just to estimate such as a product has to be consistent, risks are too high, it is inside the body or it is a new task. However, this form of visuospatial reasoning can be developed through practice, using multiple senses, encouraging it as a step in the procedure that requires prior knowledge, and for making decisions. These kinds of estimations are happening for "employees, customers, consumers, and participants in recreation ... estimation is really at work!" (pp. 243–244). Ecocultural context is a vitally relevant part of the visuospatial reasoning of estimating.

Embodiment of Spatial Knowledge

Dialling a telephone, using a calculator, and touch-typing are all examples of embodiment of knowledge of the position of numbers and letters on these tools. Pilots of planes also have spatial imagery as well as visual imagery on which they make split-second decisions (Wickens & Prevett, 1995). From Hutchins' writing and that of others who have sat with the Pacific navigators, there are embodied memories as well as mental memories assisting with the visuospatial reasoning and decision-making. In the Caroline Islands, the tilting of the head to 45° provides a kinaesthetic means of selecting the angle of inclination to view the star constellations (Worsley, 1997). This embodiment of direction is also evident when sailors can feel the state of the swell under the canoe. This may assist in positioning the canoe in reference to the island (leeward, side, or front of the island from the direction of the swell) (Bryan, 1938). In the Marshall Islands, the sailors have sea roads that are taken regularly and that take account of the swells and currents. They physically attend to the forces of nature in determining the extent of travel in a given time, so they take account of fast moving waters or winds appropriately.

PNG Indigenous peoples know of trading and cultural partners far away as observed by their long-distant trading circles such as the kula trade around the Papuan Islands, the Hiri Motu trading and winds along the south coast of PNG, the Rai Coast-Madang trade, and the highlands to coast trade. Australian Indigenous peoples also travelled long distances for trade, relationship building, seasonal adaptation, and knowing the land. The time of year, the winds, and the cycles in food production all influence these navigations together with trading goods such as food, pots, stone, oil, shell money, knowledge, and salt. Distances may be associated bodily with time taken to cover the distance. The crew of the sailing canoe respond to the boat load, swells, winds, and position of reefs (and land if visible) to adjust paddling and sail position (personal experiences). Similar skills are then applied to small "banana" (fibreglass) boats with outboard motors. These tools of movement are then part of the cultural identity and response to space and place.

Early Childhood Experiences

Nativist perspectives might suggest hard wiring like language modularity and sequencing in the brain (Butterworth, 1999; Dehaene, Izard, Pica, & Spelke, 2006). However, this nativist view of spatial sense is only part of the story about visuospatial reasoning. Yakimanskaya's (1991) social theory suggests spatial thinking is based on one's needs, urges, and motivations and that advanced thinking engenders frames of reference requiring active and dynamic manipulation of objects or activity. However, these frames of reference are culturally bound as suggested by Pinxten, van Dooren, and Harvey's (1983) anthropological theory of universalism. They gave three types of space divisions: physical, sociogeographical, and cosmological spaces. They noted all three spatial semantic categorises of near, separate, and contiguous (3 of 118 terms for the Navajo) are used by different cultural groups. Thus we need to explore the bridge between ecocultural and psychological perspectives of visuospatial reasoning.

In Chap. 2, visuospatial reasoning in early schooling was illustrated and examples of different responses were given. The chapter shows the limitations of psychological theories especially that of Piaget for which topological thinking and conservation were considered necessary before visuospatial reasoning could begin in the areas of classification, orientation, motion, and part–whole relationships within shapes. It was also established that in classrooms, students' attention resulted from both internal thinking and classroom context. Hence we began to see that a broader perspective was necessary for understanding early childhood visuospatial reasoning. Ness and Farenga (2007) provided a strong argument for visuospatial reasoning in the everyday context of block play. In spatial development, mental constructions of space are developed after the activity and are culturally bound. By analysing videotapes of children in block play, they developed a theory of how children learn to think spatially and scientifically. They observed patterns of behaviour and development of process skills and cognitive abilities that showed how children begin to learn about space and architectural relationships. As a result they presented a new, alternative way to measure cognitive abilities and development in children noting that topological thinking is not opposed to Euclidean relationships but rather the lack of some Euclidean relationships like parallel lines has not yet been established. They also counter the applicability of Piaget's theory based on cross-cultural and socioeconomic concerns (see also Dasen & de Ribaupierre 1987; Opper, 1977). Socioeconomic status, culture, and family background also impact on mathematical achievement (Ginsburg, Lin, Ness, & Seo, 2003; Pappas, Ginsburg, & Jiang, 2003; Sukon & Jawahir, 2005). Gerdes (1998) has consistently shown the rich mathematics of African groups and illustrated how they link with school mathematics, for example, in the reflections and symmetries of the women's *latima* drawings.

In Chap. 2, intention was noted as an important part of visuospatial reasoning directing attention and responsiveness. Ness and Farenga (2007) expanded on the role of intention as a search component of visuospatial reasoning. Intention can lengthen the time spent observing and attending to aspects of the environment (Baillargeon, 2004) showing that young children may have greater cognitive appreciation than Piaget contended. The link between action in say Logo programming and abstract geometric concepts is evident in school children.

> Spatially structuring an object determines its nature or shape by identifying its components, combining components into spatial composites, and establishing interrelationships between and among components and composites. (Clements, Battista, & Sarama, 1998, pp. 503–504)

Studies with small children illustrate how spatial as well as visual experiences are important in visuospatial reasoning (Greenspan, 2007; Learmonth et al., 2008). Since visuospatial reasoning has both a spatial and visual aspect, Greenspan (2007) suggests that a child who has difficulty in remembering and problem solving in a spatial environment needs more practice in problem solving in highly motivating visual–spatial situations. In some cases, the tasks need to be simpler and small, so success is achieved in navigating spaces. For example, block play provides opportunities for visuospatial reasoning through rotating and joining objects to create the playful idea. Some of the play indicates early measurement concepts such as larger and smaller in area, volume, or length. Furthermore, geometric and architectural concepts are also used in block play in making enclosed and connected spaces, straight surfaces, right angles, edges, and balancing blocks (Clements et al., 2004; Ness & Farenga, 2007).

However, Learmonth et al.'s (2008) experiments go further in suggesting that visuospatial reasoning is not merely the linking of spatial features with language. The ecology and size of a space require not only movement experience but also other spatial features to solve problems of orientation. When features in the space are closer or the space itself is smaller, young children are more able to orientate themselves than in larger spaces or with features being more distant to them. This applies to children much younger than the 6 years that Piagetian studies may have indicated. Landau, Gleitman, and Spelke (1981) noted that a child blind from a few hours after birth and blindfolded adults can be shown points in space twice which they touch and then they are able to move between them along any line, albeit not straight but as if determining gradually the position. This would suggest an embodied spatial awareness. However, Cheng, Huttenlocher, and Newcombe (2013) have reviewed the literature on navigation and reorientation in a relatively small space and suggested there are more factors involved than at first realised. There is the recognition of geometric features found in spaces with corners and surfaces meeting (with less adaptability for animals raised in circular environments). There are the landmark features and horizon features especially relevant in larger spaces. There are a range of cues and values from the environment that direct attention and encourage navigation. It seems that even unconscious visuospatial thinking is affected by the environment.

Mathematics should build on the child's sense of place from the beginning. An important role for formal education is to ensure that children explore their space and be given the tools by which to explore this space.

> Children, developing at their own individual rates learn through their active response to the experiences that come to them; through constructive play, experiment and discussion … (to) become aware of relationships and develop mental structures which are mathematical in form … about … spatial aspects of objects and activities which … (they) encounter. (The Mathematical Association 1956 Report cited in National Research Council Committee on Geography, 2006, pp. v, vi)

However, this link to home and culture may not be recognised by mathematical curricula and school systems especially for Indigenous communities.

When we look at the processes of education and the theoretical underpinnings for the educational process, there is much to be learnt from studying different cultures. There may be ideas about concepts that are foundational for our understandings but need to be made more explicit for school and early childhood education (Ness & Farenga, 2007). Furthermore, there may be different constructions of concepts that differ from those commonly pronounced in western developmental psychology that states the child constructs relative (projective) spatial concepts, those in relationship to his or her own body (in front, behind, left, right), before developing absolute (geocentric) concepts. In a small study in Bali, Indonesia, Wassman and Dasen (2007) found indications that in some cultural and linguistic contexts, this sequence could be reversed. Using two tasks they found that young children in Bali used a completely absolute (geocentric) encoding of spatial arrays; older children and adults, while also showing a preference for the absolute encoding system (coherent with the culturally sanctioned orientation system), were also able to use a relative encoding. If confirmed by further research, this would be the first demonstration of a reversal in stages of cognitive development that dominates western and hence current early educational theory worldwide, and an argument in favour of (moderate) linguistic relativity. However, Dasen and de Ribaupierre (1987) summarised a number of studies and showed that Indigenous cultures in both Africa and Australia despite a degree of acculturation favoured the spatial area and showed equivalent reasoning capacity to western neo-Piagetian testing but not necessarily in quantification suggesting child-rearing practices or ecocultural experiences strengthened their visuospatial reasoning. They also suggested that to give levels and order in development was not possible without similar ecocultural backgrounds (see also Fischer & Silvern, 1985). Thus Piagetian and neo-Piagetian perspectives and views of levels and indeed mathematics education as set out in curricula often limited our understanding of children's visuospatial reasoning.

There are other examples to be found in the literature where there is variation from the dominant western view. Harris (1989) noted that small children responded to north–south, east–west descriptions in remote desert Australian Indigenous groups including with north referencing the direction that the speaker was looking. Spencer and Darvizeh (1983) compared the route descriptions of British and Iranian preschool children. The latter group gave more vivid and fuller accounts of sites along a route, but less directional information than the former. By 3 years of age, the two groups of children were found to communicate spatial information to others in the manner of adults in their culture, suggesting that communicative competence in the spatial domain involves the acquisition of culturally patterned skills for describing space. This linguistic feature is further expanded in Chap. 4 on language.

Cultural artefacts like maps and diagrams also facilitate an individual's visuospatial reasoning. Even preschool children can acquire a sense of large scale space from maps (Clements & Sarama 2007a). This visuospatial knowledge may be restricted to certain cultural groups as indicated by studies reviewed later in this book.

Critical Perspective of Place

We often need a stimulus to think differently about mathematics whether it be linking cultural experiences with technology (Eglash, 2007), recognising non-European mathematics (Joseph, 2000), or emphasising a social justice stance supported by ethnomathematics (Civil & Andrade, 2002; D'Ambrosio, 2006; Gomes & D'Ambrosio, 2006; Tutak, Bondy, & Adams, 2011). In this chapter, I have been developing a critical approach emphasising the importance of sociocultural and, to a lesser extent, ecological aspects that impact on spatial relations for reasoning. In this section, I will expand on the critical approach and show that ecologies need to be considered as well as sociocultural aspects. Thus I develop an ecocultural perspective. This perspective is centred in the students' understanding of space from their experiences in the place (Gruenewald & Smith, 2007; Somerville, Power, & de Carteret, 2009) and their culture's systematic ways of thinking about space. In this way, space understanding is place based.

While some critical theorists note the importance of recognising the hegemony of the more dominant in society (Freire, 1992; Giroux, 1997), locally, nationally, or globally, and that this hegemony may result in disempowerment based on race or class or locality, place-based education emphasises that the ecology of place itself impacts on the learner and on education (Bush, 2005; Gruenewald, 2008). Thus ecology of place for establishing funds of knowledge with the community and family (Civil & Andrade, 2002; González, Moll, & Amanti, 2005) and the impact of those funds on the thinking of the teacher who involves the community and family in education are critical influences on pedagogy.

Ferrare and Apple (2012) suggest that the tools of visuospatial reasoning and the spatiality of ecology apply to education at different levels, that is, the size of the space. It may be the space covered by a curriculum at the policy level and how that is perceived to be enacted. It may be the issues of home-schooling, rural and urban schooling, or schooling and neighbourhoods. At the micro-level, there are the spaces within classrooms, spatial positioning and spatial arrangements for students in classrooms and the "messages" that they contain, and facilities and other aspects related to schooling, e.g. for the child with disabilities. It is also about the place of schooling. In a study in Brazil, learning was in the park for the street children while Sweden and Alaska talk of outdoor education and PNG's cultural mathematics encourages teachers to go out to the gardens and places where mathematics is carried out in activities. When working with Indigenous students in NSW, motivation was gained by looking at shapes of parts of plants and leaf arrangements, uses of the plants, and their positions in the outdoor environment. A number of studies in this region have involved mapping community areas, planting gardens with plants of significance to their Indigenous cultures, and other activities outside in the community but all with their Indigenous Elders.

Ferrare and Apple (2012) also discuss the value of having some spatial analysis tools to investigate and apply spatial theory to education. For example, social analyses and mapping tools are important to ethnomathematics and the placement of schools and the population attending the school in terms of cultural diversity. So in PNG, the urban school may use the lingua franca but a village school might be on the grounds of a particular language group and use that language. Other language groups may or may not utilise their Tok Ples language and there are varying influences of neighbouring languages, proximity of languages, and the social networks, e.g. through marriage with other language groups. As we saw in Saxe's study, access to towns or work created other linguistic changes for their own language and its maintenance. In addition, spatiality influences the connections made between a developing economy and the position (value and location—space and place) of local lands.

Furthermore,

> Bowers (2001) advocates "eco-justice" as a critical framework for educational theory and practice. Eco-justice has four main focuses: (a) understanding the relationships between ecological and cultural systems, specifically, between the domination of nature and the domination of oppressed groups; (b) addressing environmental racism, including the geographical dimension of social injustice and environmental pollution; (c) revitalizing the non-commodified traditions of different racial and ethnic groups and communities, especially those traditions that support ecological sustainability; and (d) reconceiving and adapting our lifestyles in ways that will not jeopardize the environment for future generations. Like critical pedagogy, eco-justice is centrally concerned with the links between racial and economic oppression. (Gruenewald, 2008, p. 310)

For example, Gutstein (2007) involved his class by considering the issue of a block of land that could be a play area for the youth and children of the barrio in a US city. The project involved the class in mathematics, visuospatial reasoning with walking to the place (embodying knowledge), maps, and measurement. They used their mathematics as part of their argument for the play area. Thus the sociopolitical activity gave value and meaning to mathematics (Mellin-Olsen, 1987). There is a connection between culture and place that visuospatial reasoning can enhance. Furthermore, place also links with relationships between people and the environment. Thus visuospatial reasoning is also critical for the environment and its relationship to people.

> Woodhouse and Knapp (2000) describe several distinctive characteristics to this developing field of practice: (a) it emerges from the particular attributes of place, (b) it is inherently multidisciplinary, (c) it is inherently experiential, (d) it is reflective of an educational philosophy that is broader than "learning to earn", and (e) it connects place with self and community. … Current educational discourses seek to standardize the experience of students from diverse geographical and cultural places so that they may compete in the global economy. Such a goal essentially dismisses the idea of place as a primary experiential or educational context, displaces it with traditional disciplinary content and technological skills, and abandons places to the workings of the global market. (Gruenewald, 2008, p. 314)

Engagement with experiences in space results in valuing place (Tuan, 1977). Sobel (2008) emphasises that children will develop a sense of place through different stages but the beginning is being in the place and being able to act within that place. It is our childhood experiences that give us a love of place and begin our literacy of space and place (Somerville, 2007). A sense of space as place develops through movement (crawling, bike-riding, exploring, and walking) in the place, and visual, aural, and olfactory images of places that arouse a pleasant emotion (Sobel, 2008; Tuan, 1977). The mind discerns a network of places or objects in a spatial relation and distinguishes the unique characteristics of shapes and objects. Furthermore, these experiences occur within a culture and family so that the values, relationships, and ways of thinking about the place are embedded in culture and associated language.

An ecocultural perspective is more than just engaging students; although that is important, it is also impacting on how they reason. It is this argument that the following chapters address. While classrooms and state curricula limit experience and hence reasoning, place-based education shows how visuospatial reasoning can develop when "socioecological places" are a part of "exploration and action" and developing relationships (Gruenewald, 2008, p. 318). For Indigenous communities, there is a decolonising and reinhabitation of thinking. Reinhabitation also connects and provides continuities with living in a place and utilising visuospatial reasoning in the process. It provides for globalisation and transformation of cultural perspectives as a matter of course but in emphasising visuospatial reasoning, it is possible for adequate decision-making relevant to a community and culture in the geographical space in which they live, supporting their sense of being and belonging. Decolonising thinking requires a return to cultural ways of visuospatial reasoning that maintain places and relationships and ensure that with school mathematics education, these cultural ways strengthen and extend beyond the limitations of intended and implemented curricula and high-stakes examinations. Embedded in this view is diversity of visuospatial reasoning that may come about by diverse ecocultural contexts. One way of exploring this diversity is to study cultural and natural diversity affecting visuospatial reasoning and community decision-making resulting from the reasoning.

This ecocultural approach to teaching and learning requires intellectual inquiry, practical inquiry, and emotional inquiry by the teacher and the student (Barnett, 2010; Department of Education formerly DEEWR, 2009; Pinxten et al., 1983). Ecocultural knowledge inquiry is centred in place, family, and community and is frequently undertaken in practical activity but it has a strong connection in terms of identity and meaning (emotional inquiry). Activities relevant to the community may strengthen the conceptual development of spatial concepts and indeed strengthen relationships between the mathematics of the community, students' understanding of mathematics, and other aspects of life. Indeed, this ethnomathematics is richer than school mathematics which may restrict the mathematical thinking of the user or learner. For example, the referencing of space may link to activity and associated visual imagery providing not only a symbolic and verbal representation for learning but also a visual, kinaesthetic, and episodic connection. The ecocultural mathematical identity of the student and the contextualisation of the mathematics are strengthened. The specificity to place may be closely linked with linguistic structures and the purpose for the mathematics. It is not, however, necessarily restricted or limited in its system of mathematical relationships since the principles can be applied to new spaces and shapes. Furthermore, they can be adapted to new representational systems as occurs with the nexus of school mathematics and communication with outsiders to a community. Such inquiry results from appreciating space and geometry within a global transcultural community.

Moving Forward

Rather than seeing mathematics as a body of knowledge to be learned through fixed levels of presentation as structuralists like Piaget and van Hiele suggested or curricula often suggest, ecocultural perspectives on visuospatial reasoning imply that there is not a dualism between the mind and objects but that learning is an active involvement of being and becoming enshrined in one's identity. Mathematics is not just a set of procedures or tools to be used but it is enacted through history and society by people's responsiveness to an ever changing world. Furthermore, individual and collective knowledge are "in dynamic, co-specifying and ecological unity. … Systems of mathematics can be as divergent as each of our histories" (Davis, 1999, p. 331). How then do individuals communicate this knowledge? How might ecology, society, and culture impact on visuospatial reasoning? How might ecology impact on teaching? We will consider the first question in the next chapter, the second question will permeate the rest of the book as we consider different societies, and the third question is addressed specifically in Chap. 7.

Chapter 4
Place, Culture, Language, and Visuospatial Reasoning

> *One doesn't understand concepts and 'then' solve the problems, one understands concepts 'by' solving problems. One doesn't communicate mathematical ideas 'as well as' reason about mathematics, rather, 'through' communication one refines one's mathematical reasoning.*
>
> (Elizabeth Badger, 1992)

The Challenge

In Chap. 2, psychological theories were discussed establishing that people store information that is accessible as either verbal or visual information either mentally or physically. Language was established as a representation both mentally and externally. Visual information was perceived as diagrams although the physical look of formulae, for example, can contain a pattern for the memorisation of material (Presmeg, 1986). There were different strategies for visual imagery in the mind including emerging strategies, perceptual strategies, concrete pictorial imagery, dynamic and pattern imagery, and efficient strategies. The last two were closely linked to reasoning with imagery. Furthermore, dynamic and pattern imagery may have been associated with bodily movement. However, a question arises about whether a language approach to education leads to a cognitive psychology focused on representation or does it relate to other cultural aspects or place-based aspects of learning.

Language development occurs as students learn names of shapes through family, television, pre-school experiences, blocks used in play, and in school activities where the dominant approach is to use western school mathematics. Schooling, home facilities, and language/culture influence the making of designs and arranging shapes. Everyday experiences impact on spatial language, so social inequities (Cross, Woods, Schweingruber, & National Research Council, Committee on Early Childhood Mathematics, 2009) and cultural differences for learning geometry may be substantial.

One issue is that of support for homework and another is the availability of reading material and encouragement for reading (Sukon & Jawahir, 2005).

Questions arise about whether language is a significant aspect of visuospatial reasoning and whether language changes the way one thinks about visuospatial experiences. Is it possible to have a window into this reasoning from a linguistic perspective? How do languages show different ways of visuospatial reasoning? How might this relate to identity?

From an ecological perspective, a place is a space inhabited by people with values and relationships and, as a result, the place takes a significant role in meanings and specific ways of thinking (Chap. 1) at a personal and a societal level (Chap. 3). Place is local with local ecologies but it is part of a global place. In Chap. 3, I argued for the importance of a sociocultural perspective on visuospatial reasoning and that societies influence individuals' perspectives on space and place. Language is the way in which people in a society communicate, and ideas are generated. What do we learn from language studies in terms of space and measurement understandings? Furthermore, we argued for a critical perspective of place in terms of social justice. Visuospatial reasoning is understood in terms of an ecocultural perspective permitting children to keep and build on the strengths of their place and culture.

This chapter draws on linguistic and anthropological research to assist in understanding these bases of education. It establishes how important land is to the thinking of cultural groups as portrayed in communicating in various ways. It provides evidence of visuospatial reasoning varying according to ecocultural contexts. In particular, it notes that position, size, shapes, and patterning are unique to different language groups. Language structures indicate that ways of denoting all of these aspects differ from western or school mathematics. It shows that different cultural groups value different shapes and kinds of patterns in different ways and that worldviews impact on the place of these mathematical concepts and their uses in different ways. To explore this issue, I look specifically at language for measuring space and locating in space but first I provide some background to languages.

The Role of Language

Within this critical perspective of place, Setati and Adler (2000) note the role of language in a critical ecocultural pedagogy:

> Firstly, the political and pedagogical issues in rural and urban multilingual mathematics classrooms in South Africa are different, and this *contextual diversity* needs to be recognised in language-in-education policy, research and practice. Secondly, moving between languages (e.g. English and isiZulu) is only part of the process of learning mathematics in multilingual classrooms. There are numerous, distinct mathematical discourses that require navigation at the same time. (p. 244)

Thus permitting code switching between languages and leaving the choice of mathematical register to the students in the co-learning of mathematics can "privilege the students' competence in all their languages. … code switching can be used as a mark of solidarity empowering the students in the classroom" (Prediger,

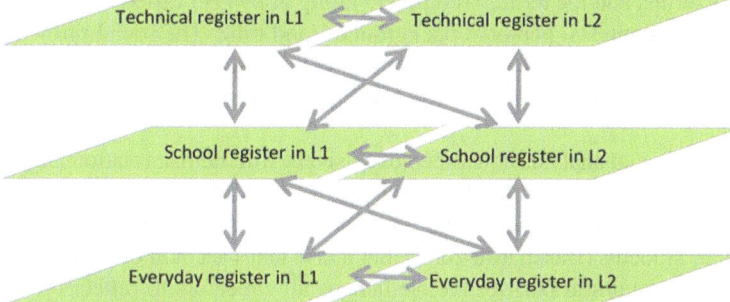

Fig. 4.1 The interplay of registers in first language (L1) and second language (L2). *Source*: Prediger et al. (2012, p. 6216)

Clarkson, & Bose, 2012, p. 6124). While informal language may assist students to understand the mathematical concepts and informal contexts permit gestures to support the language and interpersonal relations, additional language skills are needed to communicate in the school and technical registers. The interplay of registers is illustrated in Fig. 4.1.

> For mathematics classrooms, it is hence important to facilitate transitions between all these registers. It also becomes important for teachers to understand the possibilities that are afforded to students to gain deeper understanding if they are encouraged to use their languages effectively. (Prediger et al., 2012, p. 6215)

In addition gestures, diagrams, and other visuospatial representations are important in these transitions. One way of achieving transitions between registers is through a teacher "revoicing", for example, using adequate technical/mathematical language to say again what a child has said as a part of increasing the second language understanding. Revoicing a child's idea through the use of repetition, rephrasing, and expansion allows the child to be seen as the authority of the idea and an agent in the action of mathematical problem solving (Turner, Dominguez, Maldonado, & Empson, 2013). The boundaries between the contexts of the registers are necessarily fluid, so teachers can develop the registers to advantage through different activities. The register (technical L2) is important in skilling multilingual students through activities of translating from one register into another, finding and fitting registers for consolidating vocabulary, examining or aligning when registers are not aligned, explaining how to find a mathematical relation or structure in a certain register, and collecting and reflecting different means of expression within one register (Prediger et al., 2012).

Representation is not a direct correspondence but reflective of "context, functions, and social embeddedness" (Prediger et al., 2012, p. 6218). Furthermore, changing representations also indicates meaning in itself and the way it can be used and the properties it provides. For example, a graphical or algebraic expression of a straight line yields different communications. Similarly, changes in language and visuospatial representations can also explicate meaning, purpose, and techniques of the two forms. For example, Saxe's (2012) study on money showed numbers, monetary terms, and the term *fu* (complete group of plenty) varied in representation with different contexts, personal backgrounds, and time. The social purpose influences the register represented in the content field and text structure (Derewianka & Jones, 2012). Language analysis then is important for social justice in the teaching situation.

A critical social analysis concerned exclusively with human relationships fails to demonstrate ecological thinking whereas an ecocultural pedagogy seeks the twin objectives of decolonisation and reinhabitation, important particularly for Indigenous students (Barnhardt, 2007), but it also takes account of different geographies. This ecocultural perspective challenges all educators to reflect on the relationship between the kind of education they pursue and the kind of places we inhabit and leave behind for future generations and in which the learner takes initiative, makes decisions, is accountable for the results, poses questions, experiments, and solves problems, all regarded as central to mathematics education. All require visuospatial reasoning. Places are visualised. Actions in places are visualised. People reason holistically about places that they visualise (Pinxten et al., 1983).

I call this an ecocultural perspective in order to emphasise culture linked to land as embedded in relationships and language. There is a growing body of research that illustrates alternative ways of viewing position in Indigenous communities (Owens, 2013b; Owens et al., 2011; Pinxten & François, 2011; Senft, 2004b). Language and other representations reflect some of this thinking as indicated by sociocultural studies (Adler, 2002; Barton, 2008; Gerdes, 1999; Owens & Kaleva, 2008a, 2008b).

Cultural Ways of Thinking Reflected in Language

Rivera and Rossi Becker (2007) queried that culture might not necessarily provide a reason for diverging from western school mathematics because there are still individual differences in mathematical thinking in a specific cultural group. However, culturally shared language provides evidence of a widely taken-as-shared way of mathematical thinking that does show a cultural difference in mathematics that should be taken into account in understanding mathematics. Barton (2008) shows from a study of languages that mathematics is much more relative and dependent on human experience than is usually accepted. Mathematics expressed in school with English words is only one mathematical experience. There are many languages that express mathematical thinking quite differently. Furthermore, Barton shows that it is worthwhile pursuing mathematics as it is expressed in different languages. The world needs to nurture difference for its rich possibilities and creativity as well as providing culturally responsive mathematics education (Averill et al., 2009). As Rivera and Rossi Becker suggested, taking cultural ways of thinking together with school mathematical pursuits means the learner can identify culturally and mathematically, and develop his/her own ways of thinking mathematically. "A more powerful ethnomathematics program in contemporary times involves understanding the structure of complexity of cultures in ways that explain how members in such cultures are able to preserve valuable mathematical practices and might overcome those that constrain them from fully participating globally" (Rivera & Rossi Becker, 2007, p. 222). Anthropology has led the way to contested spaces where negotiation, institutionalisation, and internalisation of practice might be resisted. Thus it is important to appreciate the ecocultural reasons behind certain Indigenous practices and how they are shared, constrained, and pursued and to assist in reconciling different conceptual and practical variances.

Meaney, Trinick, and Fairhall's (2012) study pursued similar issues. They showed (a) how creative, language- and culture-rich, place-based mathematics education can develop in the unavoidable political climate; (b) how cultural background can be used to resource students; and (c) how to meet the challenges of context—community, teacher knowledge and ability, and professional development. Furthermore, Meaney et al. provide a comprehensive coverage of how one large Indigenous language group in New Zealand *Aotearoa* nationally developed its mathematics register and met the challenges of implementing education in *te reo Māori*. They provide guidelines and salutary messages for others, including smaller language groups with greater language changes as in PNG, to establish similar projects successfully. Furthermore, they support the importance of every mathematics educator recognising the importance of ethnomathematics in current school education. The language permits and encourages teachers to draw students' attention to important points through linguistic markers. Furthermore, the way that mathematical terms in *te reo Māori* were developed has resulted in assisting students to develop deep meaning. In addition, logical connectives within the syntax of the language clarify logical relationships. Their study shows the strengths of using culture and language in learning mathematics, not only on national testing regimes but also in terms of how the language and culture resource mathematical thinking and learning.

Most researchers on mathematics education in bi/multilingual classrooms have argued for the use of the learners' home languages as resources for learning and teaching mathematics (Adendorff, 1993; Adler, 2002; Moschkovich, 2002; Ncedo, Pieres, & Morar, 2002). Further, the active involvement of parents in the children's homework, often in the children's first language, can be a source of confusion if different notations and algorithms are used. If conflict between home and school exists, a suitable balance needs to be worked out. The family comes to understand the value of different approaches to learning and the implications of this in what the children do in mathematics lessons while the school acknowledges and builds on family reasoning through activity or discussion.

Children who speak a language other than the language of instruction confront a substantial barrier to learning. In the crucial early grades when children are trying to acquire basic literacy as well as adjust to the demands of the school setting, not speaking the language of instruction can make the difference between succeeding and failing in school, and between remaining in school and dropping out (Lockheed & Verspoor, 1991). The first language of a child is inseparable from his/her cultural heritage and as such deserves our recognition and support. If a child does switch to his/her mother tongue in an effort to solve a mathematical problem that child should not be discouraged (Then & Ting, 2011). Then and Ting suggested that a better policy is to recognise the value of the mother tongue, encourage its full development beyond surface fluency, but at the same time allow English or main language of instruction to become a naturally preferred medium for mathematical thought. Children taught mathematics by their parents in first language find themselves switching languages in class more frequently. Many multilingual primary-level pupils have great difficulty with reading and writing mathematics. They have poor mathematical vocabulary and can read only the simplest mathematical test, according to Then and Ting (2011).

Among the challenges for developing the *te reo Māori* schooling was that of writing mathematics in *te reo Māori* given it was predominantly an oral language (Meaney et al., 2012). In their professional development, teachers learnt to use different genres and explicitly develop field, tenor, and mode for reflection, recording mathematical activity, and conventional interaction with others. Meaney et al. provided considerable data to justify the point that not only is writing important but also the expected quality of the writing is important and possible in *te reo Māori*. The importance of revitalising the language strengthens the mathematical learning. Another issue is how to balance the need to use other terms for explaining and at the same time developing the approved mathematical register. It is important to decipher those words that easily connect from other uses to the mathematical use and those that need further translation/explanation.

Another study that particularly looked at the issue of language and geometry was carried out by Jawahir (Jawahir, Owens, Sukon, & Sunhaloo, 2011; Jawahir, 2013) in Mauritius. In his intervention comparative study, he chose not to look at written Creole but he used oral Creole for discussion and teaching (see Chap. 8 for more detail). In particular, Jawahir noted that researchers like van Hiele (1986) pointed out the importance of language in developing levels of geometric reasoning so that holistic reasoning was a foundation for later analysis, relational connections between shapes and later justifications and explanations. For example, in analysis, students may describe the properties of the shapes informally and imprecisely or they may describe explicitly and exclusively using formal geometric concepts and language to describe and conceptualise shapes in a way that attends to a sufficient set of properties to specify the shapes. Thus Jawahir considered verbal skills along with visual, drawing, logical, and applied skills. Furthermore, he noted that gestures and actions were significant in communication especially with manipulatives. Students who used Creole particularly with investigative processes of learning performed better than those using English which was a second language.

Locating, Visuospatial Reasoning, and Communicating

Structural features of language have significant implications for cognition and learning in visuospatial reasoning within mathematics particularly in terms of location or spatial understanding and processing. François et al. (2013) support this issue stating that some Indigenous languages like Athapaskan and Cherokee (Native American languages) and some PNG languages are basically viewing reality as events with substantially no noun categories.

> The structure of the Indo-European languages distinguishes between verb and noun forms. With this distinction corresponds a differentiation between things/states and operations/processes in the conceptualization of reality. Intuitively, mathematical thinking sophisticates these deep structural linguistic and cultural differentiations. Hence, the emphasis on geometric figures (with a thing-character) and their constitutive forms, on sets and their elements, on operations (of multiplication and so on) [that is] performed on entities (a number, a series, etc.). The point we want to make is that formal thinking elaborates the intuitive world view which is given in language and in folk knowledge (Atran, 1990). ... Actions in Navajo begin, stop, change or transform. The cosmos can be understood as a universe of events, rather than a universe of things. In such a view no part-whole logic of 'beings' or 'objects' and their elements [exists] ... (Pinxten et al., 1983). In Academic Mathematics, on the other hand, the very basis for formal reasoning is a part-whole logic: the world of experience is split in parts. For example, in geometry a line is defined as a set of points, or a plane is said to be a set of lines. (François et al., 2013, p. 30, pp. 29–30)

Thus language does reflect but also impacts on visuospatial reasoning and needs to be taken into account when we argue that visuospatial reasoning is impacted by culture. For example, for the Navajo, with a dominance of verbs rather than nouns, and the use of movement to describe position, reasoning is different to English descriptions that make use of prepositions to express relationships. Pinxten et al. (1983) also show that ecology is taken into account in geometry. For example, in expressing position, Pinxten et al. define this as actions related to certain landmarks evident on land that might seem to western eyes as fairly devoid of objects. Thus the description of position is unique and within a totally different system to the Euclidean western system but nevertheless rich not only as a system but also as part of a worldview of objects.

From a perspective based on western languages we might consider that space is initially referred to in terms of the planes associated with the body. These are the central vertical planes providing (a) left and right and (b) front and back. The third plane may be at our feet as the plane of the ground providing a height dimension. Such a way of referring to space is consistent with a three-dimensional orthogonal Euclidean approach that provides for pathways, areas, and volumes. The natural symmetry of the left–right plane and the expectation that one is standing in a vertical position underlie these expectations. The speaker's position and orientation are important referentials (Senft, 1997). In fact,

> One of the main features of natural language is its 'contextuality'—and it is in this context-boundness that language, perception, and cognition meet. ... Space, our perception of space, and our orientation in space are basic for human action and interaction in a number of domains- Konrad Lorenz even regards our spatial cognitive capacities as one of the roots for human thinking. (Senft, 1997, p. 2)

In many western referencing systems the speaker or the listener is considered but it is also possible to locate in terms of a third object. However, there are times when the context actually provides the meaning. For example, a ball is in front of the tree usually means the ball is between the speaker and the tree but if the tree is in the front yard of the house, it may mean in front of the tree in alignment with the generally accepted front of the house and the speaker can be anywhere. In addition, static configurations may use the way one faces but a dynamic configuration may be more about alignment or parallel relationships. Furthermore, the metaphoric and extended use of words can be linked by visuospatial reasoning (Lakoff, 1987). For example, "over" is used in a number of ways associated with position and action on a hill. Words and oppositional concepts such as "here" and "there" are very much determined by sociocultural experiences. English also changes when it is reported, so "it is cold here" is reported as 'it is cold there' (Ehrich, 1991, cited in Senft, 1997). Finally the words may also be associated with a symbolic use rather than a descriptive relationship to the object. "Over the top" refers to a person's expression that is unreasonably exaggerated. Furthermore, some words also have an emphatic purpose like the *su* in Turkish which is something that the addressee should take into account (Ozyürek, 1998, cited in Senft, 2004a). Similar emphatics are evident in Papuan languages (Tupper, 2007) and te reo Māori.

Language Patterns in Papua New Guinea

The role of language in an ecocultural pedagogy for visuospatial reasoning in space and geometry is further explained by considering the rich diversity of languages in PNG. Some PNG Indigenous languages have

> a greatly complicated verbal system, but pay little attention to the noun, lacking perhaps any system of classification or giving very little attention to distinctions of number and relationships of case to other parts of the utterance. ... (Others have very elaborate gender or noun-class systems) often involving grammatical concord with all conceptually connected parts of the utterance (Capell, 1969, p. 13).

In verb-oriented systems like the non-Austronesian language (NAN) of Kâte, the emphasis lies apparently in what happened, when it happened, and how it happened, rather than in the people or object involved or the place of the occurrence. The verb with post-, pre-, and infixes might take six English sentences to convey the same message. On the other hand, in noun-oriented systems such as Baining, a NAN language, East New Britain Province, an utterance gives attention to the persons and objects such that the action words are allowed to take care of themselves (Capell, 1969). A rarer type of language classified as numeral dominated is the Kiwai language spoken by the Indigenous people who live around the mouth of the Fly River in Western Province. As noted by Capell (1969), in Kiwai language, there is "prefixal indication of the manner of the actions—one action only, one action repeated, a number of actions together or in sequence needed to carry out the task in hand" (p. 15). Nearby are languages with "many" being determined by repeated action or increasing number of actors, so there is little counting per se and more classification of "many" or "few". These are some of the structural features that illustrate a staggering and complex linguistic and cultural diversity found among the

Indigenous languages of PNG. These structural features have significant implications for cognition and learning in mathematics education.

One early PNG study of directionals to indicate a spatial relation of person, place, or thing to another person, place, or thing was in a Narak speaking village in Jimi valley (Cook, 1967). A few examples are provided here together with metaphoric uses. Cook lists 19 directionals. *pla* refers to a high relative position on the vertical axis but also in the case of a pig, it means large and fat while it can refer to where God is thought to reside in heaven. It is also used to refer to up over a person, the direction of climbing, looking away from the person, and a long or short distance. *kalA* is vertically straight down in direction, so God came down or a place down a ridge but it can also refer to the poor condition of a small, skinny pig. It has an opposite sense to *pla. paNo* refers to the middle of something while *kora* means on the same level as something else. These examples show the metaphorical extensions and complex meanings associated with just a couple of words.

Frames of Reference for Space and Place

Position depends on frames of reference and different cultural descriptions. In general space is referred to by local and directional prepositions or postpositions ("at", "on", "in"; "in front of"; "behind"), locatives—local or place adverbs ("here", "there"), dimensional or spatial adjectives ("high", "low", "wide"), demonstratives ("this", "that"), static and dynamic motion verbs ("to stand", "to come", "to go", "to bring", "to take"), directionals (e.g. "to", "into"), and presentatives ("there is") (Senft, 1997, p. 8). These terms form deictic systems and there is a large variety of these systems across languages. In addition, languages have gestures such as pointing or raising the eyebrows to indicate position. However, the number of terms used in any one language may vary. Senft (1997) presented an argument made by others that the more the man-made spaces in a society, the smaller the size of the spatial deictic system. He gives as examples the fact that English has two terms for position ("here" and "there") but Yup'ik has 30 terms and East Eskimo has 88 terms. This is partly attributed to the man-made function given to the object associated with the position. For example, "the key is in the door" or "the satellite is in space". The locative markers of a language impose an implicit classification on spatial configurations. Indo-European categories are topological relationships (e.g. proximity, inclusion, surface contact), Euclidean notions, and functional notions concerning typical uses.

The Inuit have four different suffixes, roughly corresponding to "at", "from", "via", and "to", but they also indicate (a) if the event is at the beginning, middle, or end of the sequence; (b) expand on the position further away with an idea of "up", "down", "in", "out", and "same plane"; (c) perspective of speaker, the addressee, or some other reference point; and (d) the relative size/shape of the place partially determined by the speed and nature of the motion involved (Ascher, 1994).

Typical of studies that look at language and frames of reference (Gerdes, 1999) is that carried out by Edmonds-Wathen (2012a) using cards that have pictures with a tree and a man in different orientations (facing different ways and on different sides of the tree). A person describes the position of the man when compared to the tree. In general, frames of reference fall into three categories although the language may have a word that could cover more than one of these categories. These categories are usually described as being intrinsic, that is, the words link the items of the representation (picture or drawing); relative, that is, the words link the items to the describer's position looking at the drawing; or absolute, that is, terms such as north, south, east, and west are used. For Iwaidja (Edmonds-Wathen, 2011) on Croker Island, northern Australia, there was a second absolute frame of reference being "deep sea" (west) and "mainland" (east) but if the describer was facing west, this could also be used in a relative sense if the describer noticed that the man was behind the tree in an English description. Other languages use landmarks to denote position. Local landmarks and environmental features are also used to denote places and the position of objects. One example is the use of "west-sea-down" and "east-land-up" relevant to the geography and ecology of the Iaai on Uvea, an island in New Calendonia (Ozanne-Rivierre, 2004). (See in Chap. 5 a classroom use mentioned by Muke and in Chap. 6 the Mayan use of seas.)

Prepositions or postpositions generally provide a connection that is obvious by context or by the expected relationship between objects, e.g. the book on the table. For this reason, prepositions are frequently minimal and may or may not impact on word order. Alekano (Eastern Highlands Province, PNG) has up to 15 slots or positions for different types of words and relationships between words in a sentence. Wiradjuri in NSW, Australia, has three positional suffixes—one for being next to a person or on or in an object, another for coming to a person, and another for going away from a person which vary with the class of noun (Grant & Rudder, 2010). For example, the suffix *-gu* is added to the noun for movement towards the person or thing, that is, the person or thing is the purpose of the action (but also *-gu* is added to all the nouns for two or more objects owned by a person, rather than using "and"). Table 4.1 also shows how suffixes are used for movement away illustrating variance over type of word. In some languages, words vary with addressor and/or addressee or a third person or object as reference point and the number of people involved may also modify the words to be used (e.g. Samoan as presented by Mosel, 2004).

Table 4.1 Examples of added suffixes for movement in Wiradjuri

Word ending	Suffix	Wiradjuri base word	With suffix	English
"*ang*"[a]	-*dhi*	*ngurang*	*ngurandhi*	From the camp/home
"*i*" or "*ny*"	-*dyi*	*mirri*	*mirridyi*	From the dog
"*aa*"	-*ri*	*yinaa*	*yinaari*	From the woman
"*ang*"	-*ga*	*ngurang*	*ngurangga*	In/by/at the camp/home
"*i*" or "*ny*"	-*dya*	*wiiny*	*wiinydya*	At the fire
"*a*", "*ir*", "*n*"	-*dha*	*dhaagun*	*dhaangundha*	In the dirt
"*l*", "*r*", "*rr*"	-*a*	*gibir*	*gibira*	By the man
"*ng*"	-*gu*	*galing*	*galinggu*	To the water

Note: [a] for "*ang*" ending, "*g*" is dropped

Sometimes words vary with the use of gestures (e.g. Saliba, Milne Bay, PNG; Margetts, 2004).

Stokes (1982) also notes that for the Anindilyakwa from Groote Island in Northern Territory, Australia, the word for "come" and "go" is the same generic word requiring context for meaning, that the suffix *-manja* is used for "at", "in", "on", and "by" requiring the context to provide meaning, and that the question "where" can be asked by an adverbial for a person but of an object by an adjective.

Codrington (1885) much earlier noted that Melanesians and Polynesians have a habit of continually introducing positional suffixes, "adverbs of place and of direction such as 'up' and 'down', 'hither' and 'hence', 'seaward' and 'landward'". One deictic system that can be found in a number of Austronesian and non-Austronesian Papuan languages in PNG, in languages of the Pacific, and in Australian Indigenous languages includes varying words which refer to a place quite distant, ones that encode medial distance, and ones that imply proximity with visibility impacting on choice of words (Senft, 1997). Wiradjuri has three suffixes for "here" *-nha*, "there", *-nhana*, and "way over there", *-nhanala*. Senft's edited books (Senft 1997, 2004b) provide many linguistic examples of deictic differences from the work of linguists among Austronesian and non-Austronesian Papuan languages of Oceania, PNG, and Melanesian areas to the west (such as West Papua and Suluwase). These alternatives also affect the way people measure (see comments on Kilivila later in this chapter on measurement).

It is common for "behind" and "front" to be used for denoting persons but in different ways in diverse languages. The diversity of frames of reference in PNG and Australia illustrates particularly the difficulties of using the metaphorical "before" and "after" with numbers on the number line for various reasons. For example, when Matang (2008) carried out his study (personal observation) and used the Tok Pisin terms, he was using the words in the opposite way to that used in English. Similarly, the Iwaidja, Australia (Edmonds-Wathen, 2012b), and Walpiri in Central Australia (Graham, 1988) both used "before" and "after" in ways different to that of English. "After" in some languages can be a word used only in a relative sense so that it can also be translated into English as "before", "previously", or "after" depending on the context. "Before" in Walpiri is also used for "larger" because it has an associated time factor linked to growth. That is, a child who is "before" (that is born before) will be larger than the other child. This leads to confusion with the metaphor of the number line but it can also be easily overcome by teachers who only refer to the "number one less than" rather than "before". However, there is also a spatial orientation difficulty. In Iwaidja, a person could say a man is before the tree when the man has his back to the tree. This intrinsic approach linking the items on a diagram would be different to the English description if the tree was (partially) covering or in front of the man from the position of the describer. It could also be the case from the position of the describer of the drawing that the man was to the left or right of the tree with his back to the tree. So the word for "in front of" which has a stronger directional component in terms of the items of the diagram may be better than a term which has a different meaning in Iwaidja like "before" (Edmonds-Wathen, 2012a).

Anindilyakwa use words like "below" for a canoe on the sea but not on the land, and it is the same as the word used for "inside" the shelter in that there was a cover only "over" rather than surrounding the person (Worsley, 1997) and separating from the outside (Fig. 4.1). "Outside" is outside the jungle in a clear space and could be distant and an area explained as "from here to there" (Stokes, 1982). *Angwurn.dikirra* is for a space like a strait or between objects, usually narrow, while a confined space is a verb stem for enclosing. So the connection between lines and area is not expressed in Anindilyakwa as lines are generally associated with aspects of an object such as a spear (Stokes, 1982) and can apply to both horizontal and vertical lines which can introduce straight lines often associated with "becoming straight". Other words like "horizontal", "oblique", "vertical", "corner" would be distinguished by verbs for "lying", "leaning", and "standing upright" and the adjective for "crooked", respectively. There are "flat" objects. Shapes like rectangles are only recognised in terms of objects such as a bark painting or rectangular sail (Stokes, 1982). However, there are several ways of noting round and specifically to denote the sphere shape of the turtle egg from the other ovoid-shaped eggs (Fig. 4.2).

Compass points or cardinal points may be denoted (Harris, 1989). In addition, dimensional axes, usually in reference to the body, are used but in some cases, the position of the axes can be moved. For example, Ralph Lawton (personal communication, 2010) noted that measurement varied depending on whether the plane was on the ground or at the arms for Kilivila, Milne Bay, PNG (an Oceanic Austronesian language). Such diversity merely hints at the diverse ways of representing space verbally but also the effect of the ecocultural influence on language and visuospatial reasoning.

Fig. 4.2 Shelters in the camp, Yalata, South Australia (Owens, 1966)

Language in Comparing and Measuring space

Each language has its own way of referring to sizes. Some have comparative words such as "longer" or "more distant". For the Anindilyakwa from Groote Island, northern Australia, size is covered by three adjectives, "small", "big", and "huge" (Worsley, 1997) although Stokes (1982) notes a diversity of words for thick or fat and thin for people, animals, and things but additional ones for things. There is remarkable flexibility with the use of qualifiers such as "more" or "very" (Stokes, 1982; Worsley, 1997). There was a word for "short" usually used with a noun and interestingly the word for "foot" was also used. In Tok Pisin in PNG *centimetre* is used for the short unit. Worsley also pointed out that other taken-as-shared meanings in western education take on different meanings in Anindilyakwa language. "Empty" has a spatial sense but it implies what might be expected and while comparisons of "more than" and "less than" were measured, the approximation was adequate (Worsley, 1997). Distance is seen as the time taken to reach a place and the words for "soon" and "near" are interchangeable. However the words for "to another place" can be used for "far away" and the prefix for "rather" and suffix for emphasis are used for comparison as well as the intensifier for "more" further away. The verbs of motion have certain features to express a great distance or length of time (Stokes, 1982).

Providing a sense of size often requires narratives. In PNG, if I wanted to know how long it would take me to walk to a village, I would ask how many hours it would take the speaker (a villager) and how long it would take me (a white woman), then I would take the average. Usually it turned out to be a good estimate! Time was often used to indicate length of the track from village to village. However, the idea of an hour was not well sensed. This did not mean that they were unfamiliar with time bodily and mentally. When I was staying in a village in the mountains behind Lae, Morobe Province, women who had gathered greens from the garden would get up early enough to wrap the bundles and walk to the road head in time for the truck for the 1-h journey to sell at the market by 7.30 am. In a village in Oro Province, people without watches needed to be at the nearest airport at 8 am. They rose in the dark at the right time to prepare, knowing how long it would take to sail and paddle, negotiate the winds and swells, and arrive in time. After 4 h, we arrived at 8 am. En route, the two men worked in unison with hardly a word especially as they added paddling to the sail to round a particularly difficult point, watching the rolling waves carefully as the canoe was heavily loaded. There was clearly taken-as-shared visuospatial reasoning.

Jones (1974) argued that it was not always easy for languages in PNG to express certain concepts related to measurement. Ways of reasoning about size and making decisions in activities involving measurement in PNG are covered in Chap. 5 and implications for education in Chap. 8. Here we will discuss language patterns related to space and measurement. A study by Owens and Kaleva (2008b) in PNG revealed that vernacular words for size and other related ideas are complex. Data came from questionnaires completed by tertiary students, village observations and discussions, and linguistic records for 360 of the 850 languages of PNG. Most languages appear to be able to refer to volume, mass, area, and length but a verb may be involved implying an action on size, e.g. making bigger. There may be a limited number of comparative adjectives or very general concepts like *size*. Participants talk about size but often thought for some time before completing the word lists as there is often not an exact match between English words and their own language for meaning nor necessarily the same kind of speech pattern. For example, the word may not be an attribute word or adjective. Mussau in New Ireland has an auxiliary clause (van den Berg, SIL, personal communication, 2006). They also have the word "mother" to refer to large things. It is similar to the Huon Peninsular *awara*. In a similar way, Korafe speakers in Oro Province use metaphors for size. For example, a child is a chunk of the father or a smaller version of the father. However, they use suffixes for "bigger" and "biggest" but they will also use reduplication (repeating morphemes) which is found for descriptive words (larger and/or smaller) (Farr (deceased), SIL, personal communication, 2006).

Reduplication is common in PNG languages and is used for similarity or other purposes (e.g. continuing verb, plurals, groups like two by two, emphasis). Dobu speakers further along the coast in Milne Bay Province use the word *kaprika* for "pumpkin" which changes to *kapukapurika* for "small pumpkin" (Capell, 1943). This may indicate a different type as well as different size. Manam on Karkar Island Madang Province and Tinatatuna (Tolai or Kuanua) on East New Britain are spoken many hundreds of kilometres away but they also use reduplication. A Manam speaker gave *dadaka*, *memekei*, and *kanabibia* for "big" and *kengekenge*, *sikisiki*, *mukumuku*, *seisei*, and *bisibisi* for "little" (see Table 4.2, from the measurement study in PNG).

"Very" is expressed in consistent ways in most PNG Austronesian languages. It might be translated "enormous" and has equivalents in Tok Pisin of "mama" or "tripela". In non-Austronesian languages this term might not exist. "Larger" might be expressed by comparison of two objects with a comparison word between (order is important). Sometimes one word like "long" was negated (usually within the verb structure) to suggest "short" and a language with this structure is likely to use other paired opposites in a similar way. Other languages use a range of diminutives (linguist focus group, personal communication, 2006).

Table 4.2 also shows a number of different words in some of the languages for the same concept in English for either different objects or purposes.[1] This variation is found in Manam for several concepts, in Tinatatuna (Tolai or Kuanua) for "compare", and in Kewapi words for "little". Some variations can be explained in terms of the influence of other languages. For example, for Tinatatuna the word "to measure" includes *mak(ai)* which is similar to the Tok Pisin (PNG's lingua franca) word and the overlap of words in the three highland languages (Huli and Kewapi in

---

[1] The variations in the table may be due to different sources of data (for example, different participants writing down an oral language or the date on which the data was collected, e.g. the 1900s first contact or later records and recently collected data) but in other cases, the same participant provided the multiple number of words confirmed by others from the language group (e.g. Manam).

Language in Comparing and Measuring space   129

Table 4.2 Words related to size from a small selection of different languages in PNG

English	Fore	Manam	Tinatatuna	Enga	Huli	Kewapi
Compare	Koviga	Tongaka	Varvadaina, vandadauane, valarue, varagopina	Makande, kapakap	Manda, mandapia	Anamealapa, manda-mamea
Measure	Amakaga	Tongaka	Mak(ai), angewe, valarue	Makande, makade	Manda pia, kimagi, kemagi	Mandamea
Size (volume)	Mangawa'e	Ilo	Ngala	Kapkap, makande	Arane, timbuni, luni, taliga	Ekei-yapa andai
Big	Tave	Taila, dadaka, memekei, kanabibia, ilaba	Ngala (variations)	Andaik, yale	Timbuni	Andai
Little	Amanagando	Kengekenge, sikisiki, mukumuku, seisei, bisibisi	Kapakapana	Kolam, kokilyam, yaalam	Emene (various spellings)	Egepusi, ekei, ekesi, ogesi
Short	Alo	Tupeka	Gur tutuk, ngu ngu, tutukana	Moui, muu	Tomaki	Rundu rundai, rundsi
Long	He'elo	Salagabuli, salagauta,	Lolovina	Londe londe, londakai	Luni	Andalu (with suffixes)
Full	Puma,e	Ikauri	Buka	Tubelam, tubabpah, tuimbilam, perpertta	Toho laya, caralipa, catlope, to	Manda, rege-pe-lea, rayo
Heavy	Gunda	Moatubu	Mamat	Kend, kenda, kendepi, kendaping	Kend, kenda, kendepi	Kend, kenda, kendepi
Light	Ewasa	Malama	Papa, papanga	Yapalume, kende napinge	Hale, je-pe-pi, emene, jepeapi, yapipi, atatapi	Yapa-pea, pa

Southern Highlands Province and Enga) suggests interlanguage influence. Other words building on the same morpheme indicate relationships between the words. One particular Huli speaker noted that volume was a combination of length, width, and height (Piru, 2005). The impact of western education might be evident in this comment but only one student from the hundreds who completed questionnaires from the Southern Highlands Province noted that her family (unlike others) discussed area in terms of area units.

Some languages have an adjective for a measurable attribute that is used for all objects and others have words and suffixes for specific classifications, e.g. round objects, flat objects, people, and food (classifiers). In other cases, the attribute word varies depending on the item being discussed. For example, in Korafe *big* for fish is different to *big* for people (Farr (deceased), SIL linguist, personal communication, 2007). Other languages have words, suffixes or prefixes, or action words for different types of objects. In other cases, only certain kinds of objects may be compared (e.g. volume of stone is not compared to volume of water).

Two further linguistic features, ways of indicating emphasis and order of words, can impact on discussions about measurement activities (Tupper, 2007). Emphasis is used to draw attention to a particular point and the point could be the size of the object although there may not be a particular word used for size. Emphasis was seen as a strength of *te reo Māori* as mentioned above (Meaney et al., 2012). In the second case, the smaller object is placed before the word for the larger object in the sentence but there is no comparative morpheme, so the context of the sentence indicates that size is being considered.

Although this is only a small selection of the data from around 360 PNG languages, it does indicate that concepts of size exist and that there is a diversity of ways of expressing attributes related to size and difference in specificity. However, just the vocabulary fails to provide the more complex ways of discussing size that are embedded in sentence structure and in other ways of representing the attributes. Complicating this issue is that measurement and size may not be the only considerations taken into account in reasoning. For example, when comparing the area of two gardens people would take into account fertility, distance from the village, the number of people needing to use the land, closeness to water, cleared or currently bush or fallow, and crops. Thus size of area is only a part of the visuospatial reasoning. The visuospatial image may be much richer in terms of other attributes and features relevant to the decision making.

A further consideration in discussing locating and communicating visuospatial reasoning is the way in which a group might reach a decision about a place. In other words, it may not be a single or small number of words that locate or describe an object or person but it might be part of a larger discussion about the position or object. It is the discussion itself that can be significant to the speakers (cf. Salzmann, 2006, on disease in Mindanao, Philippines).

Implications of Language for Visuospatial Reasoning

In the previous chapter, I referred to the study of Hutchins (1995) on the navy team's use of tools including a Mercator map that required additional information to improve navigation. These tools had been developed over many hundreds of years to assist navigation and the team with specific roles required information from different people and certain commonly held and understood rules of thumb such as the better choice of visual sites for accuracy with position or the expected amount of travel distance at certain speeds.

Other historical situations illustrate the difficulties of referring to space by a spatial frame with the speaker as the centre and two orthogonal (perpendicular) axes denoted by the directions—north, south, east, and west. This static frame was an issue by the sixth century BC for the Greeks viewing the world as a sphere, so they divided the heavens into zones and the earth into five latitudinal zones (Tuan, 1977). Furthermore, reference to space also had place-based and cultural connotations. For example, European folklore linked people to their environment; for example, the north were hardy, the south easy-going (Tuan, 1977) whereas in some Arabic dialects the word for "south", where the once flourishing Yemen lay, was also used for "right", and "plenty" (Senft, 2004b). Valuing place and position is often encapsulated in the language of the cultural group. For the Chinese the four sides of the rectangle were represented by animals. Ancient Greece used planetary gods—east denoted light, white, sky, and up while west was darkness, earth, and down. For Europe, zodiac star signs were linked with patterns of farm work like the coming of rain, breeding flock, harvest, mowing, and raking (Tuan, 1977). Thus reference terms were associated with other aspects of life that required decision making.

The analysis presented in this chapter illustrates how the ecocultural context is significant in the language of spatial referencing. The language directs the visuospatial reasoning in both the metaphorical use and the significance of position/location in the culture where ideas of movement and relationships of people in a context are more important than the diagrammatic representation to be described by a viewer. This is complicated by other cultural views such as the position of the man and the woman when walking. In some cultures, the man should go before the woman or vice versa whereas in others the expected position is side by side. Similarly, women are never to be above a man such as stepping over his legs. I also suggest that a diagram which is more abstract than the man and the tree would also present some difficulty because it does not have the directional relationships clearly presented as illustrated by the Wiradjuri words (Table 4.1). This brings us to the use of maps in cultural contexts.

Maps as Representations of Visuospatial Reasoning

Location is usually considered in terms of an orthogonal coordinate system (Uttal, Fisher, & Taylor, 2006). Students may begin with locating on one dimension before using two and three orthogonal dimensions. The use of x and y axes with later developments for positions described in terms of negative to positive numbers is familiar and a basis for coordinate geometry or the visual representations of algebraic statements or relationships. Higher levels might provide polar coordinate referencing of position such as angle from north clockwise and distance from the origin or reference point. A similar idea of amount of turn and distance travelled is used in Logo geometry (see Chap. 9). Affordable programmable toys have encouraged practice and recent research shows how effective these toys are in developing spatial and pattern thinking in early educational settings (Highfield, Mulligan, & Hedberg, 2008).

Early childhood experiences in western schooling emphasise the use of prepositions like "in", "on", "inside" and words like "left" and "right". Some studies of mapping have discussed developments in primary schools indicating that responses tend to move from more pictorial representations with some indication of direction to those showing greater accuracy in terms of angles formed by non-orthogonal roads, and relative lengths (Owens, 2000a). Mapping is introduced as a plan view and in mathematics little attention is usually paid to the common use of contour lines on maps. The use of landmarks in big space is also noted by researchers (Liben, 2006) as an everyday way of giving spatial positions. There is an accepted discontinuity about descriptions of spaces that can be within a person's immediate view such as on a piece of paper and descriptions of big space.

My research with school children (Owens, 2000a; Owens & Geoghegan, 1998) showed that children can map their way from home to school from 4 or 5 years of age. Initially they note landmarks and indicate turns generally in the correct direction. As their mapping develops, they provide more detail in terms of less obvious landmarks and provide turns with right angles. Later they are able to indicate other angles, and later the proportion of parts of the land is better represented. According to Clements and Sarama (2007a) consciously self-regulated map reading behaviour through strategic map referral increased 4- to 6-year-olds' competence with reading route maps.

At a primary school in Goroka, PNG, I was observing a good teacher with her class. She had explained mapping clearly and demonstrated with the whole class participating in mapping the classroom. They had also mapped their houses. She said that some had not put an outer wall on the house but just made a map of objects relative to themselves in the house. I wondered if the "walls" of some self-help houses were not considered as part of the map of the internal space of the house and not part of the place they called home. This was the second lesson on mapping the school buildings and playground. Some had not considered the map in terms of north and some had drawn a mirror image of the school. Then the children shared their maps and discussed what was good and what could be improved with the maps (Fig. 4.3). This was a multilingual school and a mixed socioeconomic status school with a good proportion of well-educated parents. I also wondered if some of the difficulties resulted from the lack of maps in the culture and the different structure of language for location from English. I recalled the earlier work by Bishop and Lean (Bishop, 1979) in which they noted that tertiary students initially had difficulty interpreting the position of photographs of a visible structure taken from different

Fig. 4.3 Mapping the school (multilingual town school, Goroka, PNG) (Owens 2001a, 2001b)

perspectives but that they quickly learnt to read these two representations (solid model and 2D picture) and that it was indeed a lack of pictures (photographs, films, and books with illustrations) in their ecocultural context. How different might ideas be about mapping in different cultures in different places?

For the Yolngu of northern Australia, every person and every other thing is either *Yirritja* or *Dhuwa*, the two clan groupings. The division of land is dependent on the sacred sites (Fig. 4.4a). The creation of the sites comes from the dreaming creatures who created the clans who are now responsible for the sacred sites and who maintain the power by observing appropriate ceremonies and by painting, dance, and song (Thornton & Watson-Verran, 1996). "There is a metaphorical force essential for their way of life and sustaining their world" (Watson-Verran & Turnbull, 1995). In discussing and drawing the various places, a clan Elder represented the connectivity of the water flow; thus, a line is not a Cartesian mapping but a topological mapping in western mathematical terms. However, such a description (topological mapping) does not present the fullness of the representation (Fig. 4.4c). Each place has connections with activities carried out by ancestors such as a place for camping when visiting, or a place for washing cycad nuts to remove poisonous chemicals. A walking track was the Elder's responsibility and he would maintain it in songlines,[2] ceremonies, and practices. The land is represented by areas around sacred sites being either Yirritja or Dhuwa like a patchwork of nondescript shapes with the areas between being grey and not clearly delineated as the distance from the sacred site diminishes (Thornton & Watson-Verran, 1996). This referencing of space is not

---

[2] While walking/maintaining a track, an Elder would sing the song associated with the land and clan referencing the parts and points of the track as he went.

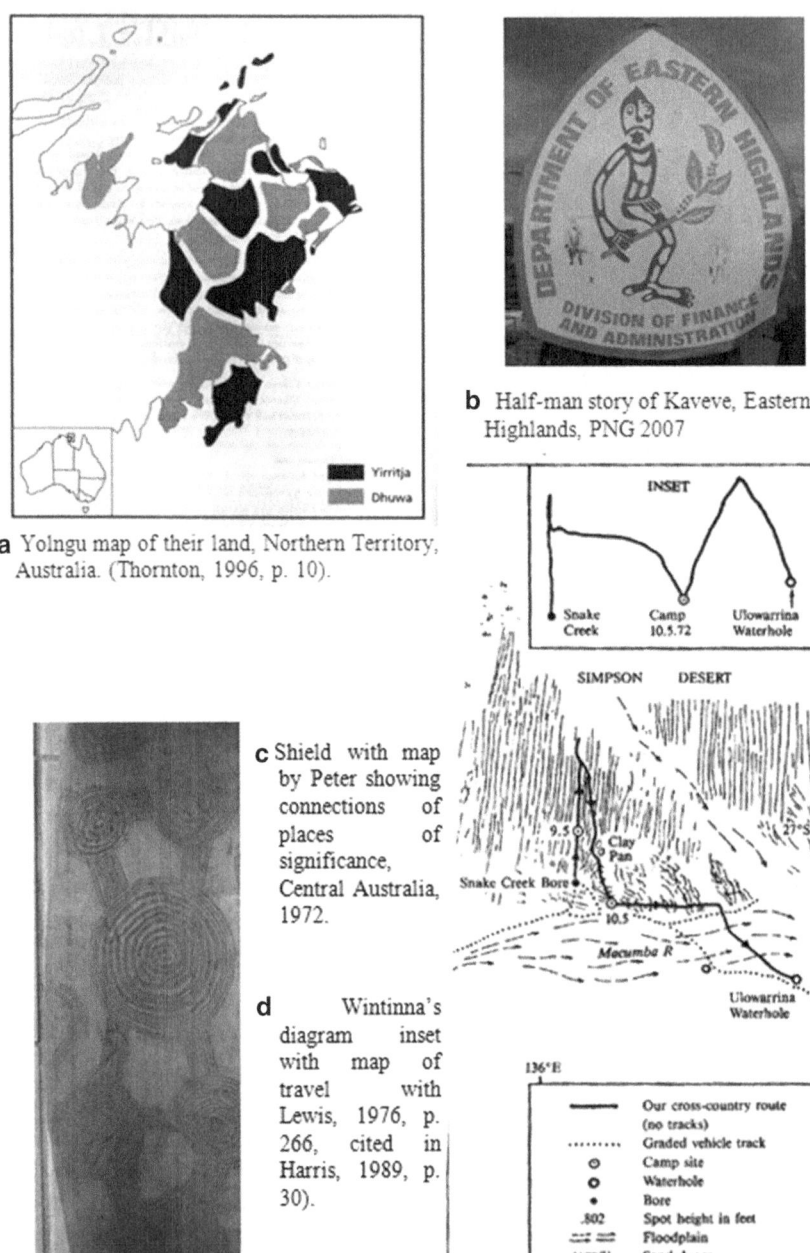

a Yolngu map of their land, Northern Territory, Australia. (Thornton, 1996, p. 10).

b Half-man story of Kaveve, Eastern Highlands, PNG 2007

c Shield with map by Peter showing connections of places of significance, Central Australia, 1972.

d Wintinna's diagram inset with map of travel with Lewis, 1976, p. 266, cited in Harris, 1989, p. 30).

Fig. 4.4 Non-orthogonal maps indicating visuospatial reasoning about place

by an orthogonal grid. Furthermore, to solve problems of space as place, people cooperate as some knowledge is known by Yirritja and some by Dhuwa.

The place where negotiations occur is roughly the shape of a stingray which buries its tail in the sand just as one of the Elders and the ancestors before him buried their spears in the sand when negotiating a peaceful solution for revenge. This is near the stingray-shaped lagoon. Thus the shape and the place are metaphors and powerful images for complex ideas. Activity is set in kin relations, land rights and responsibilities, and sacred understandings. The land is constituted by living it. The conventions of the map that the Elders drew are representations interpreted in terms of systematic relationships. The Yolngu system of spatial knowledge (which they call *Djalkiri*) is detailed and provides a means by which a person can find his/her way anywhere across the land. The structures of the various forms of representation in ceremonies, everyday living, and in the land itself locate space appropriately in the footsteps of the ancestors.

One aspect of the visuospatial reasoning for Aboriginal Australians is that the spatiotemporal entities are not as paramount as the relationship entities (Watson-Verran & Turnbull, 1995). Similarly, the travels of the ancestors in creating the landscape constitute tracks or song-lines that traverse the whole country. For the Yolngu, *gurrutu* the recursive relationships and *djalkiri* the location and their overlap form a strong mathematical structure representing ecocultural living.

People represent the position of places using dance and song. "Indigenous dance isn't just Indigenous dance—it's a map in itself, a directory of the culture behind the dance" (E. Johnston, nd, on Northern Territory languages in particular). Songs are used in many activities while traversing land and sea and for various reasons usually associated with spirituality or for rhythm of movement (personal experiences in PNG). A community project in the Blue Mountains (on the outskirts of Sydney) involves maps, pathways, and "song-lines" (Cameron, 2003). People use song-lines to maintain the connection with the route that is taken when traversing their land.

Time becomes evident in many map representations of space and place. Wassmann (1997) noted that with the descriptions and even more the map drawing (both of which are not generally required in everyday communication except with people from outside) some sense of walking the route was involved. For example, a slightly longer line represented a difficult time-consuming stretch of the track for a Yupna man, Morobe Province. Similarly, Harris (1989) provided an example of the direction and nature of walking in an Aboriginal map and discussed the meaning of maps that could be described as topological (connecting places) but deeply embedded in relationships of place and people (Fig. 4.4d).

To walk a trek in Kaveve village (Eastern Highlands Province, PNG) is to walk the story of the half-man who lived in that place (personal experience, Fig. 4.4). The story connected the place across time. In PNG, songs communicating with the spirits are used when traversing the land or remembering people who traversed the land (Rumsey & Weiner, 2001; personal experience). Hence I am vividly reminded that representations of land whether in words, diagrams, actions, or land formation embed relationships between people and between people and the land and hence reasoning about relationships.

Language is a major clue to the limitations of describing position in terms of two orthogonal reference lines or by distance and direction from a reference point. Pinxten's (Pinxten, 1997; Pinxten et al., 1983) study of Navajo concepts highlights even more the difficulties of concepts and language that are not easily connected. For example, a word like space for the Navajo referred to a saucer-like grand container. Thus the notion of infinite needs to be established in another way. With the issue of static object, the idea of a "snapshot" of motion is helpful along with the recognition that the western mathematics wants to emphasise the object that in Navajo mathematics was a spot in passing in the dynamic moving approach to place. Surface could be seen as where two volumes come together. What is quite distinct is the emphasis in western mathematics of part-wholes, of the hierarchical logic of point, line, plane, and three dimensions or of the distinction between distance and time.

There were examples in PNG mathematics of similar difficulties. However, the westernisation of curriculum meant that many teachers in their projects (see Chap. 8 for examples) were unable to focus on the mathematical thinking of their Elders and family but rather saw objects as a static object to link to western mathematics. Thus they emphasised their material culture especially built objects and final design features if they were seen as mathematics, e.g. symmetry and shapes. At best, these teachers described "deciding by eye" or "in their head" but did not have the mathematical language or connection between school and home mathematics to discuss the use of ratio or rate estimates, to note complete groups or encompassing sizes, or spiritual values of design or design making. Simply to explain as thirdspace thinking is one way forward (Soja, 2009) but Pinxten and François (2012b) outlined the depth by which Indigenous mathematical systems are more complex requiring their own unique ways of seeing western mathematics as a part of the larger mathematical complex. Furthermore, Indigenous mathematical systems are not only important but should also not be lost to the world.

Shapes

One of the issues in school mathematics is the use of labels and names for shapes as we mentioned in Chap. 2. It is important for children to describe and classify shapes and for them to recognise the generally accepted concept behind a shape name. I recall a conversation with Usiskin (personal communication, Utrecht, 2001) about the myriad of definitions for common shapes like "rectangle" and "trapezium" ("trapezoid") across the USA. Some countries have rhombus, others do not; some have oblongs, others do not. No doubt definitions are culturally determined and often without the same ways of thinking as mathematicians might claim. I have had conversations with experienced teachers about terms such as "regular", "diagonal", and "pattern" since they seemed to be using the terms differently to me. It is no wonder that Battista (2007b) noted how two boys were coming to an agreement about the use of a rectangle maker to make a particular shape in a Microworld.

Matters were further complicated because 'slant' meant not-perpendicular for Matt but tilted from the vertical for Tom. ... In school geometry, shapes are described by referring to *relationships between their parts*. ... Common-language use of the word *slant* ... refer(s) to the relationship of lines and segments to the *up-down or vertical frame of reference*. Thus, Matt's use of the word slant (to refer to the angle of the shape) was evoking a totally different, common-language meaning for Tom. (Battista, 2007b, p. 71)

The teacher's discussions with the children continued to indicate that they noticed different parts of shapes and their position differently and language continued to be an issue. When the teacher drew a rectangle with no horizontal sides next to a parallelogram with short sides almost horizontal, the children focused on the fact that these sides were not on top of each other like in the rectangle, meaning vertically above each other on the screen. However, with further manipulation and having their attention drawn by the teacher to the measures also on the screen, the students realised that the rectangle's angles remained at 90° whereas the other shape's angles varied.

The episodes reminded me of watching children with elastic loops to make shapes. One small boy made a triangle and to convince his partner, he turned it as it was a right-angled triangle, so there were horizontal and vertical sides as if that was the only way of having a right-angled triangle. "Activities assist students to start with holistic reasoning, they constantly encourage and support students' development of ever-sophisticated knowledge of the properties of shapes" (Battista, 2007b, p. 78). However, the most significant issue in regard to shape naming is the use of words for categories. In Chap. 5, I discuss further how categories of objects when counting may denote shapes in different PNG and other languages—there are different counting words for different categories.

Such contextual references are common place in visuospatial reasoning whether they limit or extend thinking. In a study on language in a collaborative classroom in PNG, Muke (2012) found that a teacher chose to assist children to know left and right by reference to places located at some distance from them in the school on their left and right. This might of course not be helpful for the children if they were in a different orientation or place but for the moment it was a reference point for them. (See above for discussions about cardinal points and left and right). It was common practice in PNG, for people to think about east and west with little thought for north and south due to their familiarity with the rising and setting sun that does not vary much throughout the year close to the equator. In temperate zones, the sun's position in the north and south is far more important in terms of its heat inside a dwelling.

Similar problems arise with terms like "straight", and "right angle" often perceived as straight ahead or vertical on the page. "Diamond" often prevents children from realising a square is a square in any orientation or it confuses the rhombus and kite names. Thus the term in Tok Pisin is particularly difficult. However, words for Euclidean shapes are not necessarily part of traditional PNG cultures and languages. The ways in which shapes are made are not necessarily linked to the Euclidean property approach to shapes. "Diamonds" are more likely to be made than squares in weaving and bilums (see Figs. 5.14 and 5.16). Shapes are associated with changes to number patterns rather than changing properties. Parallelograms are distinguished from rectangles by the diagonals being equal although every attempt is made to get angles at right angles in house building, for example. Students will provide shape names in some cases but one teacher identified "starry" as a common shape which he related to pentagon, hexagon, and octagon, all of which people knew how to make through open cane or bamboo weaving, bilum-making, and tattoos. The angles of weaving and the over–under routes of the bamboo are significant (see Chaps. 5 and 8). However, other people were not familiar with these names for the shapes as found with the shape on a bilum called naming of the *fifti toea* (a coin that is not round but not a hexagon which is the shape on the bilum, Fig. 5.16g).

It is evident that language can either hinder or be used to develop concepts about shapes or other spatial concepts. If students learn through talking mathematics, then it is important for teachers to spend time on assisting children and parents to explain and justify their thinking especially in multilingual classrooms. If there are language confusions or lack of language words then it is important to spend time on the constructing of meaning and explaining around activities that are culturally relevant (see Chap. 8 for one project attempting to do this).

Moving Forward

This chapter shows that language about space and measurement is not just associated with representations mentally as suggested by information processing psychologists nor physically as suggested by mathematicians who ignore the ecocultural context of mathematics. Language reflects an ecocultural perspective on space, place, and visuospatial reasoning not only for communicating purposes but also for visuospatial reasoning associated with place. Language gives insight into the ways of reasoning in comparing size or determining position. In the examples provided, language in words or visually is associated with communities living in the places and with communities' relationships with those places.

The differences between various frames of reference are indicative that visuospatial reasoning and decision making are reflected in language and that language and reasoning are closely interwoven and supportive of each other. With visual and verbal representations, visuospatial reasoning is extended. For example, the time needed for walking a track is more clearly portrayed in maps that reflect hard time-consuming sections of the track. However, both verbal and visual representations could be misinterpreted if the cultural and linguistic context is removed.

I return now to the diagram presented in Chap. 1 on identity as a mathematical thinker. Language is a tool for expressing visuospatial reasoning in a cultural way. Thus cultural identity and valuing that identity will be expressed in responsive social interactions and in clearly presenting meaningful relationships. This expression of culture about the land and ecology of the person, the place of the person, promotes cognitive, affective responsiveness in solving problems in a visuospatial way. Language assists the person to structure the appreciation of the environment and to use words and diagrams as tools for problem solving. As a result, a person develops his/her identity, not only as a mathematical thinker but also as one thinking ecoculturally. Visuospatial reasoning reflected in mathematical literacy is a critical part of that identity.

Visuospatial reasoning is expressed in language but are there other ways in which ecocultural perspectives are portrayed? To explore this, I will discuss visuospatial reasoning in practices of people in PNG in the next chapter and from other countries, particularly with Indigenous communities in the following chapter.

Chapter 5
Visuospatial Reasoning in Cultural Activities in Papua New Guinea

> *In real life the problem itself is at the centre and the information and skills are defined around the problem.*
>
> (Dasen & de Ribaupierre 1987)
>
> *For 'non-Western' or 'non-literate' people, 'the only approach to pattern is through seeing it and thinking through it and that is something Western societies are no longer able to do' (p. 177). Westerners, Were says, no longer know how to learn from the visual and must have verbal exposition.*
>
> (Were, 2010)

The Challenge

In Chap. 2, it seemed that tasks that might have been classified as assessing visualisation skills were in fact being done through verbal-analytic skills if they were set as a dichotomy by the researchers (see, for example, Shepard, 1971). Thus we came to question, the understanding of visuospatial reasoning or even visuospatial skills from a purely psychological perspective. I referred to some earlier studies on spatial abilities and visualisation that showed that experience or training could influence visuospatial reasoning although authors may not have used the expression, visuospatial reasoning, at the time (see, for example, Lean & Clements, 1981). In Chap. 3, I referred to other studies which considered that naturalistic experiences and sociocultural background could have a significant role to play in learning and use of visualisation. Dasen and de Ribaupierre (1987) summarised research that took some account of ecocultural contexts but their theory was largely within a neo-Piagetian framework. The issue was again raised more recently by Shayer (2003) who has been an advocate of Piagetian studies in various cultural contexts for many years. However, can we argue that difference in visuospatial reasoning can be explained by culture? In this chapter, I will consider how ecocultural context nurtures visuospatial reasoning in the everyday lives of children and adults in PNG societies.

In Chap. 3, I indicated that there are an increasing number of sociocultural studies that could be considered to have similarities to prior-to-school research in that the geometric constructions are not arbitrary but have many practical advantages (Gerdes' Foreword in Ness & Farenga, 2007). Sociocultural studies draw our attention to the diversity of mathematical ideas that are used by different sociocultural groups. Carraher (1988) found that people in all classes and walks of life are capable of performing quite complex mathematical operations provided that the context in which the mathematics is presented links with the learners' personal worlds. For example, carpenters used what school mathematics would call symmetry ideas in making designs on boxes (Millroy, 1992). They created designs centred on the boxes in culturally influenced ways. The carpenters used mathematical ideas and thinking that was tacit knowledge manifested through activities and not spoken.

The importance of recognising these visuospatial and geometric ways of reasoning expands not only our understanding of geometries but also how people actually learn to think spatially and geometrically. Equity in education will be enhanced by drawing on these sociocultural studies (Barton, Poisard, Do, & Domite 2006; D'Ambrosio 2006). Much of this sociocultural learning occurs in a place and is therefore involving the body in activity in that place. An emphasis for Indigenous communities in particular is in outdoor cultural activities which occur in a place with particular spatial knowledge and processes. However, activity situated in place begins the development of visuospatial reasoning. Such reasoning is integrated with other knowledge arising from the ecocultural situation. Such reasoning is provided with purpose in pursuit of the cultural activity which is inevitably influenced by ecology. The person is interested in reasoning through the problematic situation of the cultural pursuit.

Is it possible to show that this is ubiquitous and varied even within one country, PNG? What is the difference in this visuospatial reasoning that will establish a new perspective? This chapter will outline a number of cultural activities in various groups in PNG to see if visuospatial reasoning occurs in garden making, village arrangements, travel, and construction of buildings, canoes, woven objects, carved objects, and continuous string bags. It draws on village experiences with participant researchers in numerous places in PNG, supplemented by interviews, questionnaire data, and documents. An ecocultural perspective is evident in these cultural events. Of necessity, bodily actions together with descriptions, and the spatial arrangements of places are used to understand visuospatial reasoning. Links are made to systematic ways of thinking in terms of geometry and measurement. Furthermore, as noted in Chap. 3 and illustrated again in this chapter, cultural fluidity and adaptation of activities influence ways of thinking (Saxe, 2012; Saxe & Esmonde, 2005). This chapter gives examples of village technologies (see also prologue) that I have observed and in many cases discussed in detail with the villagers. Much of the detail is about practices that require a visuospatial understanding of measurement. The tacit knowledge that is evident in the conversations with villagers and in their demonstration of constructions is specific to certain cultural groups. There are some commonly used processes but also diversity.

Earlier Studies on Spatial Abilities and Visualisation Recognising Ecocultural Contexts

One early (1980s) study in PNG considered how cultural background influenced representations made by children with no schooling, those with no schooling but a school available, and those who went to school in the remote Jimi Valley, now Jiwaka Province (Martlew & Connolly, 1996). Their study occurred where there was little known depiction of representations of forms but there was face and body art and headdresses which were often symmetrical and colourful, woven geometric armlets, and arrow and axe bindings. The unschooled children may have seen labels on tins in trade stores or the use of pictures by health or church workers and school room items spread between scattered houses within or between villages. First, they noted that there were a number of studies in different places in the world showing differences and similarities in development of human drawings from the usual scribble, "tadpole" (combined face and body with elongated legs), to more conventional drawings with a circle above a triangle or rectangle body with stick legs and arms and then with fingers and hair (which is common among western school children). However, in some places, just an elongated shape was made with varying degrees of curves.

In Jimi Valley their careful analysis showed that there were significant differences between groups. A few unschooled younger children produced scribbles and shapes but there were more drawings specifically found in Jimi Valley with both contours and stick figures. For older unschooled children, there were more conventional figures in Jimi Valley style whereas those who went to school or were near school were more likely to produce transitional figures, Jimi conventional and school conventional figures. The unschooled did not produce the western style of tadpole (although they were aged 10). There was some interaction of gender, age, and ability on a spatial block test (introduced by tiles being placed to give meaning to the arrangements of tiles on drawings). Once the notion of representation was accepted, then students were able to produce transitional and Jimi conventional figures. A modelling of drawing soon resulted in spontaneous representational drawings suggesting minimal schooling will lead to an acceptance of representations (Martlew & Connolly, 1996). Thus we find that visuospatial reasoning in depicting people was influenced by culture and by schooling.

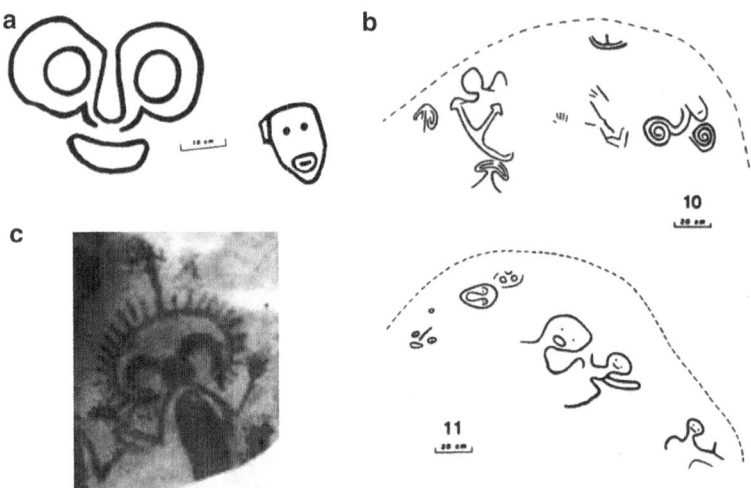

Fig. 5.1 Rock art in Papua New Guinea. *Source*: (**a**), (**b**) Tabar Island drawings (Gunn, 1986); (**c**) Snake River (1975)

Another study by Gunn (1986) on Tabar Island, one of the island groups of New Ireland, considered rock art. Most depictions engraved by pecking and abrading were exceptionally well depicted, particularly of *masalai*[1] animals, while humans were represented by faces especially with eyes. What is thought to be examples of older art include the concentric circles or spirals, some inside circles, some bent, and some said to relate to female genitalia. The rock art is mainly found around a river and although there may be links to *malanggan* art (painted on wood or sculptured in wood, rock, or coral) from elsewhere in New Ireland, there seems to be creativity, quality representations of shapes, and depiction of classification, e.g. of *masalai* images. Figure 5.1 illustrates some of the art: (a) reflects *malanggan* art but there were also free lines used for drawing noses, bodies, spirals, and bent concentric circles; (b) relates to and "tells" the story of two women who agree to meet but a *masalai* comes instead of one of the women so the other woman hides under a waterfall. The symbolism of shape is unique and illustrative of the importance of curves rather than straight lines in this culture as in many cultures. Chronology, relationships, and proximity are embedded in the art.

It is interesting to compare this rock art with the visuospatial rock-art representations from a different area of PNG, namely art drawn above bone burials in the Snake River Valley of the Buang area of Morobe Province (see Fig. 5.1c). Interestingly, the art depicted cultural stories and the features of the art are linked to perceptions and values. The Snake River art suggests a headdress on the person's

---

[1] Tok Pisin word for spirit beings.

head. Stories of who drew these figures and how they were drawn up high on the rock face changed from my first visit in 1975 to a later visit around 1985 although initially attributed to people for a specific reason and later to *masalai*.

Reichard's (1933 [reprinted 1969]) classic study of pattern for *kapkap*, the disc patterns from New Ireland, PNG, recognised the technical workings of pattern as a system of relations. These were also passed down by *maimai* to younger carvers of the shell by allowing them to observe or look on as they worked. They would not ask questions. They could also learn by remembering the image which they saw only briefly at, for example, mourning ceremonies. The size of the discs represented the status of the wearer. Some "anthropological studies see the pattern as an aesthetic medium capable of engendering person-object relations that is intricately related to mythopoeic worlds". (Were, 2010, p. 9). Gell (1998) suggests "pattern as locus of agency in ritual and everyday domains". He contends that "objects are thought-like in nature, being material manifestations of the workings of the extended mind that are able to engender social relations". The objects represent real relationships and reach into the past and into the future possibilities and aspirations. The designs are memorised. "Objects are the workings of the mind, objectified in an external form that can be displaced both spatially and temporally" (Gell, 1998). Thus the visuospatial memory associated with the creative task and the impact of ecocultural identity is strongly influencing their mathematical reasoning.

When making new *kapkap*, the maker is thinking through images and transforming images to create new understandings. To create the first circular pattern, a triangle is cut and a diametrically opposite one is cut, this is done around the circle. The next concentric circle has shapes that are smaller and the ratio between shapes in each circle is kept for the next circle. Symmetry and reduction in size (similarity) are keys to carving the *kapkap*. Central designs are often by rotation rather than symmetry so the fourfold design can be transformed into a fivefold or eight pointed design. Other modifications include straight to curved lines, negative to positive spaces, and reversals of triangles. Cultural relations are embedded in the notion of the wafer thin translucent tortoise shell connected by the string to the strong clam shell disc. Exchange of shell-money at various points, and recognition of *kastom*[2] being revived and of the intricate pattern of matrilineal and patrilineal land ownership is closely linked to the creation of *kapkap* today (Reichard, 1933 [reprinted 1969]; Were, 2010).

Kapkap also connect and cross borders (Jegede & Aikenhead, 1999), that is across the societies found on the islands from New Ireland, Mussau, Buka, Bougainville, and into North Solomons. Furthermore, the *kapkap* symbolism is modified to church *kapkap* to recognise leadership. Performances are also expected at, for example, opening of new church and there is still rivalry between clans and groups regarding how well they are keeping to *kastom*. There are headstones with

---

[2] Tok Pisin word for traditional customs.

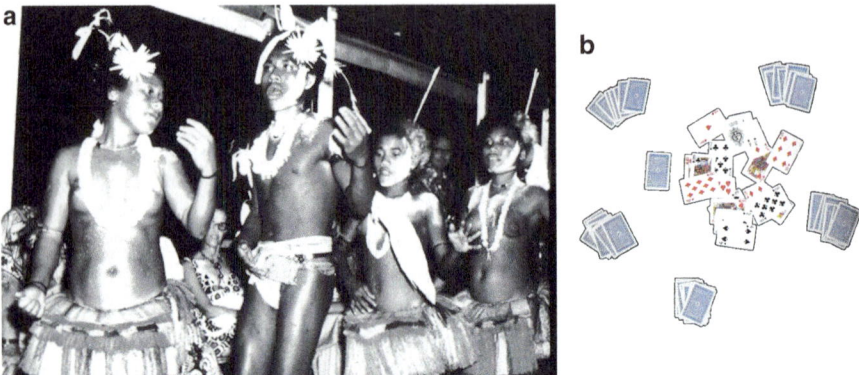

Fig. 5.2 Objects that become part of visuospatial reasoning and identity connections. (**a**) Trobriand dancers with models of discs, 1975. Note tapa cloth worn by lead male and "grass" skirts by teenage girls. (**b**) Card game Goroka

kapkap and other symbols and carved poles in churches even where Ba'hai religion was dominant suggesting a strengthening of the traditional beliefs connecting from past to present and future and a changing of spiritual life whereas the use of *kapkap* is more restricted in SDA church area (Were, 2010). Another adaptation of circular discs as symbols of leadership is seen in the Trobriand decoration in Fig. 5.2a but in this case, it has not been created in the same way or with the same meaning. Thus we see the visuospatial reasoning behind the making of *kapkap* intricately connected to culture and transcending societies appropriately.

Kula-shell from Massim,[3] Milne Bay Province, "take on their own trajectories through trading networks, across the islands, carrying the biographies of former owners as they circulate through generations and across the ocean's expanse. ... locus for thought itself—devices through which exchange patterns direct thought" (Were, 2010, p. 11). This notion was applied by (Pickles, 2009) to cards taking their own place in thought during relationship pattern changes in card games in Goroka following interpretation of the Kula trade in Milne Bay. So visuospatial reasoning has a perspective that takes account of interpersonal spatial relationships in a way that is relatively unknown in western thought in mathematics. It reflects the idea that the physical spatial interacts with the mental visuospatial reasoning of the person, an idea discussed in terms of tools and computer screens mediating reasoning (Goos, Galbraith, Renshaw, & Geiger 2003).

---

[3] Massim refers to a number of islands generally involved in the Kula trade route including the Trobriand Islands.

Earlier Studies on Spatial Abilities and Visualisation...

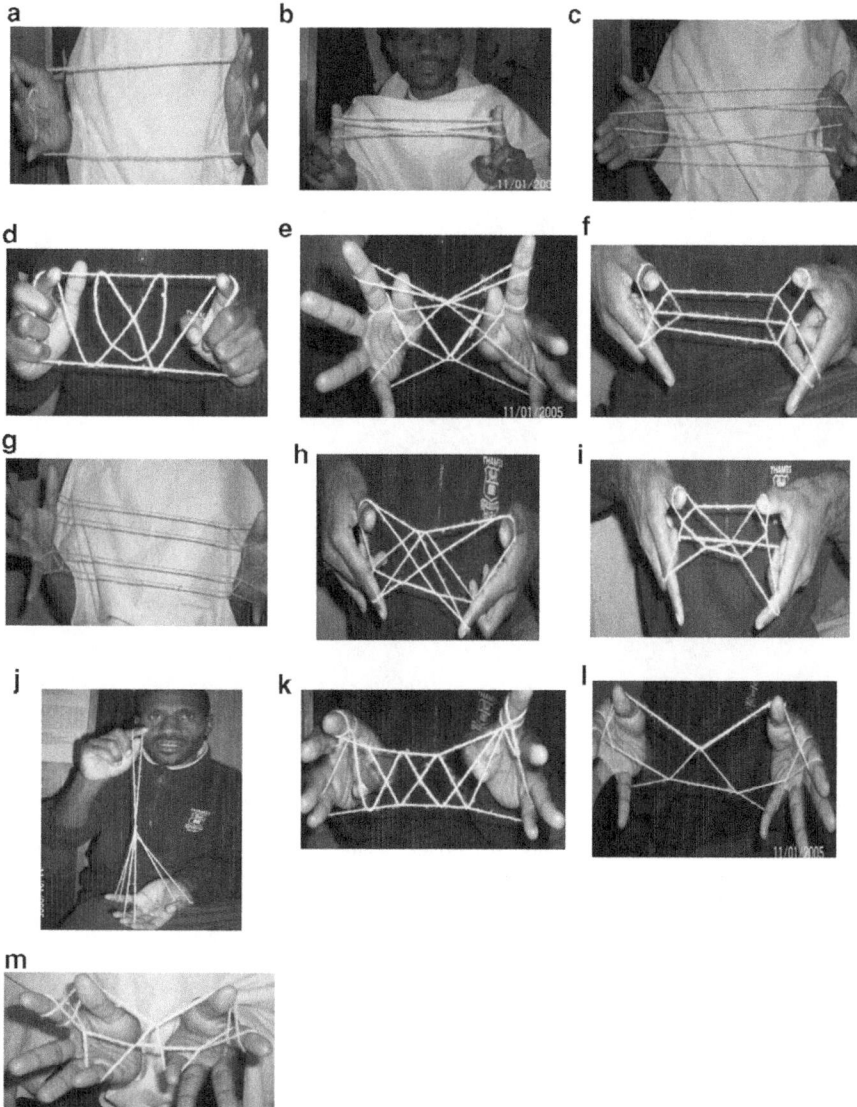

Fig. 5.3 Kambea Rambu string figures and story. (**a**) The course set for Kambea Rambu. (**b**) Passing Puri. (**c**) At Ronga village. (**d**) Wando Range. (**e**) 4 m deep well along roadside. (**f**) Cane bridge crossing Suku River. (**g**) Rushing imaginary Suku River. (**h**) Akero rainforest where wild orchids and possums abound. (**i**) Showing bird of paradise nest along the road side. (**j**) A bee heap adjacent to bird's nest. (**k**) Side view of Mount Kambea showing its exciting feature. (**l**) Showing treeless Peak of Mount Kambea. (**m**) The targeted point at the top of Kambea Peak

Fig. 5.4 Tattoo design drawn by Rea, Motu teacher at Tubusereia and copies by grade 2 children who chose which design to do from her self-made big book. (**a**) Chest tattoo incorporating tear design. (**b**) Back tattoo. (**c**) Hand tattoo. (**d**) Stomach tattoo. (**e**) Leg tattoo. (**f**) Drawing the rectangles for the basic hand tattoo. (**g**) Next stage, with others' drawings. (**h**) Proud of the finished drawing. (**i**) Lack of rulers encouraged hand drawing

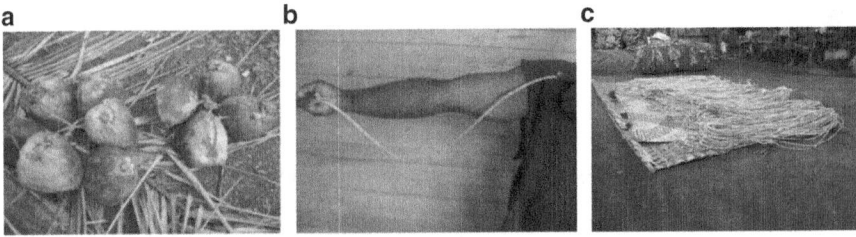

Fig. 5.5 Visuospatial reasoning in counting by Tolai community, PNG. (**a**) Nuts bundled in sets of 6 and 4 more providing words for 10 from this visuospatial arrangement. (**b**) Marking length or number of shell-money. (**c**) 10×10×3. *Source*: Paraide (2010)

Earlier Studies on Spatial Abilities and Visualisation... 149

Fig. 5.6 Triangular pattern used in planting cash and other crops

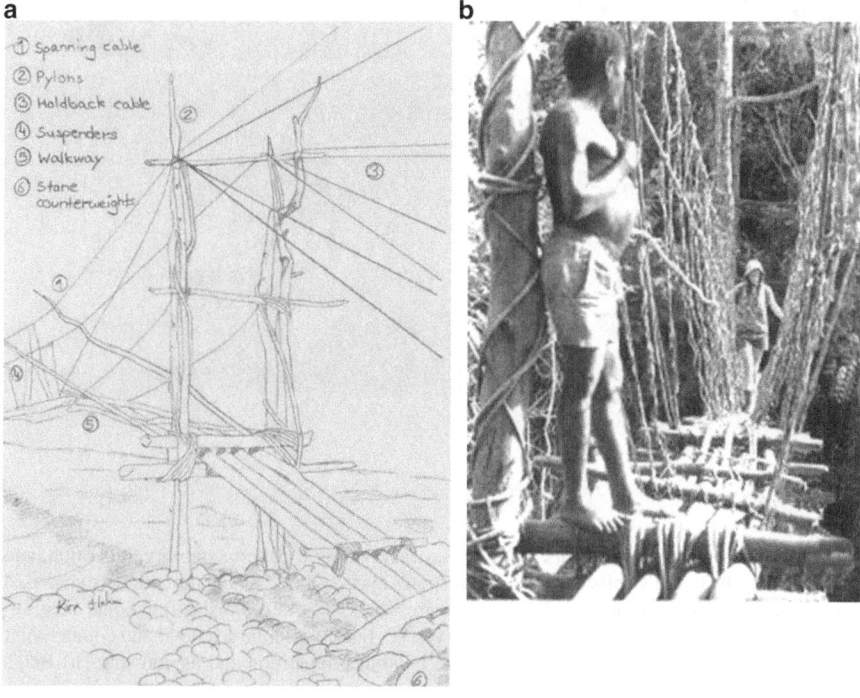

Fig. 5.7 Typical bridges in PNG requiring visuospatial reasoning to build. (**a**) Cantilever/suspension bridge that I crossed around 1986 across the tributary of the Busu between Boana and Hobu. It had been rebuilt after a flood. (**b**) Suspension bridge below Boana, one of two crossing the upper reaches of the Busu that I crossed in 1973 and another I crossed in 1983. Bridges have to be replaced or repaired when materials rot

Fig. 5.8 PNG architecture students' sculptures. (**a**) David used pattern, stability, repetition, and measurement. (**b**) Willie's holistic sculpture illustrates his use of curves and repetition. (**c**) Ian used three-dimensional shapes in his compact sculpture. (**d**) TKeps developed a functional idea. (**e**) Fred incorporates traditional decoration and curves. (**f**) Taurus used the sea devil and counterpoint balance. (**g**) Fing developed ideas from modern buildings, repetition and asymmetry. (**h**) Bell's sculpture shows traditional influences

Earlier Studies on Spatial Abilities and Visualisation... 151

Fig. 5.9 PNG gardens form area units. (**a**) Drains divide garden into areas, Kopnung, Jiwaka, 2006. (**b**) Three stalks in each mound in rows of two or three mounds

Fig. 5.10 PNG round houses. (**a**) Large, round house, Kaveve, Eastern Highlands, 2006. (**b**) Using rope tied to centre post to mark the edge of the round house. The leg at the other end of the rope is dragged to mark the circle, Kaveve. (**c**) Lower area space in a small round house covered with utensils and nuts drying. Karuku nuts drying over fireplace, Kaveve

Fig. 5.11 Rectangular houses. (a) Men's house for 25 men with four corner sleeping rooms, Kopnung, Jiwaka, 2006. (b) Making a model and showing measuring techniques, Kopnung, Jiwaka, 2006

Fig. 5.12 PNG preparing a *mumu* (Kaveve village, Eastern Highlands, 2006). (a) Heating the stones. (b) Adding the food, karuku nuts. (c) Covering with dry grass, green grass, and then soil. (d) Pouring water through a hole onto the mumu stones to make steam to cook food

Fig. 5.13 PNG coastal houses. (**a**) House with woven decorated walls, basic house on nine posts, Malalamai, Madang Province, 2006. (**b**) Placing morata on the roof of a house, checking spaces, Kela, Morobe, 1997. (**c**) Weaving walls *blind* with a desgin, Lolobai, Morobe, 1997. (**d**) Morata ready for a particular size house, Mis, Madang, 2006. (**e**) Designing new house styles, and single outrigger canoe, Lolobai, Morobe coast, 1997. (**f**) Long house for families, Kanganaman, Middle Sepik river, 1983

Fig. 5.14 Leaf and cane weaving. (**a**) Weaving patterns for walls, *diagonal, zigzag, diamond* Kopnung, Jiwaka Province, 2006. (**b**) Woven basket, internal and base layer wider, pandanus, Kepara village, Central Province, 2004. (**c**) Basket display—Yalibu, Southern Highlands trays—introduced; Buka, Bougainville lady's handbag and tray; fish traps; Sepik mask. Sepik carved man looking on. (**d**) Sepik mask , 1986. (**e**) Woven floor mat, Milne Bay, 1984. (**f**) Carry bag that men learn to make in the men's house. Note bilum with baby in background. (**g**) Hat making passed down by men

Earlier Studies on Spatial Abilities and Visualisation... 155

Fig. 5.14 (continued)

Fig. 5.15 Various carrying and storage objects, PNG. (**a**) Palm frond temporary basket for carrying food or distributing in exchanges. (**b**) Limbom basket, Wosera, East Sepik, 1983. (**c**) Karuka nut basket for trading from highlands to coast when stored nuts are dry, Kaveve, Eastern Highlands, 2006. (**d**) Mountain design woollen bilum with karuka nuts, Kaveve, Eastern Highlands, 2006

Fig. 5.16 PNG continuous string bag, *bilum*, designs. (**a**) Continuous string bags (bilum) requires string to be joined by rubbing on thigh. Women learn from each other. Bena Bena, Eastern Highlands, 1975. (**b**) Rai coast bush string being made and wound onto spindle, Madang Province, 2013. (**c**) Two-strand twisted wool bilum, Watabung, Simbu Province, 1978 with close-up. (**d**) Bilum made by Engan woman, PNG highlands with closeup, variation on the rainbow design, 2001. (**e**) Close-up rainbow design, Kamano-Kafe, Eastern Highlands, 2006. (**f**) Computer – network design made by Kamano-Kafe woman, Eastern Highlands, Province, PNG, 2006 with close-up. (**g**) Soccer ball bilum, 2005 with close-up. (**h**) Open stitch of traditional carrying bilum, 1997. (**i**) The flower pattern design from tiles used by Caroline from Sepik living in Madang, 1997. (*Source*. Jondu, 2008). (**j**) Both sides of a bush string traditional bilum, East Sepik Province, 2001. (**k**) Bilum in plastic, Tufi, Oro Province, note the use of colours and different stitches for effect and the combination of rectangles in rows to form squares, 1986. (**l**) Girl at Malalamai, Rai coast,

Fig. 5.16 (continued) Madang carrying bilum, strengthening neck muscles, being like mother, 2004. (**m**) Pig toy on cane frame with large stitch bilum and light bilum underneath, looking inside at frame, 1997. (**n**) Two different loops used for effect or for different purposes, Balob Teachers College student project, 1997. (**o**) Bilum and two kinds of shell for chest piece and status, 1997

Fig. 5.16 (continued)

Overview of PNG Material Cultures Involving Visuospatial Reasoning

Culture is influenced by ecology, and culture influences ways of visuospatially reasoning particularly through the cultural activities and creation of artefacts. In particular, visuospatial reasoning is apparent in designs. I have several books (Dennett, 1975; Madang Teachers College, 1973) and a CD (Architectural Heritage Centre, 1996) of different designs and patterns but anthropological studies provide good insight into the deep meanings of design involving visuospatial reasoning.

Some of the most beautiful examples of design in PNG (see coloured photographs for examples) are on carvings (Figs. 5.17 and 5.18) but there are also designs in weaving cane, split bamboo, pandanus, and other leaves (Figs. 5.13a, c, 5.14, 5.18b); in bilums (continuous string bags) (Figs. 5.15c and 5.16); pots (Fig. 5.19); grass skirts, headdresses, body painting, and masks (Figs. 5.15f and 5.19f); as freehand drawing on tapa cloth (made from bark of specific trees—Fig. 5.20); and in decorations that have limited life on the sides of school boards and in the sand. Woven cloth for wrapping was not part of PNG cultures. Tapa and bilum were used for coverings (Figs. 5.2 and 5.21). Bilums are in fact made from figure-of-eight type loops created from a continuous string—(see Fig. 5.16h). Creative design has led to a range of hats, clothes, and bags using bilum looping, furniture carving, arts such as pottery (Fig. 5.23), or story images (Ison, ~1986).

Fig. 5.17 PNG stone carvings and bindings. (**a**) Bindings on stone adze still used for canoe and sago working; stone axe, Highlands 1976; bindings on arrows (jagged one for men) 1975, arm band from Rai coast Madang 2013. (**b**) Rare stone carvings, Hagen, 1985. See also Fig. 5.20d

Fig. 5.18 Diverse wooden carvings from across PNG. (**a**) Bilum hook, hardwood, Palambi, East Sepik River, 1983. (**b**) Tray, Kiriwina, Trobriand Islands, 1973; Mat Gulf Province, 1997. (**c**) Carvings from Tami Island, plates 1978, bowls 1980. (**d**) Carvings from Trobriands, diversity, kwila and ebony woods, small gourd, 1975. (**e**) West New Britain—single bird with flat wings, 1975; developments to snake and birds with raised wings, 66 cm, 1982. (**f**) Trobriands, 1975. (**g**) Sepik carvings. Myth, kundu drum, male 128 cm ~1986. (**h**) Yam storage and display house, Trobriands, 1975. (**i**) Large seafaring canoe prow, Trobriand Islands, 1975. (**j**) Model of lakatoi canoe, Tubusereia, Papuan coast, Central Province, 2013. (**k**) Story board, Kambot, Murik Lakes, Sepik River, East Sepik, 1973. (**l**) Story board, Kambot, 1988. Note variations from earlier. Village woman with adze making sago. Connections of people with totems, animals, place, myths. (**m**) Carved poles of Haus Tambarans, middle Sepik River, East Sepik, 1983. (**n**) Carved logs representing all provinces, Sepik carvers, designed by Department of Architecture and Building Science, PNG University of Technology, for University of Goroka Library, ~5 m 2002. (**o**) Crocodile carving by East Sepik carver, 1997

Fig. 5.18 (continued)

Fig. 5.18 (continued)

Fig. 5.19 Pottery from Sepik River, East Sepik, and Markham Valley, Morobe, PNG. (**a**) Wosera pot, East Sepik, 1983. (**b**) "Stove" Chambri Lake, East Sepik for sago pancake plate and for smoke to reduce mosquitoes, 1983. (**c**) Sepik River pot, 1997. (**d**) Head only of sago storage pots—humour, made for tourists, Chambri Lakes, 1978. (**e**) Pot for cooking snake, Zumin, 1984; nearby Adzera village, 2006. (**f**) Mask: Tortoise shell, cassowary feather, basket, clay, East Sepik, 1985

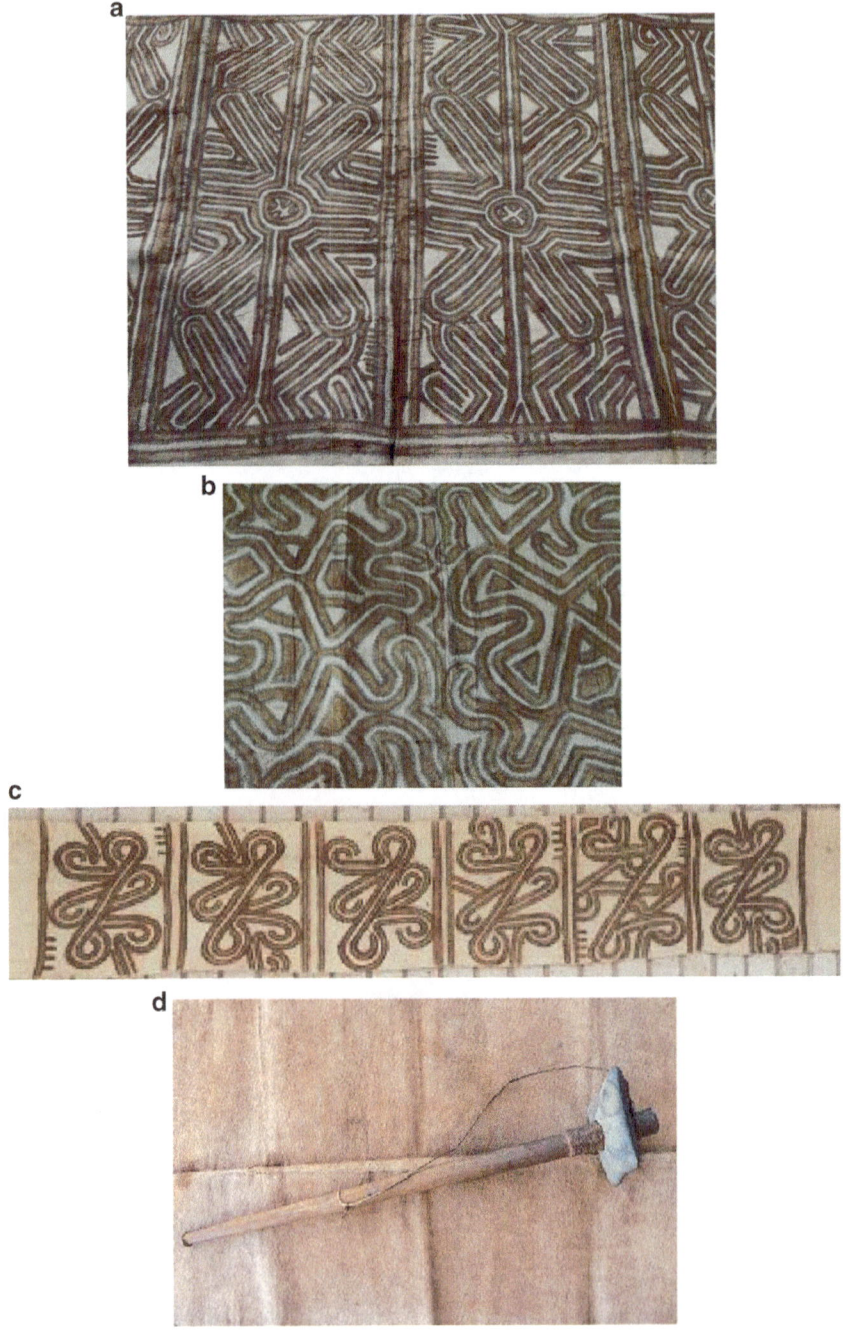

Fig. 5.20 Tapa from across PNG. (**a**) Section of tapa used as cloak for dancing. 180 cm × 110 cm, Oro Province, 1975. (**b**) Tapa cloth, 55 cm × 136 cm, Oro Province, 1976. (**c**) Long strip of tapa worn by youth and men wrapped between legs and around waist. Oro, 1976. (**d**) Anga tapa used as coat, raincoat, and hiding club, Menyamya, Morobe, 1985

Fig. 5.21 Examples of diverse dances, decorations, and displays associated with *singsing*, PNG. (**a**) Mudmen masks, *Gilitue*, preparing for cultural activity; see Table 8.1 for making of mask (John 2007). (**b**) Highland warriors with bilum skirts, painted faces, feathered headdresses, Goroka Show, 2005. (**c**) Wig and bamboo strips indicated pig kills in exchanges, 1975 (see Fig. 8.1b for making wigs from another cultural group. (**d**) Bena headgear, 1976. (**e**) Huon Peninsula kundu drums and "sails" headdress, 1978. (**f**) Henganofi tell a story of war and grief, 1984. (**g**) Anga dances at their show 1985

Fig. 5.22 PNG village activities requiring visuospatial reasoning. (**a**) Village court case seeking compensation of a pig and money. (**b**) Pig kill sharing after compensation. (**c**) Cash crops—coffee drying, tobacco, sugar cane Bena, Eastern Highlands, 1975

Fig. 5.23 PNG modern pottery

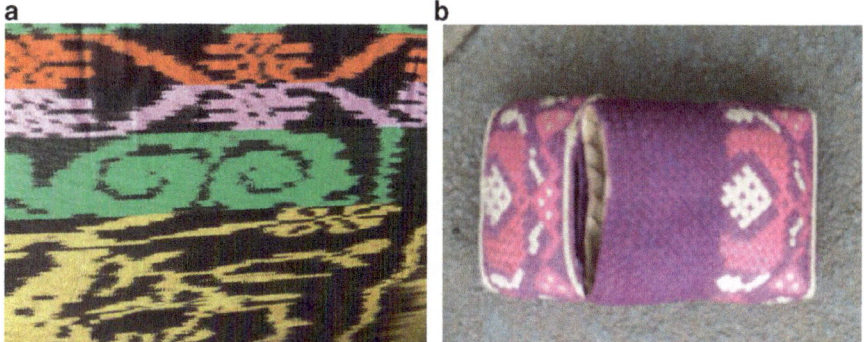

Fig. 5.24 Timor Leste crafts, ~1999. (**a**) Ikat weaving. (**b**) Small basket with fitting lid

Each cultural group that carves has different designs and patterns. Colouring is significant with carvings, pottery, and other artefacts that are painted (Figs. 5.19a, d and 5.16m). While masks can be thought of as three dimensional, the overall use especially of white paint can create alternative perspectives such as on the tortoise shell mask (Fig. 5.19f). The ends of the Trobriand tray, head, and canoe prow are related designs (Fig. 5.18b, d, e), curving intricately.

The carver in 3D makes the most of the original material. Thus beautiful Tami bowls result in perfect symmetries curving up the sides of the bowl created by the grain of the wood, or the crocodile carver considers where the legs and curve of the tail will be before carving the log. What is noticeable is the use of symmetry and asymmetry in carvings and in repeated designs (Fig. 5.20). Much of the design work is free hand but strings and sticks are also used to improve the carver's eye with symmetry (personal communication, Sepik carver, 1983). Most carving is of wood, but some is in stone both for utilitarian purposes such as axes and adzes (Fig. 5.17a) but also for special purposes including small secret objects or perhaps as ornaments (Fig. 5.17b). The following sections discuss some of the representations found on artefacts.

Visuospatial Reasoning That Connects Euclidean Geometry and Topology

While academic mathematics uses the logic of setting additional constraints on topological relations to achieve other geometries like projective geometry or Euclidean geometry (see Appendix G), curricula in schools often treat them as dichotomies. By looking at PNG design, we begin to see an alternative where visuospatial reasoning meshes ways of thinking that would be reduced by such classifications. Part of the reason for this is linked to the geometric features found in the artefacts being indicative of social identity and relationships with members of the community. Geometric design is often the abstract aspect making connection to relationships but this is connected to representational art within the same artefact. However, there is also personal expression in the design or artefact (National Museum of Australia, 2013).

In some cases, it is the act of painting that is important and not the carving per se (Hauser-Schäublin, 1996) (examples in Fig. 5.18g, j, k, l, m are painted). The white line of paint is continuous in even large complex designs in Abelam art (middle Sepik) (Hauser-Schäublin, 1996) in much the same way as the string is continuous in bilums. The artist starts at the top and leaves an end to attach the next design row. Other colours fill, interpret, and highlight the white thread. The main string for the bilum may be decorated but the white thread for the whole bilum is one continuous string in Abelam. Looking at the various bilum patterns in Fig. 5.16, there are numerous ways in which continuity is portrayed, not only in repetition but also in lines traversing the object.

Knots are often formed by tightening loops for specific purposes and according to Kuchler (1999) can be represented in carvings and formed to make wrappings (I am reminded of the beautiful bindings on house rafters in Fiji but practical bindings are evident in Fig. 5.17a). Kuchler suggested that bindings and knots are for both functional and relational senses (e.g. the living and the dead). It is the process of creating the knot or the "knot" in the carving that establishes the relationships between different objects.

The purpose associated with visuospatial reasoning varies and is evident when

> the stability of pattern is the result of the subordination of technique to the spatial framing of the image. Rather than 'finding' the design of the figure in the wood during carving, the geometric contours of the image are drawn on sand or onto the wood that is used for carving, thus allowing improved or innovative techniques of fabrication only minimal scope to influence the resulting product. It is this geometric design of a *malanggam* which is owned and exchanged in the north of New Ireland, quite unlike the New Guinea netbag described by MacKenzie (1991) which is validated in terms of its technical execution and exchanged as artefact. (Kuchler, 1999, p. 155)

The bilum itself can be transformed into a hard two-dimensional figure for men's use in ceremonial houses whereas they are left soft and bulbous for women in many areas (Kuchler, 1999).

The weaving by men of seating mats, carry bags, and hats from a coconut frond (Fig. 5.15d, e) is also common across many areas but at least in Abelam, it is the lattice that is of significance as it represents the criss-cross of moieties (Hauser-Schäublin, 1996). It is also evident in their weaving of white on dark for floor mats. The Abelam, like many groups, also use ways of "concealing and revealing", the hanging fronds at an entrance or around an area to be hidden (see Fig. 5.13f), the decorations for dance to hide the individual and to be spiritual. For the Abelam, the strips and their combinations are significant. To appreciate the visuospatial reasoning behind these, one has to look at the cultural and the environmental, ecological values. In many places, men weave the walls *blinds*[4] for houses. Men in the highlands learnt how to weave from coastal men and replaced the tightly joined wooden planked walls with

---

[4] Tok Pisin for walls woven from different materials.

them. The wooden planks were hard work by hand and the possibilities of arrows through the walls became less over time.

In Fig. 5.18m, one can see a pole with the crocodile design which was also used for ridged tattooing in initiation ceremonies (Middle Sepik, East Sepik Province, PNG). The designs on the poles prepared for the library at the University of Goroka (Fig. 5.18n) were taken from all provinces (not necessarily from carvings) but carved by Sepik carvers under guidance of architects from PNG University of Technology (Unitech) that houses the Architectural Heritage Centre. The design of the end plates of the Trobriand bowl and head reflect the front boards of the canoes (Fig. 5.18b, f, i). Interestingly, the carving of a tree kangaroo and pig from this same area stylistically are to my eyes quite different but a friend from the Trobriands on entering my house was easily able to recognise all the carvings from his island (see Fig. 5.18b, d, f). The designs of the East Sepik are rugged and often portray people Fig. 5.18a, g, k, l, m). The bilum hook is exquisitely fine hard wood; the storyboard soaked and fired. The storyboards tell important stories. The background shows the people of the village in different activities including using an adze for scraping sago. The board's story links the environment and its creatures and spirits to the people. Adzes are commonly used to carve but chisels (nowadays) are used for finer work and knives for smaller items such as the serrated ends of arrows used for fighting men (Fig. 5.17a). Shell or glass is used for smoothing the pieces, a necessity for many items like bows.

The Wosera pot (Fig. 5.19a) is distinctly different from the Chambri Lake (Fig. 5.19 b, d) and Sepik River pots—similar faces are found on lintels of the low men's houses in Wosera. Pots were important trade items in many places: Papuan coast, the Madang coast, Sepik River, and the Markham valley. All were cooked on slow fires (note the smoke colourings on the pots) so the skill of the craftsman was extraordinary given the size of some pots. Generally a coil and pat method was used. Like bilums, many traditional designs in wood and clay mark the place.

Whatever the carving or design, the spiritual culture, the artistic culture, the environment, and people living in that environment, all had an influence on the reason for carving, what it looks like, how it was undertaken, and how to share the ecocultural knowledge to others. One example recorded by Bowden (~1990) in Kwoma speaking area of the Washkuk Hills notes that men's houses are made of wood significant for men.[5]

> Kwoma art is highly stylized and for the most part abstract. ... design elements in themselves do not ... have fixed 'meanings.' Such significance as they have derives entirely from the 'meaning' of a painting as whole. ... The best Kwoma artists are never content simply to reproduce existing representations of clan totems (or other entities) but pride themselves on their capacity to modify and embellish them imaginatively. (Bowden, 1990, pp. 487–488)

Similar traits were evident in the collection of pictures by Aiyura National High School students developed to accompany a local legend (Ison, ~1986).

While symmetries are evident in the paintings that Bowden commissioned, curves dominate and the environment provides a source for content. Smidt (1990)

---

[5] Women have all married-in to the clan which is small.

discusses the symbolic meaning in Kominimung masks (on a tributary of the Ramu River and with some similarities in terms of *haus man* or *haus tambaran*[6] as the Sepik River area). Again totems, colours, and the limited number of design elements are evident but for a variety of purposes. For their matrilineal society, there are male and female masks, mostly associated with yam festivals. Visuospatial reasoning is used to create the masks which are tall, symmetrical, and attached to a basket hat. The making of different head gear in the Wantoat Morobe area was featured in one of the teacher's projects (see Chap. 8) that made links to geometric shapes for mathematics teaching.

Tapa painted in Oro Province PNG are richly curved patterns, sometimes with some symmetry reflected or rotated (Fig. 5.20 a–c). Some lines continue for at least a section of the design reflecting some of the possible cultural reasons mentioned above on continuous lines. Paints are natural colours. However, there is no regulated, stamped, or rubbed tapa designs as found in Fiji or Tonga, respectively (see Chap. 6). Tapa or other flat tree materials used on masks may be painted as stylised faces. Tapa is surprisingly widespread across Austronesian and non-Austronesian cultures but of different kinds and purposes but it does have significance [on this point, I would disagree with Hauser-Schäublin (1996)]. The softer cloth worn by young men is carefully prepared by women to fit well as a wrap (demonstrated in both Malalamai, Madang Province, and Adzera, Markham Valley, Morobe Province). It was an important cloth in dance in Oro Province (see Fig. 5.20a–c) and as a cape for covering and secrecy in Menyamya (Fig. 5.20d). In each place, the original tree, shape and size of final product after pounding flat, or amount of string from the tree were visuospatially linked in memory.

Continuous String Bag: *Bilum*

Bilums are made of fibres from different trees and some introduced sisal-type plants but commonly from the inner bark of the *tulip*[7] trees (Fig. 5.16a, b, h, j) which also supplies the supple tapa (Fig. 5.20c). When the women make the bilum, each figure-of-eight type "loop" is completed one at a time and the length of string or wool wound onto the thumb and small finger as the wool is pulled through the loop. When the length of wool is finished, they rub the start of the new strand with the end of the old one to have a continuous length. Rubbing together is the same as they do to make the shop wool strong by taking two lengths and rolling on their leg.

The string bags from PNG are traditionally the work of women and signify complex gendered meanings (Mackenzie, 1991). Mackenzie noted that for the Telefol,

---

[6] Tok Pisin: *haus man* is 'men's house'; *haus meri* is 'women's house'; *haus tambaran* is the house or building for special traditional, spiritual activities undertaken by men. They have special significance and can be quite large in Southern Highlands, Sandaun, East Sepik, and Madang provinces but many places have similar activities in their *haus man*.

[7] Tok Pisin for an ubiquitous tree with a pair of leaves from each leaf stalk. The leaves are edible and provide flavour and nutrients in food dishes.

Sandaun Province, and similar ideas occur in other groups, a bilum can be seen as our mother; it has continuous fertility. The bilum is also seen as a womb and as a sharing. It has been used for political and protection purposes too. However, there are many kinds of bilums and there are transformations in significations when used by men who are related to the maker. Mackenzie noted that generally the bilum for men is then attached to hard or flat surfaces or bound in some way. Japeth Maki (personal communication, 1997) said that if his wife from the Sarawaged mountains of Morobe "is going to make one for her husband to carry his betel nut and other personal items then it will be a small, rectangular bilum and the patterns will be associated with important relationships and meanings for the clan. This will influence the creativity of the bilum".

Different places have different purposes for the bilums although it is common to carry bilums for food and for babies, and to have small ones for special objects. For example, in some places they form front skirts (Fig. 5.21b, c) or other significant apparel for men, and sometimes for fishing. There are several different kinds of continuous looping methods for different purposes (Mackenzie, 1991). The carrying bilum (Fig. 5.15d) has a figure-of-eight type loop which can stretch easily (loops are generally spaced by a strip of pandanus). For heavy loads, they are carried on the forehead dropping over the back requiring strong neck muscles developed from childhood (Figs. 5.15d and 5.16l) but today many are worn on the shoulder.

However, in more recent times the traditional designs and bush string have been replaced, first by plastic rope bought in trade stores and then by brightly coloured wool with two strands twisted on the leg to strengthen it (Fig. 5.16). Bilum designs were associated with a particular group and even a particular woman (Fig. 5.16c belonged to Margaret Peter from Watabung; other illustrated bags were the work of specific women known to me). However, stretch is reduced in the tight woollen bilums. Designers build on their knowledge to create new patterns (see an example in Fig. 5.16d, e). These days, one woman may have learnt or developed over 20 designs and the designs pass up and down the country. By knowing the basic techniques, a bilum maker can observe another person and with few words, learn to create a new design (Lillian Supa from Simbu, interview 2008). Poilep Kamdring, a school principal, from Manus Island living in Lae, Morobe Province noted that a highlands woman showed her how to make her mountain-design bilum (Fig. 5.15d). She described its making:

> first the lower triangular sections were begun going back and forth with the one colour for the short distance. It is stepped so she would do one block less each time. There were 4 loops for each block wide and two deep. When she (the highlands woman) reached the top of the triangle she went down the side to the bottom again with the spare loop. The method here would be to look to see the blocks were of the same size but also to count. There would be four groups of four loops, then three groups of four loops, two groups of four loops and then one group of four. This might be linked to multiplication. There is a good visual representation of how the numbers increase in multiplication and can be linked to counting by groups, that is 4, 8, 12, 16, etc. Here is a natural experience of the visual imagery associated with multiplication. (Kamdring, Interview, 1997)

Caroline from Middle Sepik proudly told me how she had developed the flower design like on the tiles at Madang Resort Hotel (Fig. 5.16i). She showed how

> I started with a few rows of soft loops which I remove at the end and replace with strong opening loops. I decided to put in two then three flowers across the bilum. (Interview, 1997)

She worked diagonally too and noted, when asked if she counted, that she counted 5, 10, 15, or 20 to make up the pattern. Lilian Supa described making her bilums:

> We might need four reels (skein) to do one round so twelve all together ... To change colours, we use a number of needles, it can be up to ten or fourteen or it can come down to six or four. The soccer ball design (see Fig. 5.16g) used six needles, two for each of three colours. We twist one rope for one needle, so that's when we try to change the pattern of the bilum. ... we have the same two colours but different needles, then we stop it here and then third colour we come here and then the fifth needle sew it right down (to the corner of the bilum). The hexagons are actually made from two trapezium so the diagonal downward colour being changed will flow well. ...We sometimes say 'you must leave two or continue and make the bilum for five spaces and then you continue on with the next one' and these are some of the instructions that we give. Sometimes we say 'You use the needle and you go in and out and go down' ... The traditional kundu drum design is now sometimes called triangle today. (Interview with Lilian Supa, from Simbu Province, April, 2008)

A full description for the 50 toea design was given by Amos in an ethnomathematics teaching project (see the excerpt in Appendix D).

Some women are very creative and can imagine the pattern well. Mea Dobunaba from Central Province, a mathematics education lecturer, made bilums but she tends to draw the design first and she plans what to do from there. This she does because she has not the experience to build up new designs in her mind (personal communication, 1997).

Lengths of string are also taken into consideration in handicrafts and visualised. Mea explained that her mother was able to tell her how long to make a string for threading beads to make an anklet for dancing. The weaving in and out made a difference to the lengths and experience had taught her how long to make these lengths of string. Women sell strands of wool in the market. They take a large skein bought at the store and divide it up into manageable lengths, each about the length of a leg doubled as the foot when sitting acts as a peg. Women know how much to buy of each colour to make the bilum of their desired size and pattern.

The discussions above indicate how adaptations are systematically described in terms of enacted technical patterns (D'Ambrosio's ethno-mathe-**technical** sense). Visuospatial reasoning dominates the task as decisions are made to change colours, move to going down instead of up, and to work with several needles some of which have the same colour.

String Figures

String figures (often called cat's cradles) and clapping games are a common pastime in many Indigenous communities and played by most children in places like Australia. Clapping games involve a pair of children reciting a rhyme as they clap in various ways. They make up new ways of clapping, new orders, and new rhymes, often quite complex. They are constantly creating and repeating patterns.

Vandendriessche (2007) studied the creating of string figures in the Trobriand Islands. Like earlier researchers, he noted a series of "operations" such as hooking a piece and twisting it onto the finger. He was also able to identify "sub-procedures" and to ascertain how they were built up into "procedures" that created a design or used to produce a new design. The actual activity involves the body, mostly the fingers and hands, in transforming the loops by changing the apparent position of the loop on the fingers although this might be seen as the same topological space. Not only are the procedures repeated often in play and to while away time in the wet season, but this repetition reinforces the longevity of the visuospatial dynamic imagery in the brain. He recites the occasion on which two girls started one design but kept reaching a point where they were unsure so each time they returned to the beginning hoping their fingers would "remember" the next action. Spatial memory within the body supports visuospatial memory and reasoning. Associated with a number of objects or parts of a story, these procedures are readily repeated. Furthermore, there is repetition in sub-procedures to create new complex designs. Story rhymes accompanied some of the final products or the sub-procedures. Thus, as Vandendriessche says, these are mathematical algorithms (operations, sub-procedures, procedures) but the importance is that they are visual and spatial and ordered and remembered. In fact he associated polynomes with the various sub-procedures. Repeating procedures, recognising designs within designs, moving onto the next sub-procedure, creating new designs, relating them to cultural activities, and recalling partially forgotten ones involve visuospatial reasoning. Thus an ecocultural perspective encourages a recognition of visuospatial reasoning in this common cultural activity that indeed varies across cultures (see his references to different parts of the world).

The Senfts (1986), who also lived in the Trobriand Islands, gave further details that suggest that the string figures which are often given plant names will have other metaphorical purposes such as reference to sexual themes or sexual parts of the male or female body. Nevertheless, there is a two-dimensional shape representation of a three-dimensional object and shape comparison and classification. However, there are many different activities represented by the string figure-making such as building a house, paddling a canoe, the ebb and flow of water, a child running off to toilet, an animal or insect running, sometimes with the resultant figure moving over part of the person's body. In another the curves represent the repetition and moving circular actions of a man digging out a canoe with an adze. This dynamic visuospatial representation is complex and ecocultural. The games are for entertainment and so the connotations and actions are for enjoyment. There are also tricks that change dramatically to the final product or end it supporting the affective aspects of self-regulation in our model of Chap. 1 (Fig. 1.2). The Senfts documented 89 different figures by name and drawings (only four appeared to relate to myths, healing, or religion although there may be second meanings despite translation of the chants that accompanied the string figures) and furthermore, since string figures are made by all age groups and men and women, many could be identified by many people in the village Tauwema (Trobriand Islands, PNG) although some were more familiar to different groups of the community.

Thus we see visuospatial reasoning in creating and recreating the string figures are culturally part of the society and influenced and interpreted by their links to other people and to the environment. Even the sharing and messages embedded in the string figures or the accompanying songs illustrate these important cultural understandings associated with the different shapes and connection of shapes, sometimes but generally not, as each move is made in the learning and demonstrating of the string figures. A few of the designs connect to records of string figures made by other people in PNG or elsewhere in the world. Interestingly none were given simple names such as the number of created diamonds as westerners might call them but these were the ones often noted in other places in either the horizontal or vertical position [see further discussion on string figures in Maude and Wedgwood (1967)]. The richness of the chants with the complexity of the making of the design is remarkable even among the diversity of PNG groups.

A Kagua/Erave (Southern Highlands, Province) teacher reported on the use of a string game related to walking to Kambea Peak (*Kambea Rambu*), the highest point in the Kagua/Erave area, rising 400 m above the surrounding valleys. It is a rugged, beautiful rainforest area traversed by subsistence farmers on foot. The making of the sequence of string figures tells the story of how to reach the peak. "Rambu game is important for entertainment that pictures our great mountains, rivers, boy-girl relationships, bird's nest, bee hive … the size, shape, and features of the known mountain, rivers, and bridges". This is the story of getting to the peak. The rope is initially wavy and is knotted as the storyteller sets the scene for travel. As shown in Fig. 5.3a–m, the story unfolds. As fingers are inserted into the loop,

> the *Magawano* sing-sing group set a course to Mount Kambea [Fig. 5.3a], … the first stage called *ripiae-pare* in Kewa Language. … two first fingers (*kimala-lapo*) are placed interweave picking up segment of rope from opposite hands [b] … the second stage which is called *puri-ronga pora* (road from Puri to Ronga). The Magawano sing-sing group is traveling through Puri-Ronga road and bypasses Puri village and arrived at (densely populated) Ronga village [c]. … they set the course to Kambea Peak. After one and half hour walk they passed through Wando Range [d]… on the way they pass a well [e] … travellers gather there to quench their taste (sic) after a long walk from Ronga village to the mountain range. … The sing-sing group slowly walked down the range and arrived at Wando cane bridge (Wando Riwi-Ro) [f]. The bridge is twenty metres long that crosses over the quickening Suku River [g]. … The troop had a bathe in Suku River and continued on with the journey to Mount Kambea through a thickest rainforest called Akero [h]. In the interior of the rainforest, a Bird of Paradise nest is found on a tree along the road side [i]. One of the group member dashes towards the nest to capture the bird sleeping in its nest. Unfortunately a bee hive adjacent to bird's nest chases him out of getting to the nest [j]. The Akero rainforest itself is about ten square kilometres, contain about 9000 species of plants; over 200 are tree-size with easy crossing rivers flow from Kambea through the forest down to Suku River [k]. Most often people get off track and never make it to the top of Kambea peak. However, Mount Kambea is about 400 metres high from Wando Bridge and the imaginary mountain formed by rope is shown [l]. … Right at the top of the peak there is a target point where the Magawano sing-sing group successfully accomplishes their destiny [m]. Kagua/Erave teacher, 2007)

The visuospatial imagery and representations in the string figures are closely associated with the environment and cultural practices. Thus an ecocultural perspective for visuospatial reasoning throws a new light on the resulting figures, some of which are not generally created by others or recognised as significant steps in the making of figures. These *rambu* stories and figures vary but many people can repeat them so there is sequencing, meaning, and links between sub-procedures and resultant figures.

Weaving Patterns

In many parts of the world, people make objects by weaving. This is the case in PNG (see Figs. 5.13a, c, 5.14, 5.15a, b, 5.18b). There is a wide variety of traditional weaving for different purposes in different PNG cultures as illustrated by the figures but visuospatial reasoning is evident in the way objects are created and this has been explained extensively for cultures from other countries (e.g. the hat). There have also been adaptations such as the flat trays and trays with handles in Fig. 5.14b, c.

Straightforward weaving created the mat in Fig. 5.18b except that colour has been used to create a pattern. Twill weaving (called *diagonal* in PNG) is the weaving made by two layers of strands, staying generally in perpendicular directions to each other, having each strand going over and under more than one strand, and starting each new row a step over to one side (Fig. 5.14a, left example). When weaving, a group of strips form a design and the group of strips is repeated to repeat the design (Fig. 5.14a, b, e). Strips which go along the whole way exactly over and under the same strips are said to be in the same phase. Thus in Fig. 5.14a, the diagonal design repeats every fourth row while the middle *zigzag* design repeats every 10 rows and the *diamond* repeats every 12th row.

The basket in Fig. 5.14b has wider strips of one kind of pandanus on the base and the inner layer which forms the fundamental shape. Another pandanus is used for the outside. While the overall shape began with an available box around which the basket was woven and the natural width and splitting of a leaf into strips limits some decisions, the number of wide strips and selected width of small strips were keys for creating the dimensions of the basket as well as the design (interview with Augerea Wala and Roa Kaleva from Alukuni village, Keapara language group, Central Province). The front piece has vertical, horizontal, and diagonal lines of symmetry as well as rotational symmetry. The attractive motif on the woven floor mat (Fig. 5.14e) also has a vertical symmetry. Some weaving is carried out using leaves bent over the starting frond at 60° but at other times the weaving is determined by the natural shape of the vertical strips (e.g. Fig. 5.15a) and/or the intended outcome as seen in the mask (Fig. 5.14c, f) and the fish traps (Fig. 5.14c).

Symmetry is one of the more obvious results of designs created and often repeated by Indigenous cultures. It is evident that the artisans use systematic ways of making, creating, and recreating the symmetry. Weaving patterns are partly counted, looked at while being created but they are also spatially or physically activated. Thus it is important to talk of visuospatial reasoning in understanding the thinking. Furthermore, as mentioned in Chap. 2, there is a strong use of pattern imagery that is invoked and linked to number patterns as well as the specific visual pattern. When an Elder in Kopnung made a mistake, the other Elder took him back to the beginning row of the phase discussing how the first over or under in the row would impact on the final result (see illustration in the prologue of this book). Each were using pattern imagery and reasoning with this to make the correction. Further patterns were supplied by Kono (2007) in preparing for his mathematics classes. His work is discussed in terms of his identity to think mathematically in Chap. 7.

Tattoos

Mekeo designs are said to come from a mother Elu Auko who fell into the underground village while she dug a hole for planting taro, leaving her baby hanging in the bilum. She was taught the designs before she was returned to the village but a misunderstanding in gifts led to no further contact between the underground and above ground villages. The designs or patterns are called

mafe aka	betel nut bunch
a-u-gapa	tall tree
aipa	wild
pafa	plank
ki-u	knee
fagago	gather

The combined patterns are called:

uf u	chief's house
ipi	tapa cloth
keve	seashell

(Louisa Opu, University of Goroka project)

The Motu have a number of designs or patterns for tattoos. They include the tear drop which relates to the women farewelling and missing their men on the dangerous trading voyage in the large sailing canoes following the winds. The drawing of these designs illustrates the way in which certain partial shapes are foregrounded by the drawer in making the various tattoos which are usually made on different body parts (see example in Fig. 5.4). The children tended to draw the rectangles and parallel lines first. Resulting triangles were then coloured but in some cases children also completed the triangle (Fig. 5.4i). This is significant as the teacher had referred to this as the "half triangle" (before inservicing, for her, triangles were all equilateral). Designs are combined, for example, the drawing shown in Fig. 5.4i is repeated in a slide symmetry for the hand, reflected for a leg tattoo, and reflected on the other vertical side for a stomach tattoo and modified (Fig. 5.4d and i) or reflected and incorporating other designs (Fig. 5.4a and b). The children's copying of designs strengthened the visuospatial reasoning required to copy or create these or similar designs. The tear-drop pattern was also linked to number as a two steps, turn left, one step, turn right, two steps pattern by the grade 1 teacher teaching number patterns and combinations.

Visuospatial Reasoning with Number

First is the use of hands and feet and associated gestures. This is not unusual with Indigenous communities, for example, the Yup'ik, Alaska, USA, point to their feet as they count 11–19 and at 20 raise both hands into the air and indicate a complete circle (Lipka & Adams, 2004). Common gestures include counting off fingers as they are bent down, holding out the fist for five, slapping fists together for 8 or 10 depending on the counting purpose's counting cycle, and nodding head towards another person when "borrowing" the hands of other people especially in counting systems with a cycle of 5. On reaching a complete group, such as in a body-tally system which might mark off anything from 13 to 59 body parts, a gesture like raising both hands might be used (as occurs for "*fu*" in Oksampmin, Saxe, 2012).

However, spatial displays are also used rather than counting. These displays are for comparison especially when shell-money is used for trade and displayed rather than counted. For example, at Loboda on the north-east tip of Normanby Island in the Dobu language area, Lean (1992), after Thune (1978), writes:

> In discussing various aspects of the traditional culture such as measurement of shell-money, measurement of time, exchange of objects between groups, and so on, each of which provides a situation in which numbers might theoretically be used, Thune observes that, in each case, the Loboda invoke alternative ways of describing their world which make use of relative rather than absolute scales or in which the qualitative aspects of objects are inextricably bound up with their quantification, and thus an abstract system of enumeration disassociated from the objects to be quantified is, on the whole, unnecessary and irrelevant. For example, in the ritual exchange of yams, a group giving yams should eventually receive an amount equivalent to what they gave. In this case, it is the overall size of the total pile of yams to be given which is significant rather than the number of individual yams in the pile. The 'size' of a yam pile is recalled for purposes of repayment in terms the names of the people who received parts of it. Other categories of goods to be exchanged: pig, betel nut, store goods, etc. are treated in the same way. This form of 'name accounting' obviates the use of a precise enumeration of the items in a given category. (Lean, 1992, his Appendix on Milne Bay)

Various languages use a signifier for counting by the number in the group. This can be found, for example, with Tinatatuna (the language of the Tolai, East New Britain, PNG) even though the numerical knowledge underlying the various lengths is also well established (Lean, 1992; Paraide, 2010). The Tolai counting with fingers and toes can reach 2,000 since each number represents 100. Notches or specific fern sticks can be used for numbers between the hundreds. Quartets of eggs or corn are named with specific words whereas groups of five small animals might be counted. *Tambu*, strings of about 10–12 shells is traditional money which is counted or measured or just subitised in bundles. There are words for pairs, threes, fours, fives, sixes, and eights associated with tambu so each number is recognised visuospatially. Similar discussions on number are found in New Ireland and other places using tambu. In addition, typically Austronesian counting words are used on the Duke of York island for pairs, for example, *urua* for two pairs (Lean, 1992, his Appendix on East New Britain).

For the Tolai, different items are generally grouped in different ways, usually for ease of carrying. Thus the counting words are associated with these groupings.

For example, for coconuts, the groups of two are grouped by two to make four and then another two are added to make the set of six. For 10 coconuts, it is the set of six plus four, *a kurene ma varivi* and twelve is two groups of six *a ura kurene* or when joined as a group of 12 *altikana tanguwani* (see Fig. 5.6). Women will generally carry two groups of 12 on either side of the rope across their forehead or men with fewer over their shoulders (so they are ready to fight). For counting large numbers, 10 sets of 12 are placed together. Meanwhile, taro are in bundles of 3, 6 then 12 (Paraide, 2008, 2010) (Fig. 5.5).

For shell-money, lengths are often three times 5, 10, 12, 20, 100 depending on the purpose or purchase. Lengths of shell-money *tambu* are marked from the fingers to the mid-upper arm *a turamamalikun(u)* or hung to the shoulder *a viloai* (Fig. 5.6b), to the centre line of the body *a bongabongo*, to the other elbow *a leke* or the other hand, *a pokono/tikana pokono*. It is this last length that is then bundled into tens *a/tikana arivu* then threes up to tens and the tens of tens *a/tikana arivu* and so three bundles are *a utula mar(i)* (Paraide, 2008).

In Kilivila on the Trobriand Islands and many other languages, we find counting associated with specific classifications of objects. These classifications might be described in terms of shape characteristics, purpose, or categories with English-word equivalents like a specific food or animal or person. One of the earliest records of this was:

> In (Kilivila) the Demonstratives and Adjectives as well as the Numerals do not exist in a self-contained form, conveying an abstract meaning. There are no single words to express such conceptions as 'this', 'big', 'long', 'one', etc., in abstract. Thus, for example, there is no equivalent of the word 'one', or of any other numeral. Whenever the number of any objects is indicated the nature of these objects must also be included in the word. (Malinowski, 1917–1920, p. 41)

Thus we find interesting visuospatial cultural reasoning associated with language and number. Lean, following Malinowski, points out that there are eight groups for which different participles exist (examples are found in Table 5.1) (Lean, 1992, his Appendix on Milne Bay). (See Appendix C, this book, for full list). While more recent data may modify this information, the significance of these data are that they illustrate an alternative way of reasoning and there is no doubt that the visuospatial imagery and related concepts link closely to the categories of objects, mostly physical, indicated by the morphemes associated with the different classes. The characteristics of the objects in the classes are sometimes related to position on shapes, e.g. protuberances and corners; shape, e.g. long things; size and shape, e.g. round, bulky objects; or grouping, e.g. bundles of different types of food; and parts including large area subdivisions. These classificatory systems are also found in languages on Bougainville such as Nasioi (a non-Austronesian language) and other island and coastal areas. Often, the number word is a link to a specific bundle of food. For example, in Motu (an Austronesian language), small counting numbers (1–9) are numerals albeit related to each other (e.g. $8=4$ pairs or 4 twos, 9 is 4 pairs plus 1) and even associated with the hand for 5 (as are most Austronesian and Polynesian languages of the Pacific usually based on *lim* or *lima*) but 10 varies with the type of food bundle, that is 10 fish would not be the same word as 10 yams or 10 bananas (Tubusereia teachers, personal communication).

Table 5.1 Selection of names for classificatory groups in Kilivila[a]

Group	Participle	Category description
1	Day (ke)	Trees and plants; wooden things; long objects (canoes, sticks, poles)
	Dway (Kwe)	Round, bulky objects; stones; abstract nouns (betel nut, houses, yams)
	Ya	Leaves; fibres; objects made of leaf or fibre; flat and thin objects (coconuts, spherical containers, clothes, string)
	Kwoya (mweya)	Human and animal extremities (legs, arms); fingers of a hand
2	Kila	Clusters ("hands") of bananas
3	Pila	Parts of a whole; divisions; directions (books)
4	Kabulo	Protuberances; ends of an object
	Nutu	Corners of a garden
	Niku	Compartments of a canoe
	Kubila	Large land-plots-ownership divisions
	Siwa	Sea portions-ownership divisions with reference to fishing rights
5	Kapwa	Bundles—wrapped up (Packages)
	Oyla	Batches of fish
	Yuray	Bundles of four coconuts, four eggs, four water bottles
	(Kupwa)	(Fish counted in twos)
	(Kayo)	(Crabs counted in twos)
6	Kasa	Rows
	Gili	Rows of spondylus shell discs on a belt
	Gula	Heaps
7		Numerals without a prefix are used to count baskets of yams and numbers can be very large
8	Uwa	Lengths, the span of two extended arms, from tip to tip (fathoms)
Respondents to the Counting System Questionnaire also gave		
	(Bwa)	Short or thick solids
	(Kaula)	Groups of 20

[a]*Source*: Lean (1992) based on Malinowski (1917) with additions from Counting System Questionnaire data given by tertiary students provided in parentheses. Further work has been done by Lawton and assistants since the 1960s

Fisher (~2010) also documented the Kuruti (Manus Province) ways of counting. The frame pattern for the counting system is typical of Manus languages in that after 5, the words relate to "going up to 10". Six indicates starting up to the group. Thus $7 = 3$ to make 10, $8 = 2$ to make 10, and $9 = 1$ to make 10. 10 is *sungoh* in most of the classification groups except relating to time periods (cooking a pot of food) or lunar months. This is different to Motu which has a different word for 10 depending on the objects being grouped. In particular Fisher noted 22 different systems related to classificatory systems. For this language and its associated languages Ere, Lele, and Gele', Lean (1992) suggested there were approximately 43 groups, many having visuospatial characteristics. Interestingly, Fisher was able to give the words, at least for the round objects, fruits, and nuts up to 900,000. These are objects that are likely to be in large numbers. He also illustrated how there are different counting words for banana trees, banana leaves, stalks of bananas, hands of bananas, and individual bananas. Each grouping is significant in cultural practices. There are morpheme similarities and patterns in each of the counting systems although modified to sit with the classifier. In Chap. 3, there was also discussion of how the number "line" using body-part tallying was used and also modified referring mostly to Saxe's work together with how hands are used for representing numbers (based on our own studies).

Thus the visuospatial representations are well developed from experiences and closely linked to values associated with numbers or at least the size and nature of the group of objects situated within cultural relationships. The language indicates that particular ways of reasoning from the visuospatial representations is occurring. We can expect that a similar occurrence will be evident with measurement.

The Measurement Study

The measurement study from the Glen Lean Ethnomathematics Centre, UoG, PNG was ethnographic which is appropriate to investigate "ways of acting, interacting, talking, valuing, and thinking, with associated objects, settings, and events (that impact on) … the mental networks" that constitute meaning (Gee, 1992, p. 141). Data were collected[8] and analysed and the analysis checked with other participants' and communities' data as the grounded theory developed. Our research project required participants who were familiar with their own communities' activities and preferably investigating their own cultural practices. We were conscious of the various relationships between ourselves, the participants and the village Elders whom we interview with a participant researcher (Owens, 2006b). Villages referred to in this chapter come from a variety of environments (mountains, coastal areas, and large valleys) and language types (Austronesian, non-Austronesian, and hybrid languages). Data from 16 in-depth interviews (demonstrations, discussions, and observations with some semi-structured questions) either in the village (visited by at least two researchers) or at the University of Goroka have been complemented by questionnaires completed by hundreds of UoG and other tertiary students, and focus group data from students and linguists. The following sections represent some of the findings of this research.

Ecology is strongly evident and influential on cultures in PNG which has a huge variety of land types (albeit equatorial with a high rainfall) from islands (large and small), coastal strips, valleys (narrow and wide; upland and lowland), and high steep mountain ranges. There are coral reefs, swamps, fertile valleys, and uninhabitable areas. There are high density and low density areas and over 850 languages, that is, cultures, in PNG. People live in cities but have connections to remote and

---

[8] Data continues to be collected whenever visits are made to village communities or discussions are held with people from different communities.

rural areas where the majority of people live in bush material housing as subsistence farmers and/or hunter-gatherers. PNG cultures vary considerably, noticeably in terms of relationships between people, with the land and water, in the use of number, and in cultural activities from dancing (see Fig. 5.21) to house building (Figs. 5.10 and 5.11).

Visuospatial Reasoning About Length

Above, we saw how lengths of shell-money were compared. There are a number of examples of visuospatial reasoning about comparing and measuring lengths, and about obtaining a half and a third of a length. Common is the use of a rope to mark a length against which to compare in future or to equally space posts. Words related to length measures were discussed in Chap. 4. One informant from Mailu (Magi) in Central Province, PNG, where large canoes are still made, especially for racing made comments that are summarised:

> The things measured are length of the hull, depth of the hollow and the thickness of the hull. The specific people who carry out this measurement are men aged between 20–40. The older men either supervise or help when they are needed. … The log is measured using arm span depending on how long the owner wants it to be, for example 20 arm spans. Then it is hollowed using a hard adze. The time may vary for it to be completed depending on the size of the canoe. But before hollowing, the depth is estimated. While hollowing, the others observe and estimate the thickness of the hull and when they see that it suits the length and depth of the hull, they stop. They also measure the actual thickness of the top edges of the hull using their hand. An arm span is the combination of 9 hand spans. Sometimes the depth of the hull is measured using hand spans. This is done to be more accurate. … The size of the hull can be associated with the area of the sail as well as the speed of the canoe. All *lakatois*[9] have a longer and shorter hull [stabilizer]. The shorter hull is half an arm span from the two ends of the longer hull. This caters for the speed and size of the sail. Inaccurate measurements have caused many accidents such as capsizing, etc.

The group activity described in the making of the canoe establishes close visuospatial reasoning between boat builders. While the arm span was suggested as the means of measuring, men in other places might use strides. However, my observation is that this overall length is often not measured as it will depend on the available tree and so length descriptions in terms of arm lengths or strides are often poorly assessed. The ratio of depth and length is fairly carefully appreciated as this will limit its use in open seas. Thickness increases towards the base of the hull and feeling both sides of the hull is common practice but knocking it while the leg touched it or listening to the sound helps in determining the appropriateness of the thickness at different depths (the practice varying from place to place and between people in the same village).

---

[9] Motuan word used throughout PNG for sea-going canoes with single outrigger and usually a mast, often with a crab-sail (Fig. 5.18f).

Composite units were also discussed for the Ambulas language area, Wosera, East Sepik Province. One bamboo length called, *kama nak* is equivalent to five bamboo internodes called *ndik nak tamba*. About 5×7 bamboo length (that is seven lengths of five internodes) is equal to one garden area or *tumbu*. Other informants suggested a length of rope would be used for five steps when building a house or garden but this seemed to be a variable decision depending on the building or purpose for the composite unit. In Kilivila, Trobriand Islands, measuring systems vary for horizontal and vertical directions in conjunction with the body (Senft, 2004a) and in some cases, the position of the axes can be moved from ground level to arm level with alternative words (Lawton, 2007 and Wado, 2006, personal communications) but there is a range of unit lengths indicated particularly by parts of the arms.

Lengths that were in circular form were established visuospatially as equivalent to the same length in a straight line from various experiences. For example, in Kâte areas, Morobe Province, the circumference of the common bamboo types are well known so the lengths can be used for measuring the number of bamboo lengths required when split to cover the floor of a particular house. String is carried to the garden to get the correct lengths and the correct number. The long walk to the bamboo stand and the sustainability of the stands are motivators in cutting just the correct number of lengths (Rex Matang, personal communication). In Panim, Madang, Elders confirmed that stands of bamboo were carefully cut down after I mentioned that a group of people said they just cut lengths without measures as they said it did not matter if there were unused lengths or not enough.

Visuospatial Reasoning About Area

The concept of area in PNG is interesting because there are no traditional area units as we have in western mathematics. Only one student mentioned that she and her family used area units but they were the only ones in her village and surrounds. When people think of measuring units, they inevitably think of length units like steps and handspans. However, people compare areas and plan areas for housing, sleeping, gardening, meeting, and many other things. There are some unrecognised area units commonly in use.

Around Goroka in the Eastern Highlands Province (EHP), people speak Alekano (also referred to as Gahuku-Asaro). Land area is compared by looking at it, discussing it, and marking the boundary with tanget plants but there is no formal measurement. When making a garden close to the house, people will decide half of it by standing in the middle of one side and deciding where the half way line should go. Half the garden is left fallow.

Kaukau (sweet potato) is planted in mounds, generally with two mounds between drains (Fig. 5.9b). The fifth pair of mounds, say, might be marked with a sugar cane or other plant. The various garden sizes seem to be well established in the mind suggesting that people have a good visual image of the areas involved with the garden, halves of the garden, rows of mounds, and blocks of mounds between drains (Fig. 5.9a).

In the Whagi Valley (Yu Wooi or Mid-Whagi speakers) in the Jiwaka Province in the western highlands, Kopnung village elders discussed the floor plan of the rectangular men's house in terms of the number of men who might sleep in it and by comparison to another house. Round numbers are used, e.g. "it is for 25 men" and they refer to the area of a room as "7 foot by 7 foot" (Fig. 5.11a). The square sleeping room is visualised as sleeping around 7 men with the length for the man also being 7 foot. The prone position image seems as strong as the vertical. The rectangular floor plan is divided into three parts, roughly equal. In the middle is the area for sitting around the fire. The outer thirds are each divided into two squares for sleeping. These fractions are decided by using a length of stick from each corner of the house to mark the points with some try-and-modify techniques. Thus the 25 men's house has space for 7 men in 4 rooms = 28 men, although this calculation as such would not be made and the general size is all that is required to denote its importance. Such discussions are common place across the highlands region.

In the coastal village of Malalamai in the Madang Province (Fig. 5.13a), floor areas are decided by what space the villager wants for the expected size of the family and activities like sleeping, eating, talking, and cooking, and the extent to which he can afford to build such a house given the amount of manpower that it requires. Plans may be modified by what is available to them in the bush. People think of the floor space in terms of the number of rows of posts. These are 6, 9, or 12 posts with the base row of 3. From house to house, the space between posts is about the same. People talk of the house as half as big again when comparing a 12 post house to a 9 post house. In other language groups (e.g. at Panim, Madang), further inland where the winds are not so strong and the posts shorter, posts may be further apart.

The length of the roof *morata* (made from narrow limbom planks and sewn sago leaves) is also considered in deciding the length of the house (Fig. 5.13b, d). "The morata are about one and a half arm spans long. This will influence the size of the house and a bamboo will be used to keep this length". Other measures occur. "Short equal sticks (about a hand span) will be used to keep the space between the (morata) planks equal and the planks parallel. Trees that provide sufficient sago leaves for that house are carefully selected to provide adequate waterproofing by the roof" (Sorongke Sondo, Malalamai, field visit). Panim men said that "for a 9 post house they will select five very big sagos ready to eat and more for a 12 post house, may be nine. If it ends up being insufficient, they will get more later" (such approximation is not accepted by all places or older men in the language group). Although some people spread morata further apart, people are able to look at a pile of morata and decide if it is enough for a particular roof. Malalamai men noted that they can look at the limbom palm and know how many planks they can make for roofing. The women do the same for tulip trees and different-sized bilums made from the inner bark of these trees. In places further inland or for smaller structures that use kunai, people estimate the area of kunai needed for the roof of a certain house. People talked of the house as requiring, for example, 40 or 70 bundles of kunai. These relationships are embedded in experience and suggest that a ratio-type, comparative approach to measurement is a starting point rather than a simple arithmetic approach.

Yams may be carefully grown in rows but not necessarily. People have a practical idea of how big the area needs to be for the number of yams wanted in a house garden. Cash crops like cocoa are planted out at the vertices of tessellated equilateral triangles. Two standard bamboo lengths are used. To mark the tree holes, a straight row is first made with the two sticks. Then places for plants in the next row are marked using the two sticks to form the other two sides of the equilateral triangle. This means that plants are staggered for best use of the ground as shown in Fig. 5.6. Again the number of plants for a particular area is known from experience. They will buy or prepare the appropriate number of seedlings or small plants. Interestingly, this method is now also used for kaukau mounds in various villages across the country depending on the terrain.

Whether it is a square surrounding the mound for plants in rows or the triangles formed by marking the vertices of the triangles, these are subitised[10] area patterns that can be used in teaching about area units in school just as the space for a sleeping person can be used. Similarly, familiarity with rice bags and laplaps (a cloth wrapped around the waist like a skirt) laid on the ground, and woven coconut frond mats also form good area units by which people make decisions on area size.

What is also found is the use of lengths for determining area. The paces may be for one side or for two adjacent sides. In general, people are familiar with equivalent widths of land so the variation in length is fairly evidential of size. This applies to both rectangular and trapezium-shaped land that seems to form naturally by the slopes and clearing of the mountains. This visuospatial reasoning about area may explain the discrepancy that Bishop (1979) found, and which I subsequently confirmed in a number of interviews, but not all, with children and adults who said they add lengths at home but multiply at school without necessarily making the fuller connection of visualised area assisted by taking steps to measure at home and the areas drawn on paper at school. An investigation into why both "rules" work is important in learning about problem solving and visuospatial reasoning.

In both weaving and bilum making, there are often squares or "diamonds" used in design (Figs. 5.13a, 5.14a, b, e, 5.16c, d, f, I, j, l) and other shapes created from two smaller shapes (e.g. Fig. 5.16f, g where triangles and trapezium make up the square and the hexagon, respectively). The number of strips per wide strip in the basket in Fig. 5.15b was considered when the basket was being made. Both the narrow and wide strips create area units and the smaller squares make up the larger composite square.

Although earlier bilums from fibre in many places had stripes in different colours (Fig. 5.16a, b, h, j) or with a slightly different loop (Fig. 5.16n) to form squares, or patterns of numbers to create the design, these were also incorporated into the bilums made from plastic rope (Fig. 5.16k, l) and wool (Fig. 5.16c–g). The area of the squares might vary from bilum to bilum but were roughly equal in size across

---

[10] Know and see. In terms of numbers, children recognise small numbers of dots without counting in western communities. Other subitising occurs in Indigenous communities (Sue Willis, personal communication, ~1996) and is confirmed in my observations in PNG.

bilums for a particular group and within the one bilum. These squares provide an excellent image of tessellated squares and were relevant in the making of different-sized bilums (Fig. 5.16c). Furthermore, there was a connection between the loops in a square and that in a rectangle, often just obtained by using a wider strip of pandanus used to space the rows of loops or with increasing the number of loops to make different rectangles. Thus dynamic imagery was incorporated into shape variation but also in knowing how areas were created and covered.

Visuospatial Reasoning About Volume

Size is a general word and is often used for any filled space whether it be two dimensional, three dimensional, or one dimensional. In PNG, this word is used ubiquitously and is important in reciprocity of goods as well as the number of exchanged items. Size is often considered in terms of a group of objects. "One never gives just one thing, it has to be a group", said one of my colleagues. Generosity in giving is valued. When a packet of biscuits is being shared, no one takes one biscuit; it is always two or more. Matching of objects such as pigs will be marked by a checking of size but how this is done is very much dependent on the cultural group involved. It is not unusual to see two smaller pigs representing a big pig. People know weight well. Women in particular carry large, heavy loads in bilums hung off their forehead down their back or in limbom baskets. Men carry heavy loads on their shoulders. Gathering food such as heavy karuku nuts and bringing home garden food is common (Fig. 5.16i and 5.15).

Size is often expressed with an emphatic word such as *tripela* (Tok Pisin, the lingua franca) meaning "very big". Size is compared albeit in idiosyncratic ways in different cultural groups. Sometimes it is specific to certain objects and other times it is a generic idea. For example, words for size of rocks would not be compared with size of food item in some language groups (linguists' focus group). There is considerable discussion on this in Chap. 4 on language and visuospatial reasoning.

In school mathematics, calculation of volume is initially for the rectangular prism with the associated calculation of length × width × height to obtain cubic units. There is little looking and estimating to compare sizes or to estimate the ratio of volumes when a linear aspect is increased. However, in the villages of PNG, we find practices in which a length measure can be associated with a volume of various shaped objects. Lengths are used to compare volumes. The notion of a big pig is primarily about volume but it has mass and more importantly fat. Pigs may be carried giving an idea of mass but what is seen is volume.

Although ordering objects (e.g. the volumes of pigs, piles of food, bilums, and baskets) by sight is frequently done, people measure various lengths as a technique to help with volume comparisons. Some people mentioned that they look to see how close the belly is to the ground and others the length to the thigh. The Kamano-Kafe (EHP) mentioned the girth of the pig is measured by seeing how far apart the fingers are when the arms are placed around the pig. There can be much discussion about the comparative sizes. One participant from the Southern Highlands Province (SHP) noted that women keep the string length used to measure the girth as a record for when the measurement is needed again (in a reciprocal exchange). Another participant said that the length of each pig is marked by a knot on a long rope and this long rope is kept for later additions and for comparisons. If the length of the girth indicates the volume of the pig, then the link between girth and volume is not necessarily determined numerically but visually and it is associated with a range of other pig features. "When measuring a pig size, a small rope was used to measure the size around the pig and also the length of the pig. Then they can weigh out the cost in terms of how much fat or meat it contains" (Angal speaker, SHP). Reference to length when discussing the volume of a pig occurred across many provinces from the coast to the highlands.

Volume was also associated with parts. Kerapi, from Imbongu, SHP, noted they add up quarters of pigs (e.g. literal language translation of a *half* is *quarter quarter*) but also distribute the quarters and divide them into more parts for further distribution.

> (For bride price and land disputes, people) have different ways of arrangement, preparation and payments. However in my culture, the traditional wealth is of less significance. Upon the display of items, the length of items, traditional money, the pigs and bunches of banana are measured. They use ropes and bamboo nodes to measure the length and sizes of food stuff and pigs. One hand span (arm span) of beads, is equivalent to two smaller female piglets and half of the beads (from finger tip to middle of the chest) is equivalent to two male piglets. Although they have a digit tally counting system, they will also tie knots for each of the pieces of wealth with different ropes for each kind of wealth. They use a chain of dogs' teeth to measure the beads used for making the payments. They associate these with the length of the beads and the money displayed. (Yupna man, northern Finisterre range, Madang)

Lengths are used in conversations to make decisions about the size or volume of items and to make judgments about the wealth or value of the objects.

The width of pig fat measured in finger widths is associated with the amount of fat/oil obtained in terms of bamboo containers. Fat, a valued product of the big pig, is distilled into oil and measured in bamboo tubes. While the connection is not given a multiplicative number, experience indicates how much one could estimate. "The sheets of fat from the pig would be stripped off the membrane allowing the fat to run into bamboos. They would also boil it to get the oil. They might get four bamboo containers [of pig fat] from a big pig" (Kamano-Kafe speakers, EHP). Related to this is the use of finger widths in the highlands to describe the size of a fish, the width of the fish indicating its overall area and volume (noted by a coastal man teaching in the highlands).

"Food piles were also compared by Elders, especially on deciding greatest and smallest piles" said a Kamano-Kafe speaker and a Fore (EHP) man wrote, "People weigh (by hefting with the hands) the heaviness and lightness, [estimate the] length (long or short) and size (big or small) and finally group them in order of their size, length and weight:

- Heavy, long, big
- Heavy, short, big
- Light, long, small
- Light, short, small"

A Kumbu (or Kewapi, SHP) participant mentioned that the size of the kaukau bilum (string bag) will be discussed by up to 40 people which he compared to the 400 who might be involved in deciding the area of land. When ceremonies are held in many places, the number of pigs, bilums, baskets, and piles of food are compared. Size is only one factor that might be taken into account. For example, for the Arop in Sandaun Province, it is the number that predominates. In the exchanges of baskets and bilums of food, the type of food also matters (several villages, Madang Province). For most communities, the amount must be more than generous.

Experiments with small volumes are common in making medicines, food pastes, and colours (noted by Yupna, Yu Wooi, and Telefol speakers). Body paints are widely used and in many places colours are used for walls, artworks such as face boards of *haus tambarans*, dying strings, carvings, and pottery.

> After the designs are carved on the wood (of the door board), they used three different traditional colours to paint the designs that have been carved. The traditional colours used are; maroons called *Baagaan*, white called *Buuguung* and black colour from carbon called Amsiring. To make the paint shiny and bright some grease pig are (sic) mixed according to the correct proportion with the three traditional colours mentioned above. (Telefol, Sandaun, Onggi, 2005)

The amount of water needed for creating the steam that rises from food cooked with heated stones in a ground cooking pit (*mumu*) was a commonly given example of volume (Fig. 5.12). Familiarity with different types of mumus, different sizes, different woods, and different foods are all combined in the knowledge used by the Elders to make decisions about the volume of water. Common units were the large cooking pots or the nodes of bamboo with the full length of bamboo having usually a composite of 3 or 5 node lengths. Pits vary from round to rectangular especially if a pig or two are to be placed in the mumu. Mumu volumes are also described in terms of bundles of food soaked in coconut milk and wrapped in banana leaves for a number of coastal and island communities.

Other conversations about volume revolved around the size of the round house and its radius so the house was small enough to keep warm, large enough for the family's needs, and feasible in terms of available helpers and materials needed for construction. House building is always associated with a feast to thank the helpers. The garden must be ready for the feast and the amount of garden food, bush food, and pigs would be determined. Villagers in a Panim village noted:

> When there is a feast, then the various families will bring food. The pigs will be tied to the sticks ("stik pik") which are put in the ground, carefully indicating how many each line will get. Each group will get their share. When the pig is distributed, then all the men will check. Then a piece goes to each man and usually a coconut-leaf basket of food is used and distributed to each man in the group. Each basket has an opening about 50 cm wide. There is a rope handle to put on a pole so it can be carried. For feasts, the family would need very heavy bilums of mixed produce to share and trade. Each family would supply large bilums full of food. They would use big clay pots, make a hole in the ground for the fire and put the pots in the hole. It was important to prepare much more food than the men could eat and to prepare 5 to 7 plates full per person. Pots were made at the beach where the clay and sand could be used together whereas Panim was in the bush. They traded garden food for ground pots and never pigs. Usually 1 to 3 bilums of mixed garden food for one pot but they also took account of friendship. They would also trade eight or nine packets of sago wrapped in

the bag made from the palm. They would trade bilums of different sizes for different size pots. They might buy three or four pots at a time. (Elders from Panim in the coastal hinterland, Madang)

This description indicates that volume is a key aspect of the trading but other factors such as type of food and relationships of people are affecting the exchanges. The limbom bark basket (Fig. 5.15b) and flour sack are commonly used measures of volume for many items—sago, coffee, vanilla, and so on. Since taken-as-shared concepts develop through practice, the regularity by which a sack, basket, pot, ball, or packet of sago was used resulted in a relatively consistent unit of measure. Similar examples have been given for length (Owens 2007a). However, there are few examples of composite units for volume. One example is from Abau speakers who wrap up six sago balls in a banana leaf and pack these into the limbom basket.

Volume is important to people as part of a visual display and often as a point of reference although mass may be the real source of comparison. The visual imagery associated with some activities provides a good basis for reasoning and making comparisons. Nevertheless, other values might dominate such as ensuring more than enough is given. In actions and in discussions, there will be subtle ways of establishing volumes. In all the above examples, reasoning includes an ecoculturally influenced visuospatial consideration.

The importance of accuracy in traditional activities is determined by the individual, the available resources, effort, common practice, and purpose. In many cases, the sizes do not have to be accurate and position can be assessed by eye but in other cases, such as the positioning or removal of the central post of a house, greater accuracy is needed. Size is often only a part of valuing an exchange as mentioned above in terms of the type of food. The display of food itself is important as mentioned to me recently in a Simbu welcome display of food. Further communities make decisions about how to assess value and quality such as measuring different lengths of pigs, piling food in certain ways, and measuring lengths of mami yams, not just for size but for their other cultural connections. In most of this decision-making visuospatial reasoning is involved. The connections of experience supported by sometimes lengthy discussions are important. As mentioned earlier, it is also the relationships of the people involved in the measuring or decision-making that is important. At times, the person for whom the house is built or the chief builder whose body measurements are used in the building may provide the reference lengths. At other times the person is not signified. Sometimes lengths are kept for later events. Other times any stick or rope is used for comparing equality of lengths.

Visuospatial Reasoning About Three-Dimensional Designs

In Kaveve village, Eastern Highland, an Elder was asked to show us how he made a round house (Fig. 5.10a). He dressed for the occasion with his government badge, and pig tusks. He explained that the size of the house and its volume affected its warmth provided by a fire for which the wood had to be cut down and carried from the bush. Hence he was now sleeping in the relatively small house that he was given. However, he was also building a very big house overlooking the valley, a status house. In his description, he mentioned levelling the ground, the need for bush materials, and then, with prompting from the group of onlookers, he drew a rough circle dragging his foot. He then immediately went into the need to plant large gardens in order to provide a feast for all the helpers who might be family or others. This was a significant aspect of building a house for him. However, other Elders who were aware of the model making insisted on showing us more detail of making the circle using a central pole and a rope to drag the foot around in a circle (Fig. 5.10b), building the roof, and using steps (heal to toe) to ensure measures on rectangular houses were sound. The circle shape and its various parts were well experienced through living in the space (Fig. 5.10c). Thus visuospatial reasoning was used to determine how best to utilise the flat surfaces (at several levels including its roof space) and the three-dimensional space. Similarly, platforms are also created inside the roof spaces of houses on the coast, see Fig. 5.13).

However, young children have the opportunity to observe the construction of houses. This includes the making of walls. In many places, woven walls or *blinds* are used. If the wall is for a round house, then children see the rectangle curved to form part of the wall and another rectangular blind used for the remaining part of the wall (see Fig. 5.11). Walls and mats are rolled up and unrolled. These surface areas are created and the visuospatial images can be used to reason about rectangles and curved surfaces of cylinders (the round house walls) or of rectangular prisms for rectangular-shaped houses [Fig. 5.10b (toilet and shower house in background) and 5.13]. If vertical sticks are used to form the wall (Figs. 5.11a and 5.13e), then another visuospatial image is provided as a series of equal length sticks or planks covering the area. In other words, surface areas are made, spaces are covered.

Names of shapes often indicate their relevance and relationships to other objects. Henry Kawale from Golin, Gumine, Eastern Highlands PNG suggested shape names: for a shelter *oke*, egg-shaped *milinkalin*, and "for making the patterns of the woven walls, they use designs, patterns, and shapes given traditional names corresponding to the environment around them, *mepki*=mountain, *amil*=diamond like panda nuts". Many teachers provided shape names that often linked to nature.

Roofs

In nearly every place, discussion was held about how steep the roof should be (Fig. 5.11b). Generally it was built so that the water ran off but not so steep that the kunai grass covering (if used) would slide off. Morata was new to coastal areas, having been brought into the country by Samoan missionaries prior to 1900. However, this also led to variations such as a separate roof over the doorway, and a much steeper roof in Panim (inland, away from strong coastal winds), Madang. The spacing of rafters or morata was also carefully considered so that the roof was waterproof and each section well secured. In one house with a central apex, forming a pyramid (not very common) and rectangular houses, the vertical and horizontal roof trusses were carefully cut and positioned. Much of this was pre-planned but often pieces were checked and cut during the building phase. Inevitably, the slope of the roof which was regularly mentioned as a good example of angle was determined by a group of men, standing away from the house and telling others whether to raise the wood more or lower it until the desired angle was determined. The other end of the rectangular house was made by using lengths equal to the symmetrically opposite end. Nevertheless, the central pole or poles had to be carefully positioned (see Fig. 5.11a).

The use of equal lengths was common place for finding central points and thirds, measuring from the corners. This was evident when three elders and two younger men acting as interpreters modelled how they built a rectangular house in Kopnung, Jiwaka Province (Fig. 5.11b). The house has three sections, two ends for sleeping and the middle for sitting. First, they explained the height of the wall is equal to the height of the man's armpit while the internal wall is the height of the man's hand raised above his head. This gives a good slope for the roof as described above in terms of coping with rain and wind but also not so high, so the building stays warm. Making a model meant that this slope and wall heights could be problematic so careful decisions were made to keep the roof line sensible. The model which was smaller than the normal house had height and length roughly as expected but interestingly, the ratio of one to the other was also kept for the desired slope. There were good judgments about the lengths as modern mathematics might say in terms of trigonometric ratios for the angle of the roof. To achieve this, the inner third of the house was narrowed in the model but still roughly a third and the shorter man used as the measuring standard. Similar judgments were also made when a house was made larger than usual too (Fig. 5.11a). A careful look at this house shows the basic rectangular shaped house has rounded ends to provide more internal space and to make the house look outstanding. The roof line has been carefully planned for the modified design.

Bridges

A village (men, women, and children) had been to the city of Lae, Morobe Province, PNG to buy a tank and they were walking it back to their village, a day's walk from the road head. There is a small creek to cross at the start and we came to the creek to find the villagers had spent the night beside the creek as the creek was a flooded, fast flowing torrent. Within an hour the men had eyed the saplings and vines growing beside the track, and cut down with their sharp bush knives appropriate ones of the required length to bridge the torrent. They knew the vertical and horizontal lengths well. They also managed a hand rail and the whole village, some carrying the two halves of the tank, and ourselves crossed the river.

Many bridges are suspension bridges or cantilever bridges requiring greater understanding of loads and counterbalances (Fig. 5.7). Huge stones anchored and balanced the bridges. The bridges were made of bamboo or hardwood with vines looping under the walkway logs attached to the handrail or suspension ropes to prevent sideway slips. Heights of uprights, lengths of vines and walkways, and ramps all have to be determined. In Chap. 8, there is a brief discussion of a teacher's project that involved a suspension bridge (see also Owens, 2014).

The Ecocultural Holistic Context

Although the above sections have closely considered the various activities and how they might be linked to school mathematics in terms of familiar western concepts such as area and volume, it is salient to describe one cultural activity to see how much of the activity links with mathematical thinking and in particular visuospatial reasoning. Mathematics is holistic and integral to the unseparated parts of the activity. Mark (2006) discusses wealth and exchanges related to marriage, bearing children and their development, and Moka ceremonies of the Melpa, Western Highlands, PNG. Gifts are given at each of these times to the mother's family in establishing the relationship of the woman or child to the father's clan. An initiatory gift is returned with one of higher value. Such items as pigs, shells (six kinds), cassowary, salt, oil, decoration, bird of paradise plumes and other feathers are exchanged. The account of pigs and shell given is represented on the *owu mak*, the bamboo slats on the chest decoration where one slat equals eight shell or pigs (at least in the past) (Fig. 5.21c). The shell-money is also displayed on the ground or on the warriors as they march forward (Fig. 5.21b). On the rectangular ceremonial ground *moka pena*, there is a line of symmetry lengthwise marked by the pegs for the pigs *kung pugkl*, and the corners are marked off as places where the women can stand and sway as a group. The men will circle around the sticks in an oval shape or march up and down symmetrically on either side of the line—both groups are equal in number and number of marches up and down. There will be a men's house *manga pukglum* at the top end of the ground (Fig. 5.11a). After the exchange, young girls might dance with the men in the *murli* or *waipa* in circular patterns. There are various symmetrical designs painted especially on the face but also on parts of the body where they may signify the clan group. A pattern of equilateral triangle tessellations forming "diamonds" that tessellate are used to identify the Nenga in Mul District. Colours vary along with designs. Persons who decorate another signifies relationships.

The gifts for exchange are counted in sets of eights *engag* or *ki* (*tendta*) which is one complete group of four fingers of both hands. As the person counts, the fingers are bent down starting from the little finger of the left hand. When counting money, the suffix –*mun* is added. The fists are slapped together and then if they count in tens, the two thumbs are wiped down the lips or "closing on top of four fingers of both hands". A number such as 29 could be expressed as *engag pumb ragl pip engage pumb ragl pip wote engage pump to gul* $(8+2)+(8+2)+8+1$ or *ki ragltiki wote timbikak pukit pumb to gul* $(8\times 3)+4+1$. *ki ragl* 16 is literally "hands of two men" and *ki ragltiki* 24 is "hands of three men". *Wote* and *gugl* mean "include" or "also" while *pentipa* means "attached to" or "included" and are used as adjectives

Table 5.2 Seasons in Melpa or Hagen, Western Highlands, PNG

Month (roughly)	Language	Translation/cultural significance
January–February	ting	A word, relates to weather, wet season. Time for clearing, planting, and gardening.
March–April	owuiil	A bird name: this bird which flies along the mountain following the river down as time of the dry season is soon to arrive so work hard in preparation.
May–June	paan	A word, crops are harvested, reference made to wind that blows. Time for Moka, usually third week of June.
July–August	Puun	Red pandanus fruit
September–October	Tipaan	Wet weather is coming so collect firewood and food for the coming wet time. Pigs might start to stay in the house out of the weather.
November–December	Piill	The name of the cuscus who might "steal" the pandanus nuts that are ripe for harvest in the jungles.

Source: Mark (2006)

to describe addition or order of operations. Mark notes that there are several alternatives for each of the counting words besides the two given. He also notes that the shells especially but also other items are displayed in groups of eight. The visual displays are all part of the visuospatial reasoning. Negative numbers are reflected in the ceremony as credit or *pund*, what is not with them but in credit. Zero is reflected in *poor mon* as "finish none". The whole of the pig is *mundiimbuke*. Similarly the fractions of "half" *pukrui pentik* and "quarter" *pukrui por*[11] have considerable significance when halving and quartering the pig.

The time for moka is usually during the season with long periods of sun. Mark gives the following information on seasons. The first month in the pair has -*komum* as a suffix while the second month in the pair has -*akil* as suffix (Table 5.2).

The time of the day is denoted by "expecting light", "light appearing", "first light like light in evening with setting sun", "morning sun", "afternoon sun new", "setting sun", "time when a passer-by's face is not clear", "darkness here", "young children sleep", and "deep sleep". Thus Mark illustrates the connection between design, shapes, and time in terms of cultural activities. Cultural activities contain a wealth of mathematical thinking including visuospatial representations and reasoning.

The Importance of Ratio or Multiplicative Thinking in Reasoning

In discussing material required for different-sized houses, amounts of food required for variations in exchanges, and balance for canoes, cantilever bridges, or house features, there was always a sense of proportional reasoning occurring. If one aspect

---

[11] *Por* is also the *Tok Pisin* word for four so some borrowing is occurring. *Pukrui* is also used for a third.

increased, so did the other aspects but this is not necessarily a linear comparison. This is the case for wall, floor, roof, and grass areas although areas could be reduced to two requirements (e.g. the number and length of split bamboo for a wall or floor). The fact that a 12-post house is regarded as a half as much again as a 9-post house implies that the materials for floor, wall, and roof are half as much again. Since some lengths may be fixed (e.g. the width of the house), this makes some decisions easier. However, finding the sago trees or limbom palms that will give half as much again for morata is still part of the visuospatial reasoning built on experience.

Volumes too are considered proportionally. Even ascertaining the volume for the pig led to an increase in the number of length measures used in discussions (e.g. girth and length and height) although mostly two small pigs are taken as equal to one big pig. Volumes are increased and in proportion for medicines, paints, and for sauces (e.g. marita, a red-pandanus nut) or in other food cooking. Discussions are around comparisons with previous experiences and by eye decisions.

Architecture Students' Visuospatial Reasoning

One of my PNG studies presented as an argument for an ecocultural perspective on visuospatial reasoning used qualitative approaches to investigate the possible influences of culture on design and mathematical ways of thinking (Owens, 1999b). First-year PNG Architecture students had produced an attractive variety of paper sculptures and I wanted to find out how they were thinking in order for them to develop such unique sculptures. Students had previously made a sculpture using glue and any available materials that they wished to use (e.g. "junk", cardboard, sticks). For their second sculpture, students were restricted to using cardboard (3 mm thick) and paper that came in three colours, a maximum size for the sculpture, and the requirement glue, sticky tape, staples, etc. could not be used. They were encouraged to work directly with three-dimensional space. The students were motivated by this activity and spent many hours to solve the problems associated with building their sculptures. I wondered to what extent their cultural background influenced them? To what extend did they use their previous mathematical ways of thinking? How did they reason in developing their sculptures?

A third of the first year architecture class of 34 students[12] at Unitech were interviewed by me after completing their sculptures. The interviews were coded for themes coming through the students' comments.

---

[12] Two of the 14 students were female. Eleven were from coastal provinces of PNG, two from highland provinces, and one was from a neighbouring Melanesian country, the Solomon Islands. Two students lived with their parents on oil palm plantations away from their home province, six lived in cities, and a few lived in rural towns. Several boarded at high school and most boarded at senior high school. Comments were made about the art classes and carvings at one of the senior high schools—three attended this school. It is significant that most parents and one guardian had incomes and half had post-school training.

First, positive feelings resulted from their problem solving. This is to be expected in self-regulating students as part of their developing identity. Half the students spontaneously noted that their sculptures were pleasing them and they were particularly intrigued that they were producing such a beautiful or different or imaginative sculpture. For example, David started with the idea of a trophy (Fig. 5.8a is his final sculpture) but when I asked him what he liked about his design he said, "It is beautiful and it did not symbolise [anything]. It didn't exist and doesn't represent anything in this place, and I came up with this thing!" Willie (Fig. 5.8b) said, "The most [pleasing] thing about the sculpture which really intrigued me was about the curves and the way I made the curves". Ian (Fig. 5.8c), however, noted that his feelings influenced his sculpture in a different way. He said, "I designed it on how I was feeling at the time. I didn't want to build something big, I felt small inside, I was pressured and school was mounting on me".

Ecocultural Visuospatial Reasoning as Architects

Students noted spontaneously their imagination or their use of imagery. TKeps (Fig. 5.8d) said, "most of it was from my imagination and creativity so it gives me idea of which things to fit into each part". When describing their imagery, 30 % of the students' comments that they imagined something not relating to anything physical 50 % referred to some physical object that might have started them off but 20 % spontaneously noted that they moved away from being bound by the initial physical idea as they responded to their work and began to imagine ideas or solve the problem of joining parts together. Three students (20 %) had initially wanted to make moving objects.

Students saw themselves as creative designers of buildings. Two students specifically commented that they liked buildings in cities and that they were intrigued by them and studied them. Others commented on their personal study of designs at the National Art School or in traditional carvings. Traditional backgrounds were mentioned spontaneously as important in 60 % of cases but only 30 % directly noted that they incorporated traditional ideas into their sculpture. Their traditional background seemed to be part of a belief that their cultural experiences were valuable for creativity and were a part of their identity as a designer. Interestingly, this did not seem to be as strong with the two female students but their families had longer associations with western cultures. Cultural ideas led to the use of curves, spirals, or symmetry. For example, Fred (Fig. 5.8e) noted:

> When I started to create the sculpture, it came out of my imagination. What I learnt back at home I added onto the materials and it came to what it looks like now. … It was totally flat [cardboard] but when I started to put bits and pieces in, I think it looks like a house and I make like a house and I started to think about making house back at home … It is the spiral thing that looks like it swirls around and the woven part that makes it attractive. … like ancestors in traditional types they used to make spiral bits they just bend hard woods to make small juts and to fish…

Ian (Fig. 5.8c) perceived village life and stories as influential.

Ian: Village life, talking to elders, they were telling traditional stories and I tried to incorporate some of the traditional designs into it as well. Partly some of my sculptures, they tell people how I feel. If I am angry, the sculpture looks very scary and taunted and dull. It depends on the mood I am in.
Interviewer: Can you explain how stories are in there [the sculpture]?
Ian: Not really. Patterns on main mast and shapes on curved areas are found in my tradition. In it there are lots of zigzags.

Besides Ian, four other students deliberately attempted to transfer a traditional drawing or mask as a major part of the sculpture. Taurus from the Solomon Islands based his sculpture (Fig. 5.8f) on a sea devil traditional design and story, and Bell's mask face (Fig. 5.8h). Like Bell, others used weaving techniques for effect. Often the idea of *bilas*[13] seemed to encourage consideration of the use of colour or extra features. Taurus says:

> The part I like the most is the rising face as it expresses features of what the sea devil is like with its two big eyes, the erecting tongue devouring and the two birds in opposite, which he used to navigate the unknown seas. I like it because eyes are bent, circular, and it is out of single paper. ... It is symmetry.

Decoration had significance culturally and architecturally, and this reasoning is being vocalised. Furthermore, students readily applied visuospatial reasoning and mathematical knowledge to draw circles or to measure sizes and equal lengths. Some began with shapes like triangles, circles, and cylinders and tried to build on these. Other aspects of mathematics that were mentioned were repetition and pattern, perspective, the use of measurement to improve accuracy, the use of mathematics to develop different, more difficult designs, and the importance of three- and two-dimensional images.

For many, creating designs meant creating something original and different from what already existed in the environment or in other's work. Some wanted difference on different sides of their sculpture while others wanted some similarity in, say, opposite sides. Others strived not to have symmetry but to have balance in the artistic sense. This is noticeable in comments made by Fing (Fig. 5.8g).

> When I visited the National Art School I saw paintings and they made me interested, lines going here and there and gives me a depth of feeling I really like this and I see my own work and I can put it here and there and it looks good.... When I needed strips I measured the same length with ruler but ... I looked at it and cut them to make them look better and not too symmetrical and all the same [see parts at the top of Fig. 5.8g]. ... It was time consuming to build back structure with circles (see the repeated design, Fig. 5.8g).

Fing used mathematical ideas in several ways and linked architectural and mathematical ideas in an abstract way in thought and action:

> I was just thinking and feeling structures are just like building numbers that go on and on and structures go on and on, more like infinity or repetition or what goes on and comes back; big, small and just like that.

---

[13] Tok Pisin word for all forms of decoration.

The view of Fing's sculpture (Fig. 5.8g) illustrates how he used visuospatial repetition for effect. Repetition can also be seen in Willie's sculpture (Fig. 5.8b) and David's sculpture (Fig. 5.8a). David said:

> [The curves] are about 50 cm long and I measured it from one join to another and cut out to fit papers around, I tried it out and I saw a pattern, then I drew it and I would join it up with slits in right place.

This is an interesting comment as the physical development of the sculpture was influencing his visuospatial reasoning like Fing and many others. Furthermore, David used the slit method that was common in cultural practices.

Nearly all students made reference to symmetry, proportion and wholeness of their sculpture. One used cylinders and circles to achieve a holistic entity; others used repeated curves. As Ian (Fig. 5.8c) said:

> I don't like to make symmetrical things because I have the idea that nothing is perfect[14] so when I created my sculpture I didn't want it to look symmetrical. ... when I design it with awkward shapes, it challenges the mathematical side of me how it fits together. e.g. circular bit that protrudes, I didn't want it to protrude too much ... At last minute, I removed some of it, especially unnecessary and didn't fit picture when look from all angles, and I asked opinions from friends

Bell made similar comments about the holistic nature of his sculpture (Fig. 5.8h). Students particularly wanted the parts of their sculptures to be in an aesthetically pleasing proportional relationship. For example, Ian modified the size of his parts so they looked good together and no one shape dominated. This reasoning also related to balance. Thus they were using visuospatial reasoning as they worked with their cardboard and paper. Half the students noted the importance of either a firm foundation or some means of making the sculpture sturdy so it would not topple over. This problem frequently led to interesting creative changes to the sculpture. Willie (Fig. 5.8b) said:

> I thought I had better start with sketch of anything, I was sketching anything that came into mind of statue ... I had a sketch with base and these curves ... One problem was I had difficulty in balancing curves and they were wobbling and would not stand up straight so I put cardboards in and it was firm ...

Knowing balance was important, several students measured carefully to mark the centre of a part. This was an application of mathematical knowledge to their visuospatial reasoning. Ian used stays to hold his mast firm and circular weights on a stick to balance the high part (Fig. 5.8f). He summed up his view on foundations:

> I think it is important in architecture. If it is right then rest will stand up. ... Mainly the fitting in of joints was the bit that needed cautious work. But with patience and advice from my colleagues and the successive trials that I made, I finally made it. It was constructed in a simple manner. The solid foundation were then followed by the bits and pieces that made up the rest. For example the scalloped papers. Extra weights were forced into it on various areas requiring it so as to balance the entire figure.

---

[14] This is a cultural perspective often seen in bilums (string bags) which will have a section that is not quite the same as the rest of the bilum.

Both the mast stays and counterbalances in bush-material bridge building are commonly found in PNG (as discussed earlier in this chapter).

From an architectural perspective, functionality was not important in this design project. The students' works were considered as sculptures rather than models of buildings but TKeps (Fig. 5.8d) particularly talked about the functions of different parts of his music studio.

> I began with cardboard as a strong foundation. After setting up cardboard, I started cutting paper and bend the shapes to make the building. ... That part and four corner was apartment where musical instruments can be played and stored and top part is entertainment centre. ... It is a building not on ground floor so I want it to be a few metres high above ground level so there is a bit of cardboard as base for uplifting building. The floor level can become some sort of foot track for others coming in.

The integrity of his sculpture came from his consideration of it as a music studio so that all aspects, position and shape of parts and nature of decorations, were related to this function.

Students saw that mathematics could be used in architecture and ideas such as symmetry, repetition, balance, and relative proportion were seen as part of this mathematical perspective on architecture. In most cases, students considered mathematics as measurement and calculation. Several students noted that they or traditional workers or architects had used measurement to get symmetrical sides, to make equal slots when joining, and to help with balance. This was thought to be a practical and important use of mathematics for architecture. In one case triangular and circular shapes had to be developed from the measurements so that they would sit together.

Shapes were also considered as part of mathematics. Ian began by "playing around with common shapes" and Clive discussed the nature of a cone. The importance of estimation and accuracy were noted in terms of the purpose for which mathematics was being used. For example, they noted that in building a bush materials house in the village, estimation was most appropriate although equality of lengths as marked by the length of a stick or a piece of rope was seen as necessary. For architects, measurement and calculation improved accuracy. Ratio was seen in terms of scale but was linked to the idea of parts of a holistic structure without certain parts dominating; this was also referred to as being in proportion. One student specifically noted perspective and another the importance of plans. Two students commented that it was mathematical to be able to judge how big a structure might be when one considered the plan of a building and its height.

Interestingly, all students could suggest some type of traditional activity that involved systematic thinking. Order of events was one idea expressed by students. For example, it was needed in preparing gardens, in the kitchen, in fishing, in housebuilding (order in which sections are completed), and in using the stars. Other mathematical ideas were evident when most students referred to getting houses straight by use of ropes and sticks or by planting sticks in a line. They commented that it was a skill to get the lines straight. One specifically commented that a good builder would have a plan in his mind but the onlooker would only know what it was when they saw the finished building. They noted that good builders could vary the plans and would know how many uprights, or sago-leaf bundles would be needed for the particular house. One commented that there were differences in the ways of building houses. Sometimes the outside plan was decided and then divided up while others considered the rooms to be needed before finalising the plan. It was also noted that diagonal struts were not in all houses. However, uprights in the walls were often close together.

Interestingly, students frequently expressed themselves inadequately when they first tried to explain how they used mathematics. Clive only linked in his school mathematics when he tried to form a cone with a piece of paper. Other students referred to triangles when they meant three-dimensional shapes but others used the terms prism and pyramid although one used the term prism when he in fact had a frustrum. Nevertheless, these students had strong visual images of the shapes they were referring to in both two-dimensional and three-dimensional representation and in terms of modifying their shapes for specific purposes including dynamic three-dimensional changes.

It should be noted that this visuospatial reasoning was occurring within a problem-solving situation. About half the students commented on the problem of getting started. Most overcame this problem by drawing or cutting papers or watching others. Once started they usually kept going. Time was also a problem noted by about a third of the students. Some noted that it took time to fix aspects of their sculpture or to make the many parts.

Sometimes change in ideas and imagery came about because they had physically to make the object without glue so this problem led to invention. For example, Fing used holes and rolled paper to hold parts together and then he used this idea for effect (see Fig. 5.8g). In fact a number of students noted that the joining of papers by slotting two papers together, each with a slit half way across, was used in traditional crafts while the rolled paper in a hole was done at school for holding papers together (staples, pins, and other clips were not available to them). It was also used in traditional buildings to hold the joists for a floor although a Y shape was more likely.

The use of physical objects rather than drawings flared their imagination. However, as with all concrete materials that are used for learning, students really needed to be faced with a problem in order for there to be connections between past experiences and imagery, new imagination, relationships, holistic ideas, and application of ideas. Furthermore, students also set their own problem—how to make the sculpture more pleasing. One approach was to think of an idea, physically cut it out and try it checking its effectiveness by holding up pieces of paper to see how it would look while others drew sketches. Students commented on the difficulties of stability and joining parts together and on their thinking in order to solve the problems. Their reflection and development of solution ideas was an important aspect of their designing (see comment on pattern by David). This aspect of problem solving illustrates how self-regulating and self-monitoring of progress through the problem was essential for developing identity.

Most students spontaneously made a comment about the influence of others. Although many of them noted that they tried not to be influenced and to make something original, they saw an idea like a woven mat and decided to use it, or a friend had talked about an idea which was then used such as Japanese ikebana flowers. One noted that his friends thought his sculpture was different things and that pleased him as it was in the eye of the beholder that a sculpture should be interpreted. Others had specific help to overcome a problem like how to make joins. A couple of students noted that, although they were given comments to start them or help with a problem, what they did was still their own idea. Another said it was important for students to work in groups both in mathematics and when creating structures because students could learn best that way and give each other ideas. Thus we see how cultural context was influencing their identity as a mathematical learning in terms of the self-regulating responsive learner as discussed in Chap. 1 (Fig. 1.2).

Responsiveness During Rich Activity

The responsiveness model (summarised in Chap. 2, Fig. 2.17) was initially developed from young students solving two-dimensional spatial problems (Owens, 1996b; Owens & Clements, 1998) but it was found to be applicable to tertiary students (Owens et al., 1998) and now to students working in three-dimensional space problems. This study places visuospatial reasoning in the context of responsiveness. By being responsive, reacting to what they saw, reacting to their imagination, or the comments of others and their feelings about their work, students were being responsive. Without responsiveness, the problems would not have been overcome but more importantly their imagination and creativity would not have flowed. Responsiveness is a compound variable; its components are dependent on a balance of cognitive and affective processing. Responsiveness is the movement forward, the risk-taking of problem solving. Often multiple thoughts have to be held for consideration and action over several seconds or minutes until the context reacts to the development. In cultural activities and in projects such as the architecture students' projects, responsiveness is a movement within the aspects of Fig. 2.17 but connecting ecocultural influences and self-regulation in identifying as a mathematical thinker or architectural thinker in Fig. 1.2.

While culture was an underlying influence on students, it was clear that the students were using only minor aspects of tacit knowledge (Millroy, 1992) with both architectural and mathematical concepts and processes. It seems that explicit discussion of this cultural knowledge would assist students to develop a recognition of the social, political, class, and colonial aspects of their school, architecture, and mathematics education (Ward & Wong, 1996). Such discussions would assist students to develop their cultural heritage in their architectural design, to create culturally rich design, and to recognise the interactions between some of these less apparent aspects of architectural and mathematical education.

Decision-Making

Visuospatial reasoning also plays a role in decision-making across many cultures and activities. Architects, engineers, and builders are aware of the space and orientation of buildings, but they are also aware of the impact that their buildings have on people. Some people continue to build as they have since prehistoric or early historical times partly due to available materials and climate, but others are keener to adopt change. For example, in the villages along the Morobe coast in PNG, there are unique house styles, built out of the same bush materials that were used in earlier generations, but not following the age old designs (Fig. 5.13e). The change has come from many influences such as living close to a modern city, interactions with members of the Department of Architecture and Building Studies at Unitech, and individual expression of house builders (Owens, field research 1997). At the other end of the PNG mainland in villages of Enga Province, some houses have been built so the structure joined circular and rectangular shaped houses while others modified the rafters so they can remove the central pole of the traditional round house as groups in other parts of Enga had done (personal communication, Henry Atete, 1997). People across PNG are skilful in adapting traditional building designs by enlarging floor plans, based on the knowledge of the overall building and cultural requirements of floors, walls, and roofs as discussed earlier in the chapter. In many Indigenous communities, these requirements and plans are held mentally and require skilful visuospatial reasoning to organise building and make changes. Words may be used to create the site, but plans assist design and change.

Kaveve villagers, EHP, built a beautiful guest house with rectangular walls, an entrance way with a high square pyramid roof, a hallway to bedrooms and verandahs to access other rooms, all from bush materials and with typical village "decorations" such as tufts for roof tops. Not only are people creating mathematical designs for utilitarian reasons but also for their efficiency and beauty. Visuospatial reasoning involves the elegant and efficient reason and that of beauty and creativity.

Power, Identity, and Relationships in Architecture

The impact of culture on architecture is easily noted in the use of decorative designs in societies such as PNG. However, this may appear to be superficial compared to modern building design or construction. Ward and Wong (1996) have argued that architecture provides for basic needs and design services but should not be limited to matters of visual aesthetics. Rather architecture should address issues of power and social transformation. In their development of a pedagogy of the design process, they encouraged participants to "demythologize their [the teachers' and students'] own beliefs about architecture, education and racial prejudice and about the relationship between them".

Eisenman (1988) argued that second languages like architecture and mathematics were avoiding cultural roots and class, a view that Ward and Wong (1996) illustrate is not possible. While *Māori* carvings in buildings have special meaning, the *Whare Wananga* project (Ward & Wong, 1996) considered precise placement of building elements such as the use of the atrium as a meeting place. Another project that considered cultural roots was the design for a museum for the Mashantucket Pequot Nation for which the guiding principles were couched initially in ideas "of cultural diversity and place identity as generating principles in architecture" (Atkin & Krinsky, 1996, p. 237). Interestingly, these views from architecture are salient to the social and cultural aspects of mathematics education and education encountered in PNG.

Indigenous societies also link serious business to certain structures, as seen in buildings specifically built for ceremony and bonding relationships. This is evident in *haus tambarans* of the Sepik and Rai Rivers of PNG and the long houses of Sepik Provinces (Fig. 5.13f), Western Province and Southern Highlands in PNG. However, across many PNG cultures there are men's houses (Fig. 5.11a) as well as places for births and other female needs. Power relationships are part of the rituals and design of the place.

Indigenous communities can have complex spatial relational arrangements that educate about relationships. This is the case for the Trobriand Islanders of PNG for whom the position of houses in their villages is dependent on relationships and status (Costigan, 1995). The Sepik River architecture as well as changes and differences in settlement patterns portray visuospatial reasoning that takes account of the environment, heritage, movement from place to place, and relationships with people. Thus the shapes of the *haus man*, *haus tambaran*, who they belong to and which clans occupy space inside, and proximities are all affected (Coiffier, ~1990; Hauser-Schäublin, ~1990). "The built environment, like language, has the power to define and refine sensibility. It can sharpen and enlarge consciousness. Without architecture, feelings about space must remain diffuse and fleeting" (Tuan, 1977, p. 107). Furthermore, for many Indigenous groups, a place has its own natural features imbued with feelings (e.g. the half-man story), and this can impact on design as seen in the ways, the architecture students created sculptures based on traditional house and other designs. Finally it should be noted that places of significance can have different features depending on the space they take up. For example, the space may be small (decoration or mark of ownership or status) or of medium size like a home and this will depend on the place and culture (compare various homes in Figs. 5.10, 5.11, and 5.13).

Visuospatial Reasoning with an Ecocultural Perspective

Power structures frequently prevent the mathematician from making links outside of academic, mainstream mathematics. Links are frequently in the direction of application to the outside world. However, ethnomathematics encourages visuospatial reasoning from ecocultural contexts to school mathematics and back. A good example was provided by Adam (2010) who analysed the weaving of food covers in Malaysia, modified the mathematics and then asked the weavers if they would make these new designs. However, there were various reasons (like the cone being too steep) that discouraged them from making some designs. Nevertheless, it was possible to bring mathematician and practitioners together. In many ways, mathematicians do this in everyday work when part of a team of economists or scientists as they seek new approaches.

It is, however, difficult to describe the visuospatial reasoning that occurs within one's own head especially when this is not generally part of discussion. New approaches to teaching arithmetic by encouraging children to visualise numbers, hold in their head a visual image such as a 10 frame from which to solve a problem, and to use an empty number line to mentally calculate a solution encouraging children to describe their visualisation and reasoning. There are cultural ways of transmitting much of the cultural mathematics as is shown in the sharing of string figures and bilum making. Among the mathematics, there are

- Visual ready-reckoner like methods for ratio
- Area images that are subitised and used when dealing with area problems
- Estimates of quantities that are embedded in the cultural purpose for the objects
- Bodily movements associated with actions and the nature of the sea or mountains when sailing or walking
- Estimates of lengths of string in all orientations and in curved or straight formats or in a ball
- Bodily and visual recognition of right angles and other angles determined by extending lines in two directions

However, this is only the tip of the visuospatial reasoning and ecocultural perspectives. It is not necessarily possible or right to expand further on the visuospatial reasoning available to an Indigenous cultural group who have built up these knowledge processes over thousands of years. Suffice it to say such knowledge should be acknowledged and encouraged to be available for that community's ongoing purpose for the sake of that community (Nakata, 2011; Owens, 2013a). Nevertheless, it is important in intercultural settings and curriculum to include both content and perspectives that reflect this Indigenous knowledge.

Moving Forward

This chapter outlined some key ways in which visuospatial reasoning can be acknowledged in terms of language and discussion, comparisons, ready-reckoner ratios, prior constructions, and actions. The context for decision-making is important as illustrated by the examples related to reciprocity and the environment. Thus the ecocultural tools, purposes, and problems influence the visuospatial reasoning and subsequent outcomes. These examples support the ecocultural perspective of visuospatial reasoning in mathematical identity. They illustrate how cultural identity forms part of the mathematical identity and valorisations of the community (de Abreu, 2002).

In this chapter, there are examples that take content relevant to western schooling with an ecocultural perspective. Furthermore, the above examples also provide ways in which western mathematics and mathematics learning can be enhanced and enriched. In Chaps. 7 and 8, I pursue this issue further.

The examples illustrate how the intention of the person directs the person's attention to specific features of the environment and social and physical problems associated with living in that environment (Owens & Clements, 1998). Both visual and spatial imagery is involved in recognising the situation and comparing it to previous situations. In this mental activity is the role of practice (Lave, 1988; Masingila, 1993; Nasir & de Royston, 2013; Roth & McGinn, 1998; Rouse, 2007; Wenger, 1998). Practice provided the ready-reckoners by which many of the spatial decisions were made. The mental imagination (dynamic visuospatial imaging) was not always associated with numerical values but rather with visual images resulting from practice and experience. Practice too encouraged the embodiment of visuospatial knowledge. Nevertheless, within a new problematic situation, the person or community need to reason. Practice also provided for intuitive thought. The examples illustrate that often a community or group of people reach a decision either because the results have implications for them and for group solitary or for spatially being able to view the situation from a different perspective as in deciding the slope of the roof. Thus our understanding of visuospatial reasoning is extended to incorporate an ecocultural perspective.

While PNG provided a rich basis for this study because of the wide diversity of ecological situations and language cultures, it is important to substantiate the ecocultural perspective from other areas of the world. In the next chapter, I draw on studies from other countries, ecologies, and cultures.

Chapter 6
Visuospatial Reasoning in Other Cultures

> *What avail is it to win prescribed amounts of information*
> *about geography and history, to win ability to read and write,*
> *if in the process the individual loses his (or her) own soul:*
> *loses his (or her) appreciation of things worthwhile, of values*
> *to which these things are relative; if he (or she) loses the desire*
> *to apply what he (or she) has learned and above all, loses the*
> *ability to extract meaning from his (or her) future experiences*
> *as they occur.*
>
> (John Dewey, 1938, p. 49)

The Challenge

An ecocultural perspective of visuospatial reasoning was established in Chap. 5 through first-hand and recorded experiences in PNG. Geometry and measurement cover a broad spectrum in terms of locating, comparing, and measuring of different attributes and notions, experiencing the physical world in all its manifestations, and going beyond the physical to abstract ideas. Is it possible to capture more of the wealth of visuospatial reasoning if an ecocultural perspective is taken across recorded experiences in other parts of the world?

This chapter will draw on studies influenced by cultures in the Americas, Australia, Pacific, Africa, Middle East, and Asia. In particular we note how visuospatial reasoning is enriched by locating and experiencing in the physical world and beyond and how visuospatial reasoning is used in creativity of different people.

Early Studies of Visuospatial Reasoning with an Ecocultural Perspective

One of the earliest and continuing studies in this area was by Berry (1966, 2003, 2011). His work indicates that individual differences arising from psychological, socioeconomic, and ecocultural experiences promote better adaption in an intercultural environment. He advocates multicultural frameworks at the classroom and national curriculum level to produce mutuality and reciprocity, equity in participation, and maintenance of cultures and identities. However, in this chapter we will look at his early study of perceptual and reasoning skills that compared Scots from UK, Temne from Sierra Leone, and Eskimos from Alaska. The study was carried out at a time when quantitative analysis prevailed and relatively culture-free tests were used (he noted that socioeconomic disadvantage in lack of visual materials did impact on test results when culture was kept constant). One interesting aspect was his comparison of words and environments in which he noted that the whiteness of the Eskimo environment seemed to contribute to high ability in noticing detail and more language words (rather than English transliterations). He also attributed this higher ability to child-rearing practices in which the stricter upbringing was associated with lower scores on perception and reasoning and women's dependence more field-dependent perceptual characteristics. He noted that western education reduced transitional communities' perceptual scores. He "concluded that ecological demands and cultural practices are significantly related to the development of perceptual skills; it has been shown that perceptual skills vary predictably as the demands of the land and the cultural characteristics vary" (Berry, 1966, p. 228).

Navajo Knowledge

Another long-term study that shows the strengths of traditional community living comes from Pinxten and colleagues with a strong anthropological/sociological analysis of mathematical thinking (François & Pinxten, 2012; François, Pinxten, & Mesquita, 2013; Pinxten, 1991; Pinxten & François, 2011, 2012a; Pinxten, van Dooren, & Harvey, 1983; Pinxten, van Dooren, & Soberon, 1987). Part of Pinxten's argument for a strong, rich geometry that differs from western mathematics is reflected in the following passage, taken from a small story he tells:

> Coming out of the pass he paused to eat, and then moved on in the direction of Badger Rock. He did not see the rock for a good time, since it was hidden by the black wall of Snake Rock extending right across you, from the south to the north. It had to be followed for a long time until one could see the small pass, which was hidden behind three juniper trees. They were the only trees sitting together on the edge of Snake Rock. When one did not know there was a pass behind them, one would keep following the rock for miles, without being able to cross it. Chee had found the pass and started climbing the steep path, while Chuck barked at the herd, which was wary of following Chee up the slide. After a remarkably

short climb, Chee reached the flat top of the rock, where all of a sudden he saw Eagle's Nest, the small flat stone, only a few yards away from him. At that point, he started looking for Badger Rock. A whole range of rock formations was spreading out before him, as far as the eye could see. He stood above the canyon of his parents, unable to see anything of it beneath him. He carefully looked at the rocks in the distance in front of him and recognized Badger Rock after a while. He could walk at ease now, almost on a flat surface until he would reach a small canyon he had to cross, straight to Badger rock. When he would make a turn to the right at this standing rock called the Badger, he would walk away from the sun and reach the arroyo of Salt Water in a short time. (Pinxten & François, 2011, p. 262)

Pinxten and François then attempt to summarise the richness of Chee's mathematics for non-Navajo:

It is almost impossible to picture the real "landscape" Chee is tracking through; one is dwarfed by the enormous red mesas and other rock formations one wanders through, and no "straight" line will be definable as the shortest distance between two points one can see.
Chee works with the following elements in order to orient himself:

1. The sun; different positions function as a clock for Chee. If his spatial markers mismatch with the sun's positions, he knows he is in serious trouble.
2. Certain conspicuous rock formations are major references or markers.
3. The general topological notion of path is essential, as are the cardinal directions.
4. The need for water and feed for the animals is essential in the notion of path Chee uses.
5. Adjacency and separation are two topological notions that have an important status in Chee's movements.
6. Going up and down, front and behind, narrowness and wideness as nonmetrical spatial notions play a primary role in Chee's orientation.

If we make abstraction of most of the actual rocks and mesas, dips, and canyons, we could draw a little map to represent the major features of the orientation system Chee works with. However, that would take Chee out of his context and into the schooled context we expect from him at the expense of the home knowledge. (Pinxten & François, 2011, p. 203)

Pinxten with colleagues developed a geometry curriculum for the Navajo based on their rich cultural knowledge embedded in place and survival. This is considered further in Chap. 8 to illustrate that mathematics education can and should take account of the rich knowledges of communities. The dynamics of spatial boundaries is further elaborated by Ascher (1994). If the activity can continue on the other side, then the boundary is on Navajo land but if it has to be reversed or modified then it is a boundary with non-Navajo land.

The mountain ridge itself is an interrelated system of parts that are in motion and in the process of change. Further, the entire earth, of which the mountain ridge is an integral part, is in motion as well. The earth and sky are always undergoing expansion and contraction. … to the Navajo, the significance is the processes of which the boundary is a part and how it affects and is being affected by those processes. …space should not be segmented in an arbitrary and static way." (Ascher, 1994, p. 129)

Thus the worldview is impacting on visuospatial reasoning of the Navajo providing relationships between places. The richness of visuospatial reasoning in an ecocultural context includes mental processing of richly connected places adding to the mental and emotional "mapping" associated with visuospatial reasoning.

Pacific Navigation

Navigating on land where there are some signs such as rocks and dips is one thing but navigating on the sea requires other extraordinary skills. Some of the most intricate ways of describing place and space have come from the descriptions of navigating in the Pacific Ocean (Oceania). The following comments are based on earlier discussions by Akerblom (1968) on navigating in the Caroline Islands, Marshall Islands, and Gilbert Islands; and on the Caroline Islands by Lewis (1973); from Worsley's (1997) journalistic style of writing; and from the Penn Museum of the University of Pennsylvania (1997) whose website provided dynamic images to assist understanding. Further work on the Marshall Islands navigation is available from Bryan (1938), Davenport (1960), and Spennemann (1998). However, Hutchins (1983, 1995) provided a detailed explanation of the unique systematic visuospatial way of thinking of space used by Marshall Islands navigators. Uses of star charts in "wayfinding" also occurred in the large Polynesian routes such as from Hawaii to Tahiti (Davis, 2009; Polynesian Voyaging Society, ~2003). Islands are out of sight and without a magnetic north compass, these sailors have sophisticated and skilled ways of travelling. Some sailors travelled thousands of kilometres and returned.

In the Caroline Islands, the star positions vary over the course of the year as the earth is on a tilt. A star chart had 32 positions. Sailors know an island's position on a particular star direction and they have sea roads that are taken regularly which take account of the swells and currents. When sailors start off on a trip, they assess the strength of the swells and currents by noting how far off course they move. They then adjust their direction. According to Worsley, the star positions are less important than the sea roads but they do provide a holistic visuospatial mental map on which to superimpose the sea paths. When travelling, the stars are kept between the halyards of the mast. Around each island there are usually two concentric circles for the limits of two kinds of birds. Swells also depict the position of islands when the boat is closer. In the Marshall Islands, stick charts represent the flow of swells around islands and divide up the plane so that a sailor can tell whether the canoe is on track for the island, or when coming from the other side has reached the doldrums behind the island. Curves illustrate the routes and sail positions to take account of the currents, refracted and deflected swells. Close islands have different swell patterns. Davenport (1960) notes that more than one kind of stick chart may represent the island and sea information (Fig. 6.1b). Furthermore, there are variations between the existing charts and the information they provide. Some provide more detail, others specific travel times or distances. Bryan (1938) noted there were three kinds of charts which he associated with the whole group, a part of the group, or as a general instruction (Fig. 6.1a). Some are specific to given islands whereas others are more generic while others vary on the importance placed on either island locations or the swells.

Most important is the recognition of the sailing canoe being the reference point so that sailors give directions, along star lines including those yet to rise, of the starting and finish islands and places, whether an island or a place on the sea, to the north or south of the travelling line (Fig. 6.1c, from Hutchins, 1995). As the canoe moves

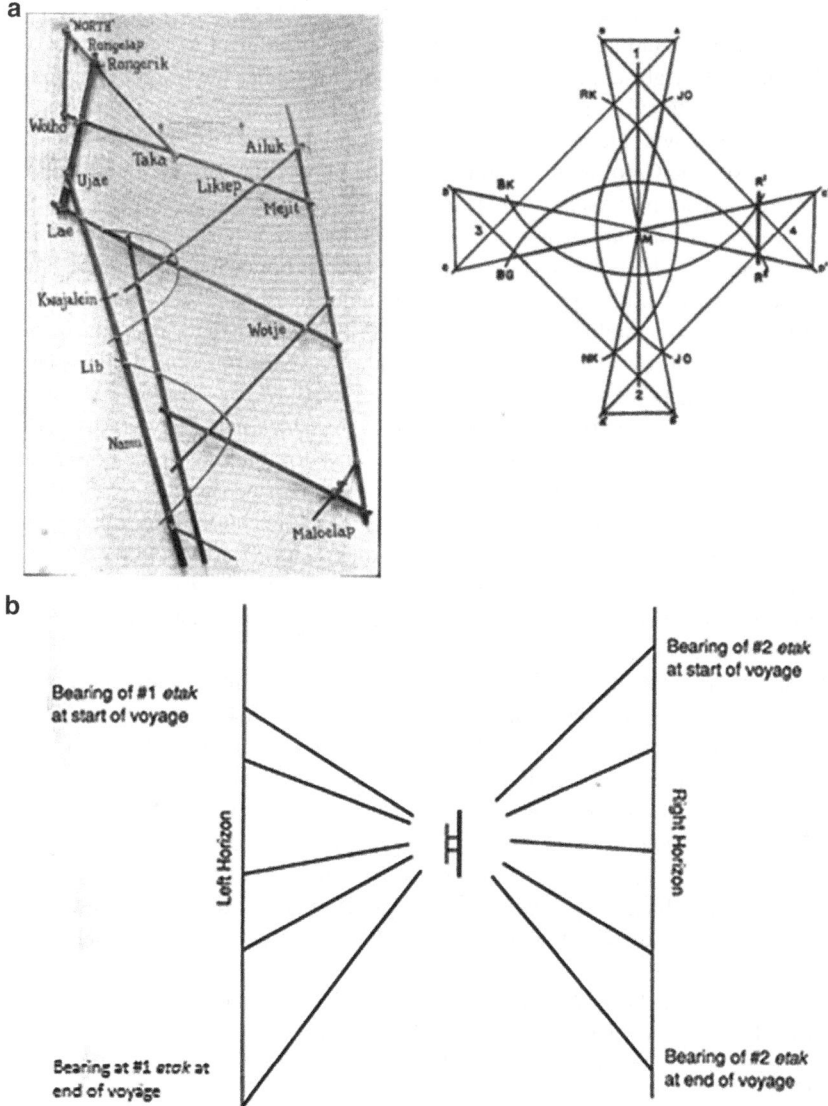

Fig. 6.1 Representations from the Marshall Islands. (**a**) Swells diagram (Davenport, 1960, p. 22). Stick chart for learning places (Bryan, 1938, p. 13). (**b**) Representation by Hutchins of the navigating system with the moving boat being the reference point (Hutchins, 1995, p. 89)

along the travelling line they know from experience their speed, time, and likely distance. Predictions are adjusted accordingly as the stars confirm position of the places in reference to the canoe. Sailors will wait for dawn and the bird flights before travelling on so they do not over sail their destination (Hutchins, 1995). Sailors have sea roads that are taken regularly and that take account of the swells and currents. Sea places are represented by a day's sail (*etak*) or by being directly south of an island, and are named, for example, from Puluwat to Eauripie, whale 1–6. The idea of a right-angle turn (breadfruit picker symbol), the trigger fish "map" also systematises the visuospatial images (Penn Museum, 1997).

The star charts with the moving stars around them provide visuospatial representations to assist the Caroline navigators to reason about "motion, relative position, and relative direction" (Ascher, 1994, p. 149). While games are used to learn the main star positions initially using stones or coral, later their position from different sea places and islands are learned. Gradually the various games become more complex. One game, island hopping, requires the names of islands on a particular star direction. Routes are combined and reversed in the games which have various nautical names. Dragging is a game in which the children give the position of places from a place which is not their home (Penn Museum, 1997).

However, these navigation systems are more than locating. They are a way of problem solving. They are elegant and effective ways for thinking in the head about position and movement. Interestingly the positional accuracy provided by these systems is embedded in the activity with an emphasis on connections and routes rather than lengths unlike scale maps in other navigation systems (Barton, 2008). The e*tak* brings together the navigator's knowledge of rate, time, geography, and astronomy and used appropriately (not as a linear measure) to determine destination. It is a logical construct or cognitive map.

Watson-Verran and Turnbull (1995) argue that the cognitive map permits the knowledge to go beyond the local even though the Marshalles used wave swells which the Puluwatans did not. It can be extended to new situations. The importance of the visuospatial reasoning is that the knowledge is used contingent with the place and the environment at the time. The vast body of knowledge can be passed on encoded in song, ritual, testing, mnemonics, group learning, connecting knowledge, and represented visuospatially with stick charts and stones on maps. The tacit knowledge and skills are also linked to the cognitive map for reasoning. The sets of information are coordinated but not necessarily in a unitising way as western measurement expects.

Ascher's (2002) thorough analysis of the stick charts and sailing techniques illustrate that the models are not intended to represent what is there like a map.

> [They] encapsulate and explain the system. When they are used by the Marshall Islanders for teaching, … they elaborate such depictions with words, but words alone would be insufficient. Particularly for dynamic systems, diagrams play a crucial role. They not only provide a way to visualize the interrelationships of the parts, but enable us to keep the entire system in mind while mentally manipulating or focusing on some part of it.
>
> The essence of an explanatory model is its simplicity. … Essentials are phrased in terms of the geometric characteristics of the ocean phenomena—the substances of the land and sea and wind are recast into points, lines, curves, and angles, and the interplay of the phenomena is recast into how these geometric aspects change and interact. (Ascher, 2002, p. 114)

Ascher (2002) also notes that the young navigators lie in the water to feel the swells and that sailors will lie on the bottom of the boat to feel the swell. These spatial imageries like the tilting of the head to see the stars between the halyards mentioned by Worsley (1997) illustrate the importance of spatial, bodily imagery given meaning ecoculturally in visuospatial reasoning.

> Numerical, spatial, and linear concepts in Melanesia, Polynesia, and Micronesia vary in accordance with distinctive physical environments and the social and cultural histories embodied in these Western assigned boundaries of the Pacific. (Goetzfridt, 2012)

One aspect that Goetzfridt mentions is the use of mnemonics for remembering star or geographic positions but in fact much of the memorisation is of a visual nature, the four points associated with a trip that forms a quadrilateral initially or trigger fish with different areas represented by the head, dorsal fin, ventral fin, and tail, together with places on the backbone for real or imagined places to assist with the navigation. The parts of the fish stylised as joined rhombus act as a metaphor for position and traversing the sea-lane. The use of fish names and many other distinctive objects or imaginary creatures also aids memory. Thus is the scheme to assist the navigator from Polowat in the central Caroline Islands to Guam in the Mariana Islands. Similarly, the parrot fish probing at the reef hole enables the travel from one island to the next until the sailor arrives back at first and catches the fish. The i-Kiribati envisaged the heavens as a giant roof with purlins associated with specific stars and through a story of travel over 150 stars could be named. Furthermore, it is in the visuospatial reasoning using trees and/or roots to discuss people in a group's relationship with the land that metaphor is important (Goetzfridt, 2012).

Australian Aboriginal Astronomy

There is growing evidence that Aboriginal Australians knew of the movement of the stars and used them for calendric purposes and other practical purposes such as measurement of distance (Norris & Norris, 2009). Across the 400 or more Australian Indigenous languages and cultures, there are some similarities in stories related to the sun and moon. One story indicates that there was a long association between tides and moon and variation in tides. This is visuospatial reasoning in large space. From the Northern Territory comes this story.

> The Warlpiri people explain a solar eclipse as being the Sun-woman being hidden by the Moon-man as he makes love to her. On the other hand, a lunar eclipse is caused when the Moon-man is threatened by the Sun-woman who is pursuing him and perhaps catching up. These two stories demonstrate an understanding that eclipses were caused by a conjunction between the Sun and Moon moving on different paths across the sky, occasionally intersecting. (Warner, 1937)

Norris and Norris (2009) suggest that a boomerang-shaped stone carving because of its juxtaposition to a man standing in front of the woman actually represents an eclipse. If so, Aboriginal Australians have provided some very early astronomy (these cultures are considered to have existed for more than 60,000 years). The stars are also used in calendars which are complex, with several seasons, and the appearing of stars rising in a part of the sky may indicate the beginning of a cold season and for deciding when to move for different food sources. The Mallee-fowl constellation (Lyra) appears in March when the birds build their mound nests and it disappears in October when the eggs are laid and could be collected (Boorong people of Victoria). Similarly, the rising of Scorpius tells the Yolngu of Northern Australia that the Macassans from Indonesia will soon arrive to fish for Trepang. The dark patches of the milky way are called the emu in the sky by many Aboriginal groups (not all), and in Sydney can also be found the engraving of an emu (feet back) lined up with sky emu at the time when emus lay their eggs. The Yolngu also track Venus knowing it is "held" to the sun but also that it only rises in the morning on a few days each year when they hold their morning star ceremony for which they prepare over some time. Though information is now lost, the engravings by the Nganguraku people on the Murray River cliff face are about the moon and sun, recording their movement in some way. The Wathaurung people of Victoria also built stone arrangements, one line being east–west, other stones placed to be in line with the setting of the sun at the equinox (Norris & Norris, 2009). Thus visuospatial representations are used for astronomical purposes by Aboriginal people in Australia as early astronomers.

Circle Geometry and Straight-Edged Shapes

Not long after I first began to explore the tile work of Portugal (Fig. 6.2) and Spain and the influences of the Middle East on these countries, I met Moustafa, a refugee in Australia from Afghanistan. He painted both landscapes and abstraction in miniature in circles. I asked him how he worked within the circle. Often he repeated his abstract design as rotational symmetry around the circle but other times he varied each sector. He first divided 360° by the number of sectors, drew a radius, and literally used a protractor to draw in each radius for the number of sectors he wanted whether that was 12 or 16. However, the next step required considerable visuospatial reasoning to ensure there was balance and no section of a sector dominated. Mostly he worked with curves carefully positioned on the sector. Colour too was carefully selected to ensure there was no dominance.

Figure 6.2 shows the basic circle constructions for developing a large range of tiled or painted spaces. Because of the interlocking spaces on the circle, one or more shapes are repeated and fill the space without gaps. Tessellations are the tilings of the same shape. The tiles join together without gaps or overlaps and with a pattern that allows the tessellation to continue in both directions. If two shapes are used it is a semi-tessellation. Some examples are shown in Fig. 6.2a but the intriguing thing is how an artist can picture the shapes within the myriad of construction lines and create another design for straight-edged tiles or painted shapes. The circle designs are not all Islamic, but for the Moors and other Muslims they contained no images.

The extraordinary hollowed patterned arched ceilings of La Alhambra (Fig. 6.2c) are the most beautiful and amazing three-dimensional tessellations that illustrate extraordinary visuospatial reasoning. Architecturally, hollowed panels in ceilings in La Alhambra and the pantheons (Rome and Paris) make them lighter but still strong.

Circle Geometry and Straight-Edged Shapes

Fig. 6.2 Visuospatial difference reflected in tiles from around the world. (**a**) Circle construction lines and resulting tessellations La Alhambra, Spain. (**b**) Firenze duomo, Italy. (**c**) La Alhambra ceiling. (**d**) Tiles in Portuguese palace, 3D illusion. (**e**) 3D tiles on cheda Thailand

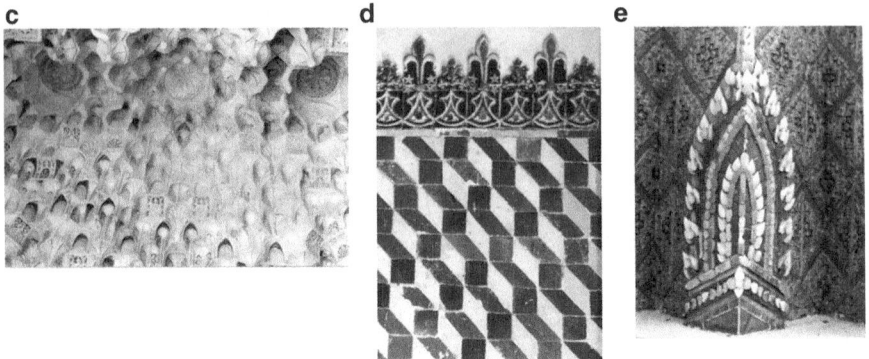

Fig. 6.2 (continued)

Domes and arches represent other marvellous architectural designs that portray visuospatial reasoning. Fine examples are the Sancta Sophia Turkey and the cleverly constructed duomo of Firenze by Brunelleschi with its internal spiral connecting the inner and outer walls and providing a means of construction. The external, more recent tiling of the duomo (Fig. 6.2b), the tiled floors of Pisa's belltower, and the three-dimensional tiling of walls of chedas in Thailand (Fig. 6.2e) and other parts of Asia must be some of the most exquisite tiling in the world illustrating a strong link between visuospatial reasoning and creativity. In modern times, the "sails" of the Sydney Opera house with its modularised construction provides further evidence of visuospatial reasoning in practice.

Creative Designs Across the World

Strong images from the Pacific are the sand drawings from Vanuata. These continuous curves are carefully crafted and associated with cosmology. The order in which each part of the figure is drawn indicates a strong overall visual image together with memory of order but the exactness is remarkable as shown in the figure from Deacon and Wedgwood (1934) shown in Fig. 6.3a. A strong sense of reflective symmetry is evident. However, Goetzfridt (2012) emphasises the associated story encouraging a recognition of the emplacement of myth (Rumsey & Weiner, 2001).

Ascher (1994) points out that some designs are repeated curves that may come from rotations or symmetries as shown in Fig. 6.3b and which she associated with some of the *kolam* prepared by Tamil Nadu women (e.g. Fig. 6.4b) although they also have other symmetrical reflected designs that can be incorporated in large designs (one or two reflected symmetries) (Ascher, 2002). Interestingly she notes that the Siromoney have used the Tamil Nadu designs in computer rules such as in Logo, albeit in more straight rather than curved formats. It is noted that Ascher divided the *nitus* sand drawings into those with even-degreed graphs and those with

Creative Designs Across the World

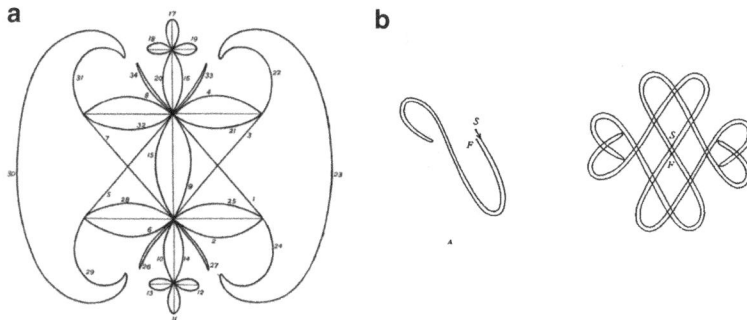

Fig. 6.3 Creating designs in the Pacific. (**a**) One of at least 91 sand drawings recorded in Vanuatu (Deacon & Wedgwood, 1934, p. 148). (**b**) Rotational and reflective symmetry in creating sand drawing using a base tracing, note the start (s) and finish points (f) (Ascher, 1994, p. 53)

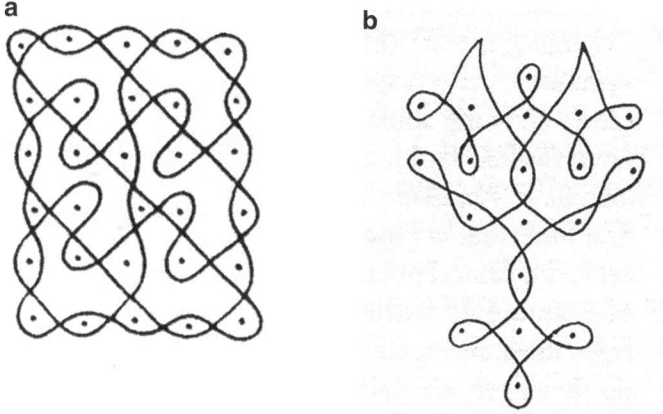

Fig. 6.4 Rows of dots used as markers for the curved lines to create the design. (**a**) Tshokwe (Central Africa) *Sona:* "the marks on the ground left by a chicken when it is chased" (Ascher, 1994, p. 42). (**b**) Tamil Nadu *kolam*: nose jewel that is embedded in other *kolam* (Ascher, 2002, p. 65)

a pair of odd vertices, by the systems used to trace them as well as by the presence or lack of visual symmetry created by the tracing procedures. The *sona* decorations on embroidered cloth or sculpted wooden objects, masks, woven belts, hats, and other objects produced by the Bushoong for the Kuba exchange system of mainly Angola are copied as sand drawings by children (Ascher, 1994). They can also be analysed in similar ways. The men of Tshokwe in West Central Bantu area create similar continuous line designs for significant cultural purposes. These are topologically analysed by Ascher (1994). The Tshokwe place dots in rectangular format to divide the space and provide the structure (Fig. 6.4a) but size may vary, and in most cases the vertices are of degree 4. (Some with odd vertices are actually being joinable outside the rest of the structure.)

Fig. 6.5 Fijian tapa, 1977

Many studies of the symmetries of the beautiful Polynesian strip patterns such as those found in New Zealand *Aotearoa* have been made but visuospatial reasoning is evident in some of the designs described by Ascher (1994). Decisions were sometimes made so that colour usage did not follow the carved symmetry but rather created a new idea in terms of symmetry loosely described as "juxtaposing one symmetry with another" (p. 170). Importantly while the ecology has influenced the designs,

> the harmonies, balances, rhythms, symmetries, and asymmetries ...[are] related to and expressive of the structures that underlie the Maori belief systems. Complementarity, the relatedness of pairs through difference, and symmetry, the relatedness of pairs through sameness, are seen as organizing principles in much of Maori myth, religion, social life, and economies, ... [including] the male-female complementary relationship, ...the world of humans and the world of the gods. (Ascher, 1994, p. 171)

Tapa is often decorated with symmetrical designs. In Fiji, stencils or stamps are used to repeat a design as individual blocks together with lines in regular patterns (Fig. 6.5) while in Tonga a board *kupesi* is prepared tying on curved and straight sticks into different patterns, using different triangular forms but also a plant motif with curved lines. Once the repeated pattern is rubbed directly on to the tapa placed above the kupesi and shifted along, then it is carefully painted along the rubbings. This is a group process so the making together of the *papa koka'anga* with the repeated pattern built along it using coconut sticks, then rubbing and painting together is an important part of pattern and design relationships. In Fig. 6.6, various designs with rotational and translation symmetry are shown. Finally different colourings are applied. Finau (Finau & Stillman, 1995; Stillman & Balatti, 2001) linked this work to matrices for different translations to make a connection to

Creative Designs Across the World

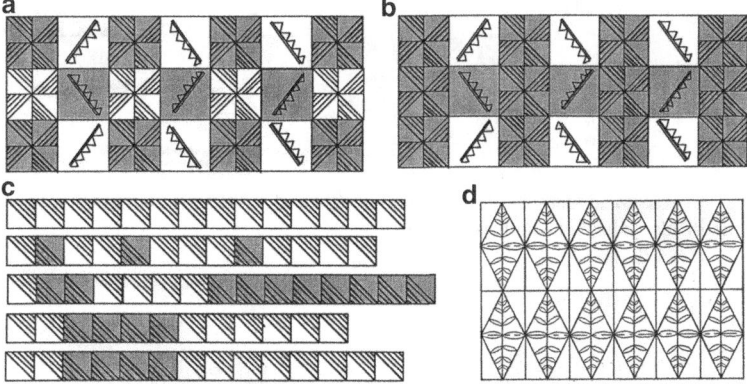

Fig. 6.6 Tongan tapa designs. *Source*: Finau and Stillman (1995). (**a**) Repeated rotated design with symmetries and a "perfect" colouring. (**b**) Repeated rotated design with symmetries and "imperfect" colouring but still attractive. (**c**) Various patterns incorporated into the repeated motif design. (**d**) *Kupesi* board coconut fronds bend to form plant designs

academic mathematics and to a variety of growing patterns in the background colourings or otherwise, all of which are pleasing and provide for creative variation in the tapa. As in PNG where group connections and group activity are just as important in the house construction or design, so it is in the creation of these Pacific designs. Much of the tapa of PNG is unpainted but in Oro Province, free-hand designs with some repeated curves are common (Fig. 5.20).

Symmetry is one of the more obvious results of the designs created and often repeated by Indigenous cultures. It is evident that the artisans use systematic ways of making, creating, and recreating. The following, discussing embroidery and weaving of the Hñähñu: the Otomies, Central America, is typical of most Indigenous artists in which detail, complexity, and cultural significance in terms of relationships ecoculturally are important.

> First, to become a traditional Otomi weaver is a life long process. A common situation is one in which there are several generations involved in creating woven or embroidered projects. That is, there is often a gathering that includes girls, mothers, aunts, grandmothers, and so on. Second, Otomi weavers and embroiderers must keep track of many counts of threads, and must make precise measurements. They must know the entire design from memory. The impressive part of the weaving or embroidery process is that the artist typically is not using diagrams for the patterns, nor is that person using a ruler to measure distances. A third consideration is that the products finished by the artists (as is the case in many traditional contexts) often have important cultural significance. Finally, it is important to mention that many kinds of weaving and embroidery designs are very time consuming to make. It is not unusual for traditional textile artists to take several weeks to make a work of their art. (Gilsdorf, 2009, p. 91)

Symmetries are evident in the decorations from southern Africa. Significant is the spatial embodied aspects of visuospatial reasoning in creating the designs. The *litima* designs

Fig. 6.7 The artisan and learners' designs. *Source*: Cherinda (2012). (**a**) Artisan's basic mat. (**b**) Highlighting with colour. (**c**) Another colouring

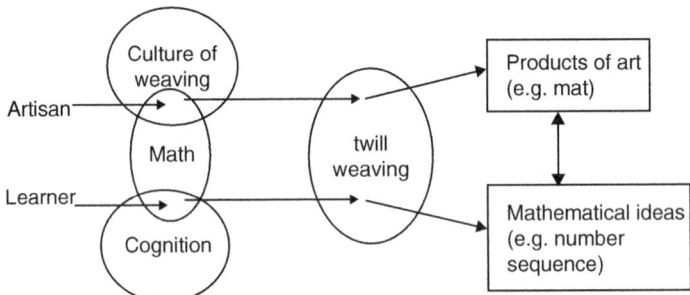

Fig. 6.8 Cherinda's diagram of ethnomathematics in terms of the mathematical learner. *Source*: Cherinda (2012)

> function as an extension of human action, and echoes the structure of the body. A painter, when drawing on the ground in explanation or preparation for mural painting, often draws with both hands. She begins at the top of an imaginary vertical, and the resultant forms on either side of this are simultaneously realized and are mirror images of each other. (Changuion, p. 35, cited in Gerdes, 1998)

The litima designs consist of a basic square that is repeated or reflected in the vertical axis and then a further two squares below are repeated or reflected in the horizontal axis. In other cases, the change is also accompanied by a reversal of the dark and light areas. Many designs involve curved lines but there are also straight-edge designs. Despite the amazing diversity, each is beautiful, technically well done and well integrated in terms of the overall four-square design forming a new design unit. For example, some curved lines do not finish at a corner but join to a curve on the side of the next square.

Cherinda (2001, 2002) noted that colour can transform the same pattern of weaving (a design motif) into different appearances (see Fig. 6.7a–c). He provides us with a way forward in connecting artisan's work to that of the mathematician as shown in Fig. 6.8.

> The premise is that mathematical objects (ideas) produced by the learner in the process of weaving… can be applied for stimulating new mathematical ideas and for producing new patterns in the weaving of the mats as well. … It is the instances where I say that both are 'weavers', versus both are 'mathematicians'. (Cherinda, 2012, p. 932)

Importantly, Cherinda details the visuospatial reasoning of students. For example, he notes that one student said that you count over and under to copy a line of weave, that you look at what is there overall to know whether it is a good look or not. You look at what is above or below. To start a new phase, the student isolated the first line of the weaving to copy. This is an important selective attention but also inductive reasoning. Cherinda noted that the student was able to create his own design, albeit by combining ones he had probably seen. These were created out of his mind. Another student too was able to count and reverse rows to create what they called beautiful designs. This affective aspect may or may not suggest recognition of symmetry as Cherinda claimed but it does indicate how affect is a critical aspect of self-regulating for a mathematical identity developed from an ecocultural perspective (Cherinda, 2012).

Cherinda perceived the connection between artisan and mathematical learner as bringing their own contextual knowledge to bear on their weaving and thus producing both inquisitive artisans and learners creating design in their own right (Fig. 6.8). Further argument on the process of learning visuospatial reasoning in schools is elaborated in Chap. 8.

Weaving often creates three-dimensional objects. Adam (2010) worked with weavers of food covers indicating the ethnomathematics of the artisans and the visuospatial reasoning evident in deciding what modifications were appropriate. Adam provided some suggestions that she had generated from advanced academic mathematics. Weavers rejected one that would be too steep but considered one that lengthened the food cover. Gerdes showed how some African weavers took a flat piece of woven mat and folded it to produce a cone shape container. The connection between flat surface and final product is not the same as the western view of surface area of a cone but there is embedded a cultural perception of size and purpose and strategies for modification. Many cultural groups make fish traps (Fig. 5.14c), storage containers (Fig. 5.15a), baskets (Fig. 5.14b–d), and masks (Fig. 5.14c, d). The visuospatial reasoning in producing a slightly smaller woven container compared to the woven lid is evident in the Timor Leste container (Fig. 5.24b) or vice versa in the Buka basket (Fig. 5.14c, bottom). This is not just a counting exercise nor a slight difference in length but one in which experiences provide visual images of what is required and what happens during weaving. A similar reasoning occurs with ikat weaving from Timor Leste (Fig. 5.24a) where the lengths of thread are carefully determined so that the tie dying of sections of the skene of threads means that in the weaving process, the colour starts slightly before or after the previous row. A similar reasoning occurs with making a box and lid from a sheet of paper. The sheet for the lid will be about 6 mm shorter on adjacent sides (Owens, 2006a).

Geometries are not necessarily linked to number as we found in PNG where visuospatial reasoning did not always occur with number and measurement although in some cases of gathering and counting food items there was a strong connection (see Chap. 5 and, for example, Paraide, 2010). Weaving and bilum making patterns may be, but are not always, counted to create designs. The visuospatial design imagery and subitising are often sufficient for the maker.

However, there are examples from ethnomathematics studies that show the visuospatial representations support numerical ideas especially if numbers are associated with spiritual values. In Mayan cosmology, following the diagonals down and up a mat of three divisions to count 1–6 results in each row equalling seven which was associated with God in divine power and the resultant diamond pattern is linked to the significance of the rattle snake. Priests traversing the steps of a temple also formed the diamonds that are associated with all of life in Mayan culture. The diamond was positioned representing space with blue diamonds for the seas on either side of the land in middle America, and for a sun on the sea from which it comes with a trajectory to the other sea. Like many Indigenous cultures, the rising and going down of the sun provided key reference points.

> Through the development of a sacred number system using mats with divine patterns, Mayan people possessed a sophisticated geometric and numerical creation story of their universe, whose first record is related to sacred numerical values. Numbers, symbols, and words could direct the priests to corresponding numerical values. A study of Mayan practices demonstrates one use of an ethnomathematical—global perspective. Ethnomathematics serves as an academic counterpoint to globalization, and offers a critical perspective to the internationalization of mathematical knowledge through attempts to connect mathematics and social justice. (Rosa & Orey, 2007)

It is important for ethnomathematics to take up the issue of some amazing mathematics existing in Indigenous cultures especially in terms of design. Their particular mathematical processes including visuospatial reasoning are to be valued, respected, and should be part of mathematics curriculum and education for the rights of learners as it is so important for their cultural and hence mathematical identities.

Settlement Patterns and Shelters: Place, not Just Position

In the Pacific, the patterns of settlement vary. However, in much of Polynesia ranking, symmetry and position of housing is evidently significant to the people. In Palau, the long history of symmetry in the meeting house and in the village gave visuospatial representation to power in decision-making processes (Wickler, 2002). In Chap. 5, the organisation of Trobriand Island villages was mentioned; the Trobriand Islands are relatively unique culturally in PNG in terms of hierarchy. In the Sepik PNG, other settlement patterns were evident. Land in PNG communities, like the Yolngu map (Fig. 4.4a) is often in patches, not only for sacred reasons. It may be due to intermarriage, status, the need for a variety of soils and environments such as fertile gardens, forest resources, grasslands, or food gathering resources such as swamps. Housing may be found on the various lands or even over the water. Hence the sections of housing at Tubusereia PNG over the water connected by a walkway, sections on land nearby and further away could belong to the one family group.

Place markers for different sacred sites evident on Lamotrek Atoll, Yap, Micronesia provide another aspect of geometry (Metzgar, 1991, 2004). Such points indicate lines and enclosure, a basic aspect of geometry in schools as well as traditionally. Similarly markers occurred on boundaries in most highland areas of PNG and in Hela and other areas of PNG drains demarcated the boundaries. The straightness and slope of drains is another example of visuospatial reasoning (Fig. 5.9, Edmonds-Wathen, Owens, Sakopa, & Bino 2014). Bellwood's (1979) historic study drew on examples from all parts of Polynesia. He concluded that ecological factors including the nature of the environment (often relatively poor soil on coral bases), people living in that environment, social and political factors determine individual settlement location in terms of clustering and dispersal.

In west and central Africa, Eglash (1999, 2007) found that there was intentional design with social meanings mapped onto the scaling architectural patterns of houses. Circular houses in circles of circles and rectangular houses in ever-diminishing rectangles occurred. However, he found these scaling patterns, that he associated with fractals, in a wide variety of designs, for example, textiles, paintings, sculptures, hairstyles, and religious symbols. The depth and richness of culture behind these is "lost" in summarising. However, one thing that Eglash has provided are useable tools for creating patterns on the computer and this will lead us into the next chapter. For example, he used the hairstyle cornstalks to assist in engaging Afro-American youth, and fourfold "beadwork" for creating a number of Native American designs. He notes:

> The presence of four-fold symmetry in Native American design is not a trivial geometric feature; rather it provides deep cultural connections spanning many facets of life, from religion to astronomy. Moreover its mathematical implications go far beyond that of reflection symmetry, allowing exploration of processes ranging from transformational geometry to iterative computation. This is just one example of the more general need in ethnomathematics to expand from a focus on static images to include process-oriented frameworks that illuminate design in the making, and that offer students a creative medium they can appropriate, for the purpose of expressing their own mathematical and cultural ideas. (Eglash, 2009)

In Brazil, de Castello Branco Fantinato (2006) found that belonging to a place and showing this by countering the expected way of driving on the road made it difficult to assess by observation people's spatial knowledge and skills. Furthermore, women felt the quality of a garment to decide if they could buy it when unable to read the price. Thus in her brief report on the study, she illustrated how ecocultural, especially from a living and socioeconomic situation, impacted on people's visuospatial reasoning to solve mathematical situations.

In Chap. 5, the role of visuospatial reasoning related to house building established a strong case for culture and ecology being considered in discussions on visuospatial reasoning. In the USA, a similar situation occurs. Partly related to prevailing winds and other factors, the Sioux tipi has three poles each planted at the vertex of the equilateral triangle (this triangle is not hard to achieve with two equal sticks as we saw in PNG). The long poles are tied at the top so the rope attached at the top falls to the centre of the triangle. While Rosa and Orey (2012) point out the significance of the centre culturally, they do not make it clear whether the centre is actually found by using the medians (lines from vertex to middle of the opposite side) or whether this centre is estimated or determined by the three equal poles tied at the top of the tipi and the rope falling. The centre is for the fire for cooking, heating, and burning incense in the centre of holiness. It seems that their use of etic (outsider perspective) and emic (insider or empathetic perspective) ethnomodelling could be simplified by talking of an ecocultural perspective of mathematics, especially visuospatial reasoning. Sioux setting up their tipi is one further example to add to the collection discussed in this book. Thus the links between ethnomathematics, ethnomodelling, and school experiences are seen to exist and be valuable. They are embedded in understanding visuospatial reasoning from an ecocultural perspective.

Moving Forward

One principle considered by the Alaskan universities for cultural competence is that the curriculum "respects and validates knowledge that has been derived from a number of cultural traditions" (Alaskan Universities Council, ~2012). It then becomes evident that visuospatial reasoning taken from an ecocultural perspective, especially from Indigenous communities must be a consideration of curriculum. Can this be achieved in schools and in teacher education? In the next chapter, the arguments for the importance of this perspective for mathematics and mathematics education are presented. In Chap. 8 programmes that take an ecocultural perspective on education are illustrated showing the role visuospatial reasoning plays in the development of space, geometry, and measurement.

Chapter 7
The Impact of an Ecocultural Perspective of Visuospatial Reasoning on Mathematics Education

> *At its best, schooling can be about how to make a life, which is quite different from how to make a living.*
>
> (Neil Postman, 1996)

The Challenge

Over the years there have been differences in the way researchers have viewed ethnomathematics (Shirley, 1995). Historical studies and studies of large societies (e.g. India) have referenced and evaluated non-western mathematics in terms of western mathematics rather than just referring to difference. This is evident in books and papers like those of Joseph (1991, 2000) who had grounds for emphasising the non-European bases of much mathematics to counter the Eurocentric view of mathematics. Another group of studies have looked at the mathematics behind the products of culture and made links with the western mathematics. These include studies by Eglash (2007), Fiorentino & Favilli (2006), Vandendreissche (REHSEIS-UMR7219, 2005; Vandendreissche, 2007) and Gerdes (1998, 1999). Some of these studies have used anthropological approaches and mathematical modelling to describe, for example, kinship relationships. This is evident in studies showing reciprocity and recursive patterns as shown in the Garma Project for the Yolgnu people in Australia (Thornton & Watson-Verran, 1996) or Eglash's (2007) studies on self-similarity. However, other studies by Saxe (Esmonde & Saxe, 2004; Saxe, 1991, 2012), Wassmann (1997) and Dasen (Wassmann & Dasen, 1994a) used both anthropological and psychological approaches to illustrate the social cognitive psychology of knowing. This approach may result in assessment tasks that are itemised and not necessarily part of thinking in an actual cultural activity. Another group of studies have focussed on equity and expectations in learning in context including studies by Civil (Planas & Civil, 2009), de Castello Branco Fantinato (2006), Knijnik (2002),

Owens (1992b, 1999b), Restivo et al. (1993), and Voigt (1985). Some of these studies of situated cognition (Lave, 1988) such as Carraher's (1988) work on street vendors or Millroy's (1992) work on carpenters have led to the idea of explicating tacit knowledge (Aikenhead, 2010; Frade & Falcão, 2008) as a key to enriching western mathematics with that of other cultures.

There is an overlap in ethnomathematics theory-building and work on languages in mathematics education (Adler, 2002; Barton, 2008; Clarkson & Presmeg, 2008; Setati & Adler, 2000). In a study of language, culture, and mathematics education, Matang (2008; Matang & Owens, 2014) has shown the main difference between learning in the vernacular, a Creole (Tok Pisin or Papua New Guinea (PNG) Melanesian Pidgin English), or English is in terms of language of formal instruction. The children in vernacular schools spend at least 80 % of their classroom time learning to read and write in their own mother tongue unlike those in the Tok Pisin and English schools. The higher performance by children in vernacular schools is due to longer length of time spent by children in learning early number knowledge embedded in the counting number words. The digitally counting systems (e.g. Kâte), automatically reinforce the idea of composite units assisting children to construct larger numbers through the use of cyclic pattern numerals (i.e. 2, 5, 20) that are physically expressed through the use of fingers and toes, and hands and feet. The measurement study highlighted visuospatial reasoning in mathematical activities related to distance, volume, and area (see Chaps. 4 and 5; Owens & Kaleva, 2008a, 2008b).

Having considered in depth the value of ethnomathematics in terms of visuospatial reasoning in various place-based situations and the ecocultural nature of these activities, it is important to take a critical look at what this means for mathematics, mathematics education, and mathematical identity. Each of these will be addressed in this chapter.

It is often stated that over the ages people developed mathematics and expressed their ideas in various ways, in language, diagrams, and actions. People shared their ideas about representation of systems and gradually a fairly dominant European mathematics was developed (Menghini, 2012). In some cases, mathematical ideas such as the relationship of sides of right-angled triangles were noted by many ancient cultural groups. However, it is also known that there are complex systems of mathematics that have often remained outside the European-influenced school systems such as Vedic mathematics. This book reinforces the importance of recognising diversity in mathematics and the creation of valuable mathematical ideas by sharing often unwritten mathematical processes and systems. By strengthening the importance of visuospatial reasoning in mathematics, I explore the richness of mathematics beyond the arbitrary symbols used in relationship and logic mathematics. It helps to establish the value of collaborative thinking in visuospatial reasoning much as Fermi[1] suggested as a means of problem solving.

While other authors have emphasised the importance of a consistency between culture and school for strong identity, I explicate how this occurs in practice.

---

[1] A collaborative estimate is made to solve these problems. For a brief explanation see http://www.edu.gov.on.ca/eng/studentsuccess/lms/files/fermiproblems.pdf

The model of identity as a mathematical thinker draws on ecocultural identity and establishes a strong mathematical identity by virtue of strong visuospatial reasoning being accepted in the learning. Data from some ethnomathematics projects in PNG and other places will be used to illustrate the challenges and values of this perspective.

Part of the purpose of encouraging an ecocultural perspective of visuospatial reasoning is to improve education especially in this important area of visuospatial reasoning. There is a justification for both an emphasis on visuospatial reasoning and an ecocultural perspective in mathematics education. Many of the limited conceptions and lack of visuospatial reasoning can be attributed to the lack of links between ecocultural backgrounds and the mathematics of the classroom. By interpreting school mathematics in terms of the students' background mathematics, it is possible to engage students in improved spatial sense and visuospatial reasoning.

In essence this chapter will develop the arguments presented and justified in the previous chapters to show the importance and impact of an ecocultural perspective on understanding and valuing visuospatial reasoning. This chapter will bridge to the next chapter where an ecocultural perspective is established as significant in terms of Indigenous cultures and transcultural/multicultural education and evidence of visuospatial reasoning in practice in these contexts is given. Let's first turn to the impact of an ecocultural perspective and visuospatial reasoning on mathematics itself.

Impact on Mathematics

Mathematics can be considered as a way

> to explain and understand the world in order to transcend, manage, and cope with reality so that the members of cultural groups can survive and thrive (through) techniques such as counting, ordering, sorting, measuring, weighing, ciphering, classifying, inferring, and modeling. (Rosa & Orey, 2012, p. 3)

Visuospatial reasoning has a strong role to play in mathematics when people use techniques that are strengthened by visuospatial representations and skills and reasoning to make decisions, no matter how simple or complex. As Amos (2007) said about the many PNG women past and present who make bilums, "when making this bilum a lot (of) imagination is involved with vitalizing (sic) of the design or pattern". This imagination in visuospatial reasoning brings life to the design.

Rivera (2011) argued that visuospatial reasoning was a legitimate way of thinking in mathematics. For example, π can be visually verified by initially measuring the length of the diameter of a circle with a piece of string and then showing that its circumference is slightly more than three times the diameter by tracing three copies of the string on the circumference. This is not a difficult task or unknown in PNG where round houses are common and people need to obtain materials for the walls (they also use six times the radius as the circumference, Fig. 5.10). Phi, φ, whose exact value is $(1+\sqrt{5})/2$, is another example of an irrational number that could be

Fig. 7.1 Visuospatial reasoning about ratios represented by irrational numbers in western mathematics. (**a**) Extending the house, half as much again. Malalamai, Madang Province, PNG. (**b**) Two ropes to form a right angle in PNG (Yamu, ~2000)

visualised by obtaining the ratio of the length and width of a golden rectangle[2] and whilst this shape does not seem particularly valued in PNG for deciding shapes of houses, the golden rectangle does occur. For example, when a house that has 4×3 posts has half as much more floor area than the rectangular floor of the house with 3×3 posts (Figs. 7.1a and 5.13; Malalamai houses discussed in Chap. 5), then this is approximately in the form of the golden ratio (≈1.66). Visualising such ratios requires inferring particular relationships between parts. Rivera (2011) made the point that every square root of a non-perfect square number could be depicted by taking the length of the segment corresponding to the hypotenuse of the relevant right triangle. In some places in PNG, this was visualised in practice. For example, where rectangular houses are generally of the same size, the diagonal length as well as the side lengths are well known in steps and in visualised lengths of ropes or saplings. In one area it was recorded that the villagers keep a rope with three knots (one at each end and in the middle obtained by folding exactly in half) and the diagonal rope with two knots (Fig. 7.1b). These ropes are used to form right angles (Yamu, ~2000).

While these are specific cultural examples, a general argument can be established. Culture may be seen as a way of life in its entirety for a particular cultural group or society while mathematics, on the other hand, is a systematic problem-solving method purposely developed to solve the everyday problems of the existence of its members. Mathematics education can be seen as the processes and organisation that enable cultural knowledge whether that be in school or in the family to develop and survive. The type of content knowledge is determined by the existing conceptual-knowledge frame of a particular cultural group including developments from various groups over time. This is further refined by the individual needs of learners as a prerequisite requirement to becoming an effective everyday problem solver.

An ecocultural perspective suggests that learning is not only about personal accomplishments and growth but also about the person's function in a community's activities and how they develop identities as mathematical learners and thinkers (Greeno, 2003). In particular, concern is for the interaction of learners with the ecocultural context—people and resources. Key to development of visuospatial

---

[2] For a brief note on the golden rectangle and associated golden ratio, see respectively http://www.mathopenref.com/rectanglegolden.html; http://www.goldennumber.net/golden-ratio/

reasoning is the opportunity for students to engage in constructing meaning of concepts and problems by engaging, trying ideas, visuospatially representing ideas mentally and physically in a way that they contribute collaboratively to each others' communications verbally, materially, and visually. Ways of doing this need to be developed within the community of learners in an ecoculturally appropriate way. Thus there is no dichotomy between school learning of procedures and practical or applied procedures but rather the learning draws on the ecocultural experiences and approaches to develop the mathematics throughout the curriculum or learning sequences and to develop abstract mathematical reasoning. Visuospatial reasoning, explaining, and justifying are keys to this in terms of learning concepts and what it means to carry out these mathematical processes. It is likely that there will be a development from intuitive reasoning to developing more sophisticated and integrated ideas and proofs where the oral and visuospatial precede and guide any later written reasons and proofs. Language plays a key role in these oral and shared visuospatial reasoning as well as what the community accepts as appropriate.

One of the purposes of mathematics is to solve problems and an ecocultural perspective encourages mathematics to be appropriate to the ecological and cultural context. Navigation required the use of felt motion, visual stars and sea, a memory and application of the star maps, and a mental ready reckoner in visuospatial terms of the distance travelled in a period of time. These complex pieces of information formed the inputs into the visuospatial reasoning pursued by the navigators. Through practice, judgments over time made such decisions more reliable but I claim the reasoning was visuospatial although numbers may have assisted in minor ways such as counting days. There were representative maps held in the mind on which islands, sea places, and stars could be placed but if the navigator was at the centre then other places were dynamically moving. Dynamic imagery is in fact an important part of visuospatial reasoning not only in navigation but also in other activities. For example, a person adjusting the position of posts, rafters, and other parts of a house, a bilum-maker growing a pattern towards the desired connecting shapes on a bilum (continuous string bag), or a person taking the next step in a string design (cat's cradle) will all use visuospatial imagery even if they have used counting or measurement to support their imagery. Similar imagination occurred with knotting, fishing, and carving.

Another significant type of visuospatial reasoning was the recognition and use of patterns. When the Elders were discussing the weaving of the diamond, they reasoned about where to start the row, knowing from past experience and in the example that they were working on what the pattern of overs and unders would result in the colour turning up in the right place to continue the slanting line of the diamond (see picture in prologue). Other patterns occurred, for example, in one place where a specific plant marked every fifth pair of *kaukau* (sweet potato) mounds. The marker may have had other purposes such as a food, shade, or for decoration. However, the marker may not have been used all the time.

Another critical part of the patterns that may not have been as regular as we might expect from a western education are those embedded in the mental ready reckoners that matched certain spaces with others. These links were established through experience in the ecocultural context. For example, they occurred when men knew how much kunai grass to cut to cover a house whose floor area was half as much again as another one, or when women said they would need two *tulip*[3] trees to make the 30 balls of string for a fishing net, or the Elder went from describing the size of his house to that of the size of the garden needed to provide a feast for all the helpers required to build such as big house (Fig. 5.10, Chap. 5, Owens & Kaleva, 2008a). These forms of reckoning were substantially in terms of related visual imagery associated with the place and aspects of the place. Like other tools in the visuospatial arena, decision making was based on best practice or probability of success. Experience largely informed the development of the mental reckoner and of its use in decision making.

One aspect of visuospatial imagery embedded in ecocultural contexts that has been given little attention in western mathematics is that of spatial imagery resulting from physical involvement in the activity. The ability to make decisions without necessarily explicating the reasons has been a part of air pilots skills but it is also a part of knowing the safety of a structure being built out of bush materials, the effectiveness of an arrow or a trap or the balance of a canoe. Such visuospatial reasoning does not seem to have an equivalent in school mathematics. Odobu (2007) from Manumanu village, Central Province, PNG provided a number of examples illustrating the spatial aspects of visuospatial reasoning derived from cultural activities (Fig. 7.2).

Odobu provided similar examples of how people make sectors of circles especially in sharing and house building, make decisions about volumes (mostly thought of in terms of liquid and food for containers such as clay pots and baskets), and tell time especially for the Hiri trading and other activities that cover long periods like weeks and parts of a day rather than hours, minutes, and seconds. Odobu matched each aspect of geometry and measurement with typical examples from school textbooks which used PNG contexts but not necessarily from one place. These were similar to Jannok Nutti's (2010, 2013) examples in which Sámi contexts were used but the basic curriculum is western as discussed in Chap. 8. Throughout each of the discussions and examples given above, there is a bodily movement associated with the visuospatial reasoning. Odobu was less clear about the size of a kilometre and how that related to a hectare as this was school mathematics and his school experiences had not associated this space with cultural embodied spatial experience. Nevertheless, his ecocultural mathematics was strongly associated with visuospatial reasoning—something that was commonly found in the teachers' projects (some referred to in Chap. 5 and others in Chap. 8). His concepts were strongly established for ecocultural mathematics.

The question remains about whether visuospatial reasoning has the so-called logic of mathematics. I contend that it does when used by skilled people. Even the use of 'by eye' decisions in achieving a straight wall or a right angle between the walls of a house have logical connectors. There is logic in knowing that looking from further away along three or more sticks or people provides better for deciding

---

[3] Tok Pisin for two leaves coming from one leaf stalk; inner bark also used for tapa.

There are no clear methods of calculating areas. Area is a matter of practical experience usually done by comparison. People know what areas look like and so they can construct them through activities. For example, constructing base of a house, door of a house, base of a canoe, canoe floor surface, clearing a piece of land for garden and so on.
The sides of surfaces are done in hand-spans or leg-spans.

For instance, one leg-span forward and one-leg span across forms a space like this.

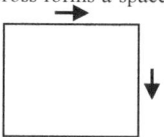

Unequal leg-span or hands-spans give a shape like this

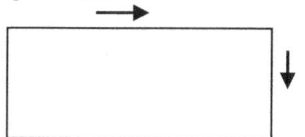

One can form many different shapes in the similar way. The shapes may represent many different objects in traditional cultures.

The measure of the area bounded by towns, cities or plantations are too large to be measured in mm, cm, or m. The area of large dimensions are measured in metres (sic) or hectares

Calculate the area of a triangular coffee plantation in hectares.

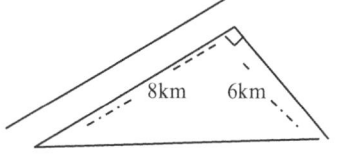

Parallelogram is a shape obtained by tying two longer sticks and two shorter sticks end to end and then skewing them to the right or left. Similarly, skewing a shape formed by four equal sticks, you form rhombus. Parallelograms do not have a name in culture. However, these shapes tend to tell people that house is going to fall.

Do activity done by people to resemble this shape. Example: Two people walking in the same direction. They are 'w' apart. One walks a total of 'a' distance, while another walks 'b' distance. The former walks longer than the latter. If the distance between paths these two people take remain same (w) and two paths are parallel than area formed is a trapezium.

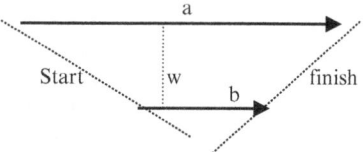

For the circle, kundu, base of highlands kunai house, clay pot, cut end of round object like log. Measures are made by comparison. The designer could use the hand span, a piece of rope or visual comparison. Sometimes a designer of the object thinks about the contents it could hold say more food to feed larger family.

To draw a circle using a traditional method, hold with two hands a piece of rope, holding one end of the rope firmly on the ground. Move another end around to form a round shape. You are forming a circle.

Fig. 7.2 Excerpt from student project on traditional measurement and shape construction (Odobu, 2007)

the straightness of a line or corner right angle. The decision making in itself is logical as seen in the following situations:

- Looking from two or more directions for positioning the central pole of a house
- Noting which slopes for a roof are sufficient for run off

- Knowing that the closeness of the *morata* roof material depending on the length of the leaves makes a difference for waterproofing
- Knowing whether roof poles are close enough for tying on kunai thatching
- Preparing baskets of food that are large enough to satisfy an acceptable exchange
- Mixing materials in the right proportions to make a good paint, soup, or pancake

Hence I argue that an ecocultural perspective on visuospatial reasoning is mathematical and it enhances mathematical applications, ways of reasoning, and decision making.

Impact on Understanding Mathematical Learning

By taking the idea of situated cognition as a beginning premise, we can then begin to explore visuospatial reasoning in terms of ecoculture. For example, to what extent may experiences related to weaving impact on the reasoning of students about spaces and shapes. This may be quite different to the way a western student from an environment with television and structured blocks (such as pattern blocks with specific shapes) might both develop intuitive reasoning, conceptual reasoning, and purpose for reasoning. Furthermore, cultural values may become widespread in a society leading to reasoning from the cultural values. De Abreu refers to this as valorisation.

> Cole's [1998] version of cultural psychology has as a strength an emphasis on tool mediated action rather than pure cognition. Emphasis on action enabled the recognition of the heterogeneity of psychological processes of groups engaged in distinct social and cultural practices. It also allowed recognition of the distributed nature of human cognition. Thus, the action of a person does not need to be situated in one mind, but rather in an activity system. … However a weakness of Cole's type of approach is that it does not yet provide a satisfactory account of within-group diversity (de Abreu, 2002, p. 174)

There is no doubt that the impact of various cultural tools used for mediating thought may indeed produce diversity within the group if values are considered. What one student or their family values may impact on the degree to which a particular tool is used. This is exacerbated by the many cultures that may be impacting on students and in some cases producing dissonance in thinking. In particular, perceptions and apprehensions of visual representations or reasoning from visual representations may take quite distinctly different perspectives. Owens and Clements (1998) illustrated this by a scenario in which two people are viewing a house and while one views it from the perspective of being situated historically at a specific period and notices features related to this perspective, the other views it in terms of its suitability for their particular family. Again we see a link between diversity and the more psychologically embedded notion of attention which is a key to visuospatial thinking and reasoning.

In her thorough analytical review of papers presented at conferences of the International Group for the Psychology of Mathematics Education, Presmeg (2006) noted that education to encourage visualisation was critical for quality visualisation in learning and mathematical practice. To "fold back" to a visuospatial basis as Pirie and Kieren (1994) purported for the learner to then go forward with deeper understanding is a critical perspective to take on learning. Interestingly, one of the studies reviewed by Presmeg that illustrated this point was about a plumber having limited strategies but the studies discussed in Chap. 2 (Owens, 1999c, 2005) were about classrooms in which visuospatial reasoning improved with activities involving spatial problems.

Informal mathematics learning takes place outside the formal classroom environment but within the boundaries of a particular cultural group's everyday activities and worldview and subsequent epistemology. Such learning has the advantage that it is not only familiar to the learners, but importantly, it provides the necessary contextual meaning to many abstract mathematical ideas and concepts taught in the formal classroom. It is also applying locally derived mathematics to solve the problems encountered in the particular ecology and culture. For example, in PNG, utilising the Indigenous knowledge within the formal classroom will not only enable the learners to construct meaningful mathematical relationships, but also provide an opportunity for interactions to occur between the learners themselves, as well as the teacher in explaining the many contrasting differences that exist across different cultural groups. The students can be learning in their home languages and using the words that hold powerful concepts embedded in the languages and using the wisdom that is thousands of years old (Trudgen, 2000). The teacher can learn from the discussions if they are teaching cross-culturally and learning the language and culture. They can enhance school mathematics with the funds of knowledge available from forming a community of practice with the families of the students (González, Moll, & Amanti, 2005). There are advantages for the learner if formal education takes seriously the notions of informal education, and in particular ideas that arise from ethnomathematics.

The Context for the Current School System

Most school systems are dominated by a financially powerful and controlling government machine. These systems often stipulate a curriculum which is to be taught although the means of controlling the actual classroom practice may vary. In some cases, external examinations and testing regimes provide high stakes for successful completion. These examinations are creeping into lower years of schooling. The examination items can control the curriculum in practice. In some cases, examination items are as close as possible to the curriculum intentions given the restrictions of the type of testing (usually restrictive paper-and-pencil testing) and the facilities for marking and assessing. Nevertheless, in spite of the examination regime, the curriculum generally remains dominant especially in the lower years of schooling although it may be interpreted by a textbook for teachers and students. The curriculum is generally developed by people in central government offices working under guidelines from governments with varying degrees of input from the mathematics education researchers and teachers in the classroom. The impact of the government's agenda may be less controlling in less authoritarian states but systems with neoliberal tendencies, consumerism and globalisation as westernisation may still prevent the diversity of mathematics developing (Atweh, Barton, & Borba, 2007). The centralised curriculum may prevent full potential in any system of mathematical reasoning for the student from a minority, with a first language background different to the official school language, from a class or gender group that prevents full opportunities for learning in some countries, or in a country that has been colonised.

Current schooling is strongly influenced by views of geometry and research into mathematics education that looks mainly at early geometry learning centred on naming, classifying, and defining shapes that were of interest to Euclid, the Greek mathematician two and a half thousand years ago. Some researchers mainly see geometry as a way of strengthening the area of proof (Mariotti, 2006). Visuospatial reasoning is seen mainly as a way of reasoning spatially as part of mapping or locating (Clements, 2004) and later graphing in algebra to reason about relationships (Meira, 1998). However, visuospatial reasoning has a role beyond positioning as indicated by both NRCCG's (2006) and Shah and Miyake's (2005) comprehensive studies.

Space and geometry curricula in the primary school tend to focus on the shapes introduced in Euclid's geometry. These begin with shapes like triangles, circles, rectangles, and squares. Curricula could begin with three-sided, four-sided general shapes and then name the subsets (Dreyfus & Eisenberg, 1990). Objects are also given names like cylinder, prism, and pyramid. This emphasis on shapes and on classification is reflected in the commonly emphasised levels as outlined by the van Hieles (Battista, 2007a; van Hiele, 1986) based on their studies of secondary school students. These have formed the basis of many curricula (Halat, 2007; Owens & Perry, 1998) and many research studies (Owens & Outhred, 2006). A look at these studies will indicate the approach taken to geometry education.

Many studies have used van Hiele's levels to indicate how students learn to name, classify, and define shapes. Researchers have suggested that the development from Level 1 (recognition based on global perspective of a figure) to Level 2 (analysis of the properties of figures) is not straightforward. This transition can be understood in terms of students' responses based on a realisation that aspects of a figure are important (identification of features), an attempt to document more than one feature, and grouping of figures based on a single property (Pegg & Baker, 1999; Pegg & Davey, 1998). The transition from Level 2 to Level 3 is also problematic (Clements & Battista, 1991). Researchers focussed on class inclusion (for example, squares as a subset of rectangles) as a distinguishing feature of Level 3 (Currie & Pegg, 1998; De Villiers, 1998; Matsuo, 1993). However, Matsuo (1993) suggested students' classification of a square as a rectangle seemed to depend on the property that they focussed on. This might suggest alternative perspectives on classification in which sociocultural experiences should be taken into account. Students might use known definitions or definitions they develop from perceiving certain properties (Shir & Zaslavsky, 2001, 2002) or procedural definitions that develop from constructing shapes, for example in dynamic computer environments (Furinghetti & Paola, 2002) or simple thin elastic. It is at this point on constructing shapes and creating definitions that sociocultural experiences may lead to totally different perspectives on shapes. For example, beginning with weaving, curves, or paths as the dominant basic feature of various cultural groups discussed in Chaps. 5 and 6, different properties and definitions may result in quite different systems of geometry.

Impact on Mathematics Education

Geometry is grasping by exploring the space in which the child lives, breathes, and moves (National Council of Teachers of Mathematics, 1989, referencing Freudenthal). In this statement, geometry is firmly situated in the space and place where the child is. The child explores this place both from the exploration based on his own perceptions and intuitions and from the influence of those around him. Through exploration, relationships of objects, their positions, their similarities, differences, and features are established and generalisations drawn.

If you watch young children play, they will tell stories and establish relationships and roles. They may use trains with personalities, dolls, animals, cars, lumps of wood, or themselves in imaginative or adult roles to tell their stories. They are imagining creatively and talking their own stories in their own language. Despite intonations and long sentences, it is not generally standard school language. They will draw in any adults willing to participate by a few recognisable words or gestures. These imaginary activities will involve some form of mathematics such as the object that represents food being placed on a plate and held horizontally to bring it to another person or to a toy for tasting. It is clear that children have a strong sense of horizontal and vertical developed at least from experiences with gravity. Ness and Farenga (2007) noted that young children have a sense of location in space that has horizontal and vertical axes but also movement forward or back. Their world is not just that of topology as Piaget mentioned with spaces inside and outside or with proximity being considered as a significant aspect of the space being explored by children. Furthermore, such imaginative play occurs in a space but that space is a place to these young children. As a result, the space takes on meanings that go beyond the simple geometric relationships. These meanings will become involved in the reasoning that the children use in play.

Early childhood contexts are mainly families (often extended) and in some cases institutions such as prior-to-school settings in which they may have contact with only a few other people on a regular basis. Children are developing relationships with significant others whose interaction with them is often dominated by cultural mores about these relationships, what is considered appropriate ways of permitting children to learn and what are appropriate content for them to learn. This includes the spatial experiences provided to children and the various representations associated with these experiences. It will influence the way in which children can interact with the spaces and objects around them.

Language is a dominant aspect of these contexts. Multilingual contexts can both enrich but also hinder development especially when the child fails to learn any language well (Clarkson, 2009; Valdés, 1998). Values are a strong influence in determining the selection of experiences that significant others provide for their children (Walden & Walkerdine, 1982). These are particularly relevant in experiences that may differ by gender but they also impact on the dominance of labels for shapes in western education home and school situations. They are also important in the value put on place and the visuospatial reasoning associated with valuing place. The sociocultural background of the child will be influencing the way in which these relationships are described and perceived. Geometry is the abstraction of these relationships which are then used for further exploration and development of ideas. Geometric reasoning then is "the invention and use of formal conceptual systems to investigate shape and space" (Battista, 2007a, p. 843). However, "formal" may be specific to a cultural group.

Visual and language knowledge are combined to provide an understanding of shape and space agreed upon together as people interact over time from an early age (Clements & Sarama 2007a, 2007b). Descriptions of specific experiences may be pragmatic arguments but they are interpretations of mathematical activity and social construction of concepts (Patronis, 1994). For example, a class of 16-year-olds took adjacent angles whose sum was a straight angle and showed that the bisectors of the angles were perpendicular, relying on angle measurements for "proof". Even in dynamic geometry environments, students continuously move from "spatio-graphic geometry" to "theoretical geometry" when elaborating a proof. The student uses the figure to make conjectures or to control results, then shifts to using definitions and theorems, then goes back to the figure and so on (Laborde & Capponi, 1995). This approach appears to occur early. Students apply different cognitive actions such as attending to features like "it's pointy", to decide on prototypical images (such as the equilateral triangle for triangles), to dynamic changes like sliding or pushing images to transform into another shape (Clements & Sarama 2007a, 2007b; Lehrer, Jacobson et al., 1998; Owens, 1996b). This pragmatic activity-based approach to arguments is closely akin to that found in Indigenous communities who discuss their house construction to make decisions. No single person is expected to make the decisions or to have a too restrictive rule-bound approach to the geometry involved (villages in PNG).

Tessellations can be found in many cultural groups. These may result in designs that involve Euclidean shapes but they are frequently developed from circle constructions as in Middle Eastern art (Critchlow, 1992) or weaving (Cherinda, 2001, 2002). The progression of western students' knowledge of tessellations is not well understood (Callingham, 2004), with the exception of an array of square units (Outhred & Mitchelmore, 2004; Owens & Outhred, 1998). There is a development based on visualising the coverage of the area with tiles which results in seeing the structure of the tessellation. Callingham (2004) reverted to using the van Hiele levels as a way of describing students' understanding of tessellations. Most students could describe an array of squares giving the name, informally or more technically describing the array and explaining the transformation to make the array. For other shapes students were at the visualisation level and could only recognise and name shapes. Whether this was a sociocultural result or not was not explored. Nevertheless, the studies referred to in Chap. 2 show how students can develop a sound pattern imagery understanding of tessellations while studies of weaving indicate an alternative understanding of tessellations.

The South African curriculum has emphasised development of visualisation as critical for geometry in the early years (Kuhn, personal communication, 1998). Another programme that moved away from the van Hiele development was provided by the *Count Me Into Space* studies (NSW, Australia, Department of Education and Training) which emphasised developing strategies for investigating, visualising, and describing rather than just classifying shapes. Nevertheless, the same shapes such as triangles featured strongly in these programmes. Transforming shapes was seen as a strategy for exploring shapes and visualising rather than as a way of using rotation and reflection to prove congruence, as many curricula do (Gutiérrez, 1996).

A large study illustrates this interaction between formal school and the child's intuitive learning. Appropriate classroom experiences were designed

> around children's everyday activity related to (a) perception and use of form (e.g. noticing patterns or building with blocks), leading to the mathematics of dimension, classification, transformation; (b) wayfinding (e.g. navigating in the neighborhood), leading to the mathematics of position and direction: (c) drawing (e.g. representing aspects of the world), leading to the mathematics of maps and other systems for visualizing space; and (d) measure (e.g. questions concerning how far? how big?), leading to the mathematics of length, area, and volume measure. (Lehrer, Jacobson et al., 1998, p. 170)

The emphasis here is on activities that are discussed in terms of using form, wayfinding, drawing, and measuring which are all seen as everyday childhood experiences. These are universal activities but practiced differently in different societies (Bishop, 1988; Dehaene, Izard, Pica, & Spelke, 2006). Space and geometry at this stage consists of concepts such as dimension, classification and transformation, position and direction, maps and other systems, and measuring which become both the geometry and the tools for exploring space. Again each of these appears to be universal but within each there is latitude for cultural difference and difference over time.

Processes for exploration in early geometry include inventions of ways to represent space, conversations that fix mathematically important elements of space such as properties of figures, argument, and justification around activities that involve manipulative tools or images, and narrative around what learners have done in stages, what they know informally and intuitively, and what they then own as part of their mathematical knowledge (Lehrer, Jacobson et al., 1998). Such learning requires increasingly more sophisticated investigating and visualising; describing and classifying (Owens, McPhail, & Reddacliff, 2003). The link between visualising and investigating and geometry is visuospatial reasoning.

There is no shortage of studies that have now taken cultural competence of teachers seriously in education (e.g. Averill et al., 2009). We discussed this earlier in reference to Indigenous education especially in Australia in supporting the role of Elders (see Chap. 3, Owens et al., 2011, 2012). The Sámi in Sweden established a Sámi

Handicraft School for adults to specialise in the various crafts such as knives with handles carved from reindeer horn for men and traditional clothing for women, and ways of thinking in their culture supported by revitalising their language. At the same time, Sámi schools for children were established "to pass on norms, values, traditions, and cultural heritage" (Jannok Nutti, 2013, p. 58) with teachers who were fluent in at least one of the Sámi dialects. The Sámi Handicraft Centre, supported by the museum, also established a programme to bring to schools. With model reindeers (on wheels) and sleighs, the traditional equipment included ear-marking tools (each reindeer is marked by the owner's geometric mark), ladles from beech tree boles used for drinking, reindeer skin pouches for coffee that folded as the coffee was drunk, ropes, and easily transported *lávvu* (cone-shaped tent).

In the school hall, the children simulate a trek to the reindeer and set up camp learning about

- The lightness and minimal space needed for the trek
- How to tie different knots for the shapes
- Sizes, spaces, and uses of the different artefacts

The children learn to recognise different geometric marks by making them on "soft" (recycled foam) ears attached to the model reindeers. Adults and children learn how to cut to size the desired clothes and shoes, how to sew them and how to make the important decorative patterns. In practice, reindeer herders and their families learn to track the reindeer and to map their routes, and are able to lasso their own reindeers and place them in their own corral when the reindeers are herded in summer. From the hundreds moving around quickly the skilled reindeer herders can recognise the markings on the skins of the calves of their own herd. This is a quite extraordinary visuospatial skill. The Sámi have their own approach to mapping based on the north, the rivers, and the reindeer routes. Much of the skill of the Sámi is in the ecocultural perspective taken to visuospatial reasoning around activities that relate to geometry and measurement (based on two personal visits, several oral presentations by Jannok Nutti, and Jannok Nutti, 2008, 2010). Jannok Nutti provided a summary of her earlier research Jannok Nutti (2007) into the mathematics of Sámi reindeer herders and handicrafters as follows:

> there are several conceptions, for example different names for reindeer herds based on the approximate number of animals. Unusual reindeer, for example animals with distinctive colours, function as support in counting or approximation of the wholeness of the herd. This is because reindeer herders easily recognise and identify this reindeer and if some of them are missing the herd is incomplete. The number of branded reindeer calves was counted by making marks on a wooden stick, by saving part of the ears of the branded calves, or by making notes on a piece of paper. Locating was made possible by well known objects in the natural environment, by the wind, or by rivers. The cardinal points were based on the landscape, the rivers, or lakes and the valleys around them. Body measurements were used. Depth of snow and water was measured with a stick or a rope and body measurement units, or with the help of the complete body. Distance was measured by the time it took to walk, by sound, or by sight. The concept of *beanagullan* is an example of a unit of measurement of distance. *Beanagullan* can be translated as the distance at which a dog's bark can be heard. Eight seasons divided the year and time was regulated by heat, light, or seasonal activities. The designing activity involves designing of buildings and artefacts. (Jannok Nutti, 2013, p. 61)

Each of the mathematical activities given in this paragraph involve a considerable amount of visuospatial reasoning as well as skills, and most relate to geometry or measurement of the environment, a cultural activity, or ecological response. Thus an ecocultural perspective of visuospatial reasoning is well exemplified.

Jannok Nutti showed from an action research study that:

> teachers changed from a problem-focused perspective to a possibility-focused culture-based teaching perspective characterised by a self-empowered Indigenous teacher role, as a result of which they started to act as agents for Indigenous school change. The concept of 'decolonisation' was visible in the teachers' narratives. The teachers' newly developed knowledge about the ethnomathematical research field seemed to enhance their work with Indigenous culture-based mathematics teaching. (Jannok Nutti, 2013, p. 57)

The self-empowerment of teachers who had to creatively make use of Sámi cultural knowledge and through those experiences develop students' mathematical knowledge came from working together and their interest to attend seminars about Sámi mathematics, mathematics education, and Indigenous education. Thus the notion of ecocultural context becomes important in the self-regulating, affective learner directly, and through the social competencies of the learner and ecocultural identity (Fig. 1.2). Furthermore, the teachers were responsive and affect was a part of their developing mathematical identity:

> The teachers' active engagement, and visions of culture-based teaching and its implementation were central. They tried to rediscover or reinvent Sámi culture in a mathematics school context. The concept of "rediscovery" led to joy and dreams, but also to mourning for lost knowledge and made the concept of "mourning" visible. (Jannok Nutti, 2013, p. 69)

It seemed that the Sámi cultural theme lessons with ethnomathematical learning were more productive than providing standard textbook type problems with a Sámi context. Teachers who followed this latter strategy were concerned that they needed to teach the students the national curriculum for them to become independent. Furthermore, there is other evidence to suggest that using cultural activities to teach mathematics can result in improvements in national assessments (Lipka & Adams, 2004; Meaney, Trinick, & Fairhall, 2013).

Other ecocultural situations have also illustrated the importance of visuospatial reasoning in mathematics education. Lipka, Wildfeuer, Wahlberg, George, and Ezran (2001) illustrated how to introduce elastic geometry, or topology, into the elementary classroom through visuospatial reasoning using intuitive, visual, and spatial components of storyknifing[4] as well as other everyday and ethnomathematical activities.

Tacit Knowledge in Visuospatial Reasoning

Frade and Falcão (2008) discussed the issues of making implicit knowledge explicit. We can generalise to say that people have a sense of area (tacit knowledge) developed through sleeping, gardening, and house building in particular. People are able

---

[4] For an example, see http://aifg.arizona.edu/film/storyknifing

to use this idea of area to make judgements such as the estimated amount of material needed for a house of a particular floor size. Many participant researchers referred to the pacing of (the length of) a garden as a measure of a garden. However, people would visualise a garden by knowing its length. Some visualised the number of kaukau mounds, others visualised a garden with a common width. Similar comments could be made about floor plans and roof areas. The static environment provides some mathematical examples whereas mathematical thinking occurs during the process or activity. By making these points explicit, teachers can reduce the discontinuities in knowledge and hence build a firm basis for school mathematics.

In PNG's measurement study, quantities were provided on numerous occasions to indicate amounts but these were frequently indicative of approximations or possibilities like round numbers are used in western societies. Dehaene et al. (2006) noted a similar effect among the Mundurukú speakers of the Amazon, South America when they mentioned "five" or "a handful" to refer to displays of five up to nine dots or using "four" or "a few" when five dots were presented.

Mathematical features such as shapes of bilums, pigs, holes, and houses were not mentioned in reference to volume; they were assumed by sight. However, length was seen as an important "rule-of-thumb" way of determining volume. For example, a length of string or part of the forearm was linked to the volume of a bilum, or the girth of a pig to its volume and hence its mass. Nomographs and ready-reckoner tables are possible equivalents to these mentally stored Indigenous knowledges. Nevertheless, the lengths that were referred to in describing a house did recognise the basic shape of the house. Thus radii for round houses or lengths of the sides for rectangles were mentioned together with heights.

One should also note the sophisticated ability of people to estimate needed amounts, for example, water for *mumus*[5] or garden areas. In each case, a good sense of comparative rates is applied, based on previous experience rather than on a mathematical calculation of volume. For this reason, comparisons are frequently made and confirmed by a group of people when payment or decisions involving sizes of mass or volume are made. The visual reasoning dominates over the numerical reasoning although numbers will be called upon to support a discussion. In other cases, the number rather than the size will dominate so long as items are roughly equal in size.

If we turn to the issue of engaging students in school mathematics,

> in reality, many students do not see the need to learn school mathematics further adding to barriers of meaningful learning of mathematics as many of these formal mathematical methods are viewed by students to be inappropriate in solving many everyday practical problems at hand … Ethnomathematics, unlike the school mathematics, is both context-relevant and problem-specific thus provides the necessary linkage between the everyday cultural practices of mathematics and the teaching of school mathematics. … Recognition of students' ethnomathematical knowledge also increases their self-esteem, which in turn increases their performance on school mathematics. (Matang, 2001, pp. 2, 4)

D'Ambrosio (1990) also raised the relevance of mathematics and the importance of self-worth as significant for mathematics education and aspects of tacit knowledge

---

[5] Food cooked in the ground using hot stones (see Fig. 5.12).

in visuospatial reasoning. A greater appreciation of the thinking behind out-of-school mathematics and school mathematics will bridge not only the conceptual-knowledge barriers but also the motivational barriers to learning. The teacher is no longer the sole transmitter of knowledge but knowledge resides within the community and community knowledge is valued. The teacher is a learner in the community of the contextualised classroom. As D'Ambrosio said in 2004 "ethnomathematics is the backbone of mathematics"[6] without which mathematics will not stand up and be of assistance to society.

Around the world, including PNG, curricula have become more proscriptive in the last 15 years. Little attention was paid to ethnomathematics. However, current mathematics curricula under reform began the process to include a "wide variety of rich problems that: (a) build upon the mathematical understanding students have from their everyday experiences, and (b) engage students in doing mathematics in ways that are similar to doing mathematics in out-of-school situations" (Masingila, 1993, p. 19). The details and analysis of PNG research provide sound evidence that ethnomathematics should be taken into account in curricula development and implementation. Mathematics teaching that is contextualised and concept oriented (rather than procedural) implies that teachers must incorporate students' ethnomathematical knowledge into the planning of learning experiences. "In the long term this will not only make mathematics to be a meaningful and reflective subject but relevant to solving everyday problems found in a complex and an evolving technologically-oriented society" (Matang, 2001, p. 7).

Language and Concepts in Mathematics Education

Concepts are established by language so it is important to recognise the range of ways by which groups indicate measurement attributes. The diversity and uncertainty by which speakers provided words for the commonly used school terms of volume, mass, unit, and composite unit (Chap. 4) indicated that most communities need to consolidate their Indigenous knowledge and determine how best to refer to these ways of thinking and acting in their language and then to appropriately link to school mathematics either by a clause, phrase, or single word. The mostly oral languages are rapidly changing and being overtaken by Tok Pisin (the main creole). In areas with long contact with English around Port Moresby, the language Motu has many transliterations (English terms sounding like Motu words) for mathematical concepts. This was also prevalent in Maori in New Zealand *Aotearoa* prior to the establishment of *te reo Māori* terms for mathematics (Meaney, Trinick, & Fairhall, 2012). Linguistic ways of comparing vary (Smith, 1984). For example, there may be a limited number of comparative adjectives or very general concepts like size.

---

[6] Keynote address: International Congress on Mathematics Education 10, Copenhagen, Denmark, 2004.

Other languages have a wide variety of terms for "smallness" or "largeness" which was evident in our PNG data (see Chap. 4).

In each situation, the use of visuospatial reasoning will assist the community to determine what might be a good way of describing school mathematical terms in local languages. For example, if the visuospatial mental image for a preposition indicates one idea, how might it be modified for another related concept. If number words are used for different groups of objects, then how can the language provide alternative abstractions for school arithmetic. Alternatively, as you will find in the discussion on elementary schools described below, only the treasures from the language are used for mathematics rather than taking a whole language approach. In other words, the strengths of cultural practice associated with language that provide alternative abstractions or related abstractions for school mathematics might be used. An alternative abstraction might be the importance of "heaps" in spatial arrangements. Another abstraction might be the idea of "a complete group". In measurement, the idea of estimate might be strong as well as ratios for comparative lengths, volumes, and areas. The idea of comparing and measuring with a unit might be evident even if the word for unit still needs to be derived.

There is no doubt that there are culturally different concepts that are in some ways more complex but abstracted. The number for deciding equality may always be associated with quality. Yet it is used consistently. Number might be an initial approximation for equality when volume is the main equaliser. This is evident in exchanges based on pigs. Exchange systems are complex yet well established mathematically without reliance just on number. Does this go beyond Davidov's ideas of abstraction (White & Mitchelmore, 2010) in which concepts in mathematics associated with different objects or constructs are eventually abstracted to a general mathematical term such as angle? It seems that western abstractions will dominate discussions on this issue due to global influences. However, it is important to recognise that there are equally important worldviews and abstractions in different ecocultural situations.

Impact on Mathematical Identity

The main purpose for taking an ecocultural perspective to both mathematics and visuospatial reasoning is its impact on the person. The studies on mathematical identity cited in this book relate to adults and teachers. However, if a teacher has developed their identity as a mathematical thinker from an ecocultural perspective, then it is likely, as Fig. 1.2 suggests that the teacher will influence the students in a similar way. For example, the lecturers at UoG had a significant impact on their students—the preservice and in-service teachers (see examples in Chap. 8). The teachers in turn were preparing classes taking an ethnomathematics, often ecocultural, perspective as they made many references to the landscape, the environment, and people's ways of living within the environment.

Kono (2007) illustrates the impact of culture on his planning for teaching. Interestingly, he developed a number of mathematical ideas that were not from a standard textbook. First his cultural identity is expressed:

> Mathematical concepts and principles are involved in most of the handiwork of the indigenous people of Papua New Guinea. These works are often overlooked and hence alienate the thoughts of mathematics learner. This makes the learner to think that mathematics has a foreign origin and has no relevance in our socio-cultural contexts.

Then he illustrates the weaving patterns including the "three up—three down [*diagonal*], chevron, V [vertical reflections of each other] and block" (similar to the *diamond*). He notes the lines of symmetry of each (0, 1, and 2) and says

> Students should be told to look at the pattern of the weaves rather than the shape of the complete work. The symmetry lines are being darkened. Instead of asking the students to draw what is in the textbooks the teacher can encourage the students to improvise strips of materials such as papers, bamboos, ... to create patterns for themselves from weaving'

He suggests they could do these in other subjects if necessary. He then continues to produce some mathematics which is original to his own exploring of the visuospatial representations of shapes. His report continues in Fig. 7.3.[7]

It is evident that the teacher was developing his mathematical reasoning from the visuospatial representations established through cultural learning and about which he was able to reason in an original fashion. These ideas were not covered in his own education.

One aspect that was strongly developed in the different teachers' projects was the use of visuospatial representations in village objects (houses, traps, bilums, carvings) and village activities (selling food, playing games, imitating parents in hunting, making objects, building houses, even arguing) to recognise mathematics. The teachers often said the mathematics was not necessarily recognised by the Elders as mathematics. The teachers only thought of mathematics as school mathematics up until they developed their projects but they continued to think in terms of the school syllabus in terms of topics. However, they now extended their conceptualisations to incorporate traditional or contemporary cultural ways of thinking and doing. Pepeta (2007) espoused this for his Enga community in Hawks' land[8] around Wapenamanda, noting his cultural identity. He listed the activities he then developed from two kinds of houses (round house for men and rectangular house for women) (Fig. 7.4).

He also provided a description for a game *bras flaua* (like Happy Families) associated with algebra, and woven bands associated with shapes and corresponding angles. It was very common for teachers to apply the ecocultural representations to school geometry shapes but behind this was knowledge of how the shapes were made. For example, how the surface area of the cone roof of a round house is covered with kunai grass, bundle by bundle, or planks placed vertically around the

---

[7] Some modifications to the text were made to describe angles rather than to use letters on the diagrams. Some diagrams have been omitted.

[8] He referred to his totem and his ecocultural links with the land.

The lines can be constructed in two ways: one, by drawing the lines to meeting the corners, and two, by drawing any perpendicular lines from the centre of any side meeting the opposite side right at the centre … .
For a regular hexagon

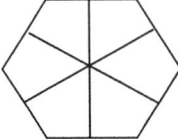 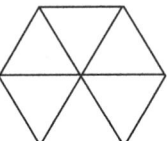

(a) Lines subtended from sides.   (b) Lines subtended from corners

If you should combine the number of symmetry lines, you will notice that the number of lines of symmetry in any regular polygon is equal to the number of sides of that polygon. Therefore we can write as

$$N = L \quad \text{Where } N = \text{number of sides of a polygon}$$
$$L = \text{number of symmetry lines in that polygon.}$$

Example: For a pentagon, $N = 5$ $\therefore L = 5$, Nonagon, $N = 9$ $\therefore L = 9$ and so on.

The number of sides is the equivalence of the number of bamboo strips that are required to make one of these shapes. This also holds true for any equilateral triangle or a regular quadrilateral. Ask your students to try it out.

Rotation and total interior angle of a polygon

This section will give any mathematics teacher an insight into his/her approach of teaching geometry that the normal way of finding the total interior angle is not the only mechanism of determining solutions. Lines of symmetry can also be employed in any polygons to determine the total interior angles.

As in the example on regular hexagon, there are six (6) equilateral triangles.

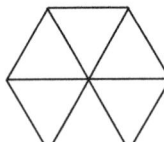 However, in an equilateral triangle the measure of each angle is 60° two angles = 120°. There are six sides so 120° x 6 = 720° the total of the interior angles. Or 60° x 12 (total angles formed at the corners) = 720°.

In another dimension (sic), say the symmetry lines are perpendicular to the sides. In a pentagon, for instance. Pentagon has five symmetry lines. The lines are drawn perpendicular to the sides to meet at the centre. Then, angles at the centre sum to 360°. Since all the angles are equal and there are five angles, denote $5x$ for all five angles, and hence $5x = 360°$. $\therefore x = 72°$. You can be able to find obtuse angles of the triangles because the other two angles are right angles.

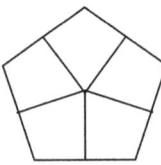 Hence 360° − (72 + 180)° = each obtuse angle, that is 108°. These sum to 540°. These are the angles of the pentagon. If we isolate one of the shapes in the large pentagon, it looks like a kite. But all the kites are similar. Therefore it can be deduced that a pentagonal shape is formed by a full *revolution* of only one kite. Any teacher should be aware that as long as any two angles in a kite are known the others could be calculated by *rotating the* kite for one revolution. This idea can be applied in other regular polygons too.

Fig. 7.3 Cultural identity reflected in mathematical identity (Kono, 2007)

> The traditional way of doing things especially designs reflect back the wisdom and knowledge of the people though there was no knowledge of mathematics known to them as today but mathematics was also used in the types of activities they performed every day.
>
> In the traditional society of PNG, adults take the responsibility for teaching children about the cultural values and resources. The children learn by watching adults by trying things out by correcting where necessary. Children spend lot of their time with the same age group. They would build, make-believe, houses and gardens, pretend to hunt birds and pigs ... They acted out feasts, dancing, making *moka* (*tee* in Enga), building houses, *bras flaua*, selling of food stuff, making bilums and string bags for different purposes, belts, paying bride prices, and imitate the older people in many different ways. The girls also imitate their mothers working in the gardens, carrying bags, washing babies. These are the shapes which are identified from the two traditional houses.
>
> | Shapes: | What part to be calculated: |
> |---|---|
> | Triangle, | its area, base, height and the angles. |
> | rhombus, | its sides, height, area and angles. |
> | rectangle, | its area, width, height and the angles. |
> | semi-circle, | its circumference, diameter and radius. Sector, area. |
> | cylinder, | its height, area of the base, radius, volume and diameter. |
> | circle, | its radius, diameter, area and the volume. |
> | cone, | its area of the circular base, height. diameter and radius. |
> | square. | its area, angles, height and the width. |
> | Sphere | its area and the radius. |
> | Trapezium | its area length and the height. |
> | Cuboid | its area, volume, height, length and width. |

Fig. 7.4 Examples of built environment linked to school mathematics (Pepeta, 2007)

wall or the sticks arranged in the frame of the roof of the rectangular house making triangles and trapezium (Figs. 5.11 and 5.13).

Cultural identity and recognition of mathematical knowledge as having a basis in culture were effective drivers in producing quality mathematics. Much of this can be credited to self-regulation for the following reasons (Fig. 1.2; Owens, 2007/2008, 2014; Wilson, 1997):

- The projects were on topics that the teachers selected themselves
- They applied their own goals
- They explained the mathematics
- They solved the problems of sourcing the details of the activities and connecting them to school mathematics
- They structured learning environments for their school students
- They evaluated their successes in their conclusions

Affective aspects were revealed by the degree of engagement with the task, portrayal of ownership of the mathematics in the culture, their imagination to prepare examples, resilience in problem solving, and the quality of their reporting.

Values about improving social cohesion if students found mathematics more relevant and less students might become "rascals" (criminals). Hope and aspiration was evident in some of the projects prepared at UoG.

There was a synergy between ecocultural mathematics and representations of school mathematics. Teachers valued the abilities or mathematical processes of their ancestors and Elders but their understanding was enhanced by school mathematics. The teachers recognised the importance that an ecocultural pedagogy had in terms of learning and developing their students' sense of worth.

Moving Forward

By taking an ecocultural perspective on visuospatial reasoning in mathematics, especially geometry and measurement, the meaning of mathematics is extended to incorporate more of the cultural ways of thinking mathematically. Indigenous cultures in particular have strong spatial experiences associated with mathematical concepts. Visuospatial reasoning is embodied in activity, often in a group situation. Furthermore, the few examples provided in this chapter support the view that cultural mathematics has a stronger basis for learning than school mathematics, creating visuospatial representations, and ways of thinking to which the learner can fold back during problem solving. Furthermore, despite concerns by teachers for students to perform on the national stage, it is clear that the learners were highly motivated by culture and that cultural identity could be harnessed for mathematics through an ecocultural approach in geometry and measurement education. However, can this extension of mathematics to value ethnomathematics address the social justice issues and the global issues of education for Indigenous communities? The next chapter delves into how an ecocultural perspective on visuospatial reasoning for geometry and measurement assists learning for the Indigenous student and the curriculum.

Chapter 8
The Importance of an Ecocultural Perspective for Indigenous and Transcultural Education

Kay Owens, Marcos Cherinda and Ravi Jawahir

> *Give a child a set of mathematical problems to be solved in ten minutes and graded for accuracy against the work of others, and the resulting performance may be dismal. Put the same child in a situation in which the problems are made meaningful, the same mathematics is used, and the solutions matter, and the child's performance can soar.*
>
> (Kilpatrick & Silver, 2000)
>
> *In the mathematics classroom…an ethnographic approach can give valuable insights into education and knowledge technology… towards how knowledge of pattern is generated and reproduced as carrier of thought of a particular kind.*
>
> (Were, 2003, pp. 25–26)

The Challenge

While there has been considerable work on visualising to assist children to learn early arithmetic (NSW Department of Education and Training, 1998; Wright et al., 2006), and some work on using patterns to begin early arithmetic for Indigenous Australians (Warren, Cole, & Devries, 2009), there is still a significant challenge for this book. An argument is now presented that revolves especially around the ecocultural perspective of education for visuospatial reasoning in geometry and measurement and how this reasoning assists students in problem solving and in maintaining a cultural identity as part of their mathematical identity.

Lehrer, Jacobson et al. (1998) suggested that quality instructional design required input from researched models of student thinking, classroom-based collaborative research, parents as partners, professional development workshops, and teacher authoring. Parents who are enabled to recognise their ecocultural mathematics will influence curriculum if they become partners. In their study,

> teachers went from posing tasks in isolation to developing sequences of tasks that provided children with opportunities for progressive elaboration of core concepts … on wayfinding, mapmaking, and Logo … In particular teachers increasingly emphasized representational fluency; children invented or appropriated multiple forms of representation. (Lehrer, Jacobson et al., 1998, p. 176)

Children reasoned from physical movements to recognise equivalent transformations and to generalise from their actions. Such questions as "do you think this is true all the time?" encouraged students to visualise and reason visuospatially. For example, in deciding whether a particular rectangle was larger than another, one child folded the thinner rectangle (6×2) in half and measured the other one (4×3). From this, she generalised to claim the 1×12 rectangle could go into the first one too and proceeded to fold it into four and show how the strips covered the 4×3 rectangle. From professional learning to developing substantial resources for teaching from a cultural basis, a strong identity and community of practice developed in a series of workshops in which student reasoning was the focus. The keys were building on children's informal knowledge, promoting their invention, having classroom conversations for understanding, and teachers orchestrating the curriculum tasks and tools. These young children's visuospatial reasoning was an entrée to conjecture and proof.

However, the challenge is to provide examples related to visuospatial reasoning from various cultural practices. The weaving boards developed by Cherinda are a well-researched example of an ethnomathematical activity in practice (Cherinda, 2001) along with many examples from origami (but not covered in this book). I have provided several papers on using just paper and string to teach mathematics as a result of my transcultural experiences (Owens, 1996a, 1998b, 1999a, 2001b). There are other examples within the realm of paper-folding which link to other African practices (Gerdes, 1999). Other examples will come from object making, travel and position, and various topological visuospatial reasoning activities.

In effect, much of the change towards an ecocultural perspective comes from face-to-face discussions between teachers and community members in order to develop a hybridity of thinking that will authentically link the ways of thinking of the community and the ways of thinking for a global education (González, Moll, & Amanti, 2005). In the future, students will be facing new and challenging problems for which a maintenance of ecocultural ways of thinking can provide a sound basis.

Continuities in Education Between Community and School

Many studies have considered the transitions and continuities between the contexts of learning mathematics (Owens et al., 2012) but the forces of globalisation work to homogenise mathematics curricula and negate the differences (Atweh, Barton, & Borba, 2007). Nevertheless, there are studies that have recorded mathematical activities incorporating ethnomathematics into formal mathematics (Eglash, 1997, 2007) ensuring that diverse mathematical thinking is not lost in school. These include studies by Cherinda (2001, 2002), the University of Goroka coursework projects (1995–2008) (some of which are referred to in this book under their authors' names), Gorgorió, Planas, and Vilella (2002), McMurchy-Pilkington and Bartholomew (2009), Meaney and Fairhall (2003), and RADMASTE (1998). Some of these studies have resulted in actual curricula recommendations (Jannok Nutti, 2008; Lipka & Adams, 2004; Litteral, 2001; Pinxten, van Dooren, & Harvey, 1983).

Providing a culturally appropriate curriculum is a challenge for small communities as found in PNG (Litteral, 2001). Nevertheless, some efforts are being made to ensure that mathematics of small communities are used in education as determined by the education reform (Matang, 2008; Matang & Owens, 2006; Owens & Kaleva, 2008a, 2008b; Paraide, 2003). Similarly in Alaska, funding has been provided for the establishment of culturally relevant mathematics units. Lipka and the Yup'ik community have established units of work that begin in the activities of the community. Students might build models of fish racks or discuss the importance of locating eggs on the island. They have shown that students' attainment on mathematics has increased significantly (Lipka & Adams, 2004).

According to Clarkson and Kaleva (1993) one crucial reason for failures in implementing curricula that consider cultural aspects is an unwillingness by many experienced teachers to change their ways of presenting mathematics lessons that have become routine for them. Hence change in the classroom is difficult, leaving aside the bigger political difficulties inherent in any educational change. Successful implementation of any curriculum reform in the classroom focusing on culture will depend at least on (a) the role of the teacher under the reform, (b) teacher beliefs and values about the new curriculum reform, (c) teacher background knowledge in mathematics (Matang, 1996), and (d) involvement of community in the school curriculum and learning activities (Department of Education Employment Workplace Relations, 2009; Yunkaporta & McGinty, 2009). We would argue that these four points must be addressed.

One way to meet this change is for teachers to begin to utilise students' mathematical spatial experiences, which will be set within their own cultural experiences (ethnomathematics), gained from everyday encounters (their informal education) as the basis to teach school mathematics (Trudgen, 2000). We believe this will result in students becoming active participants of the information-sharing process within their classroom. Clearly in this scenario the responsibility of the teacher changes from one that is imparting knowledge in an authoritarian manner to one that seeks to create a learning environment that promotes meaningful and interactive mathematical discussions not only between teacher and students, but also between the students themselves. This no doubt will include the students telling the rest of the class mathematically related family stories (Matang, 2003; Matang & Owens, 2006; Owens, 2000b; Trudgen, 2000; Yunkaporta & McGinty, 2009). Similar suggestions which could be followed have been made in relation to mathematics and language (Clarkson, 2009).

In a comparative study of transcultural education in four countries: Sweden, Australia, PNG, and Yemen, I concluded that variation in the contexts highlighted the following themes in teaching:

- Aspects of cultural context relevant to mathematics
- Meeting language differences in different ways
- Maintaining culture in different ways
- Teaching in a cultural context
- Teaching mathematics in a cultural context
- Having an emphasis on national values
- Using national language appropriately
- Developing context-specific strategies for diversity (Owens, 2008)

If the impact of students' informal education outside the classroom, and the notions of ethnomathematics are to be taken seriously, then teacher education programmes must also change. Pre-service teachers will need to be given the opportunity of in-depth investigative studies of the mathematics content knowledge that they will teach. This mathematical background will give them confidence to approach the teaching of mathematics within the immediate socio- and ecocultural contexts of school students, as has occurred in the Luleå University programme in Sweden (Johansson, 2008), the secondary education projects in PNG (University of Goroka Students SMAC351, 1998–2007), and the elementary school project on improving mathematics education (Bino, Owens, Tau, Avosa, & Kull, 2013).

Language-Based Activities in Multicultural Classrooms

Children in multicultural preschools assist each other very slowly to speak in English but factors such as the table mix of languages, the amount of English understood and spoken at that stage, the child's choice of playmates, the children's personalities, and non-linear rates of learning, influence the progress made by children (Fassler, 2003). One recommendation developed from this study was that teachers need to spend time talking with each group of children.

There are a number of language issues that arise in terms of visuospatial and geometric concepts. Words (morphemes) like verbs for actions and words like nouns for observable objects such as blocks are best used so students may learn to compare, estimate, and measure before they appreciate attributes of shapes. Even so, words that sound similar need careful pronunciation and experience e.g. for English "side" and "size"; "estimation" and "evaluation", "triangle" and "rectangle". Words may have two meanings—mathematical and general e.g. English "area" means a place in general usage but the measurable space inside a 2D boundary in mathematics (Owens, 1996b). Difficulties arise with what accompanies the idea implicitly e.g. length when referring to volume (Owens & Kaleva, 2008b), not dissimilar to the linear scale on a measuring jug but requiring considerably more visuospatial reasoning from experiences. Some concepts have many constructs and representations. For example angle and fraction require considerable language and experience for students to grasp. Difficulties arise with prepositions as discussed in Chap. 4 on deixis but numerous hidden meanings in mathematical expressions may be overcome by using multiple language patterns (Davis, 2009).

Oral work by both teacher and students needs to be slow, purposefully repeated but not rapidly or loudly repeated. Oral work is often facilitated if teachers use whole class discussion, followed by paired or triple group discussions before individual thinking or work. This should then be followed by further sharing in the small group and then in slightly larger groups prior to any further whole class discussion so that all children have a chance to listen, talk, read, and write. Teacher's and other students' sentences need to be simple. Students should be allowed to code (language) switch. Code switching may occur between every day and more precise mathematical language; between dialects; between lingua francas and home languages and various national languages. Switching languages assists learning if knowledge is constructed rather than kept as unrelated ideas. Students of the same language group may talk in their own language to explain to each other while students from other groups can try to explain as their gestures are often beneficial in helping another student. Students help each other and speak more mathematics. Gaining understanding by playing with words and using bilingual facility will assist in constructing meaning. For example, a grade 2 student with English as a second language was able to develop the meaning of "bigger angle" by realising it did not mean "sharper" but the opposite "opening wide" (Owens, 1996b). The words were switched for the different visuospatial imageries through reasoning and listening to and watching the teachers' explanation and the actions of other students.

Visuospatial Reasoning in Metaphors: Selection and Relevance

Many metaphors are used in mathematics and teachers need to explain the metaphor. Teachers should avoid using western colloquial metaphors if these are not used in everyday talk in the community. However, there are often rich metaphors in cultures that can be used in mathematics. We have already shown that the slope of a roof is a good introduction to work on angles. Visualisation (images in the mind from action on objects and pictures) provides a major context for oral learning. It requires students to make explicit in words what their learning is in addition to their learning by observation.

Difficulties with a new language do not mean students have a lower ability to reason and furthermore the metaphors and experiences that can be drawn on by students are all the more critical for mathematics education. Students should be exposed to a range of tasks and not just easy ones and they should problem solve and speak about mathematical concepts to enhance the transference of skills and metaphorical language. Visuospatial representations either in drawings, gestures, or metaphors are important. Language games assist students to speak and listen to mathematical terms e.g. drawing what is read in the school language or speaking this language to explain the drawing. For example, Murray provides a range of activities for teaching children with English as a further language (Murray, 2011).

Some mathematical areas are often stronger in other cultures e.g. circle geometry in Arabic art than in a western school culture, topology in Navajo or string designs (Vandendriessche, 2007). Furthermore, estimation skills (not necessarily matched by language) are often strong in Indigenous, subsistence cultures, and among artisans (Millroy, 1992; Owens & Kaleva, 2008a). These can be important for the planning of curriculum. However, teacher education also requires addressing.

PNG Secondary Teachers' Ethnomathematics Studies

In Chap. 5, I discussed the mathematical thinking involved in different bridge-making experiences in PNG (Fig. 5.7). In Yambi's UoG (2004) project, he drew a bridge (Fig. 8.1a) and related a number of school mathematics problems to the bridge. In particular, Yambi related the making of the bridge and the bridge diagram to assist students to visualise the context, the triangles, and the trigonometry required to solve the story problems, thus assisting the students visuospatial reasoning for trigonometry (see Owens, 2014). Yambi also used the wave of the vines holding the platform to the hand rails as an example of the sine wave. While this might not quite fit a sine wave in practice, it was a wonderful way of modelling a sine wave (see Fig. 8.1a). Another teacher, Imasa from Pindiu, Morobe, also linked the sine wave to the curving of the rope used to bind split bamboo to two sticks to form a platform for a bandicoot trap.

Another teacher from Hela Province, PNG, noted that the framework for making a wig as shown in Fig. 8.1b was similar to a parabola (Piru, 2005). This allowed Piru to encourage students to plot points and explore parabola on Cartesian coordinates. Piru emphasised the importance of accuracy in measuring so the wig was a snug fit for the dancer's head. They used a small unit of length called the *ki*, related to the width of the finger. (See a wig from another cultural group in Fig. 5.21c.)

Martin who speaks Magi (Mailu), Central Province, PNG, illustrated the connection between the traditional making of fishing nets from bush string and the tying of knots to form squares (Fig. 8.1c). He noted measurement of both length and area (in Chap. 5 we noted that the idea of an area unit was not common but there were examples in culture of area units that could be used). The squares could be larger for different fish. He also mentioned other links to geometry such as parallel and perpendicular lines, upper and lower limits (lines), cylinders, and two-dimensions to three-dimensions in making a net for the weights (mostly stones). Paraide (2010) discussed Indigenous knowledge in terms of currents, swells, and fishing weights for baskets and nets and links between culture and school mathematics.

In the five stones game (called knuckles in some countries) played in Finschaffen, Morobe Province, PNG, and many other areas of PNG, the stones are seen in a visuospatial arrangement by David (2007) although scattered. From the arrange-

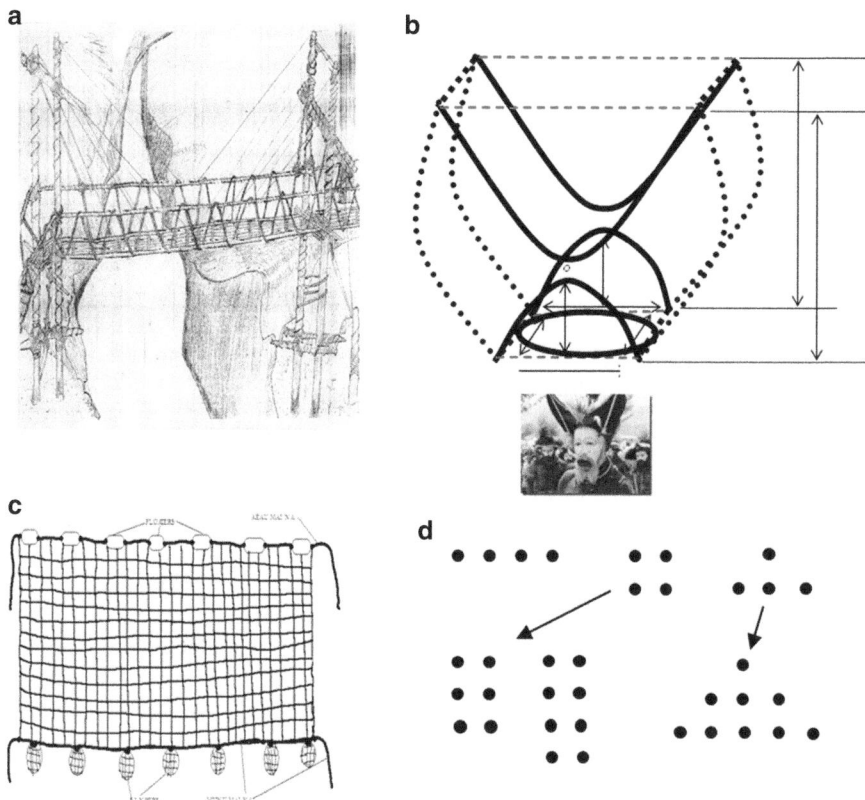

Fig. 8.1 Examples of visuospatial reasoning in ethnomathematics for secondary schools. (**a**) The vines from the handrail to the walking platform provides a metaphor for sine wave (Yambi, 2004). (**b**) The wig shape is outlined by the *parabolas* and *dotted lines*. The *head circle* is marked (Piru, 2005). (**c**) Fishing net ecocultural visuospatial representations (Martin, 2007). (**d**) Visuospatial arrangements in the five stone game: first line illustrates how stones are to be swept up (singles, twos, three); second line illustrates how rectangle and triangle numbers can be extended to give new patterns (David, 2007)

ment of the stones in pairs and a triple further number patterns can be generated (Fig. 8.1d). There can be a step-by-step increase in the size of the square (as well as rectangle) and a triangle formed. She provided questions such as:

1. What shapes can be formed with four stones?
2. What other mathematical properties can be formed with four stones?
3. When a square is formed. We can increase the size of this square by adding more stones. How many more extra stones do you add to increase the size of the square to its next biggest size?
4. What do you get after adding those extra stones in Question 3?

and she encouraged students to draw tables to see the patterns and then to form algebraic statements.

Figure 8.2 based on John's (2007) report illustrates how the teachers were linking ecocultural mathematics and visuospatial reasoning to school mathematics. Each of the points provided by John in this table was supported by example problems for the students. He chose the making of an Asaro mudmen mask (see Fig. 5.21a). While the syllabus and textbooks gave some examples of the topics related to typical PNG experiences, the project encouraged teachers to be creative and use visuospatial reasoning themselves but related to ecocultural contexts. As Kono (2007) said,

> this project only serves to give some clues to practicing or prospective mathematics teachers to be resourceful by incorporating cultural activities as concrete examples in teaching mathematics rather than abstract borrowed western ideas.

Julius (2007) is typical of the way in which the ecocultural aspects were integrated into their mathematical presentations (Chap. 5, Figs. 5.13c and 5.14a, c give background details). Figure 8.3 is from Julius' report. Further examples are found in Owens (2014).

These teachers were perhaps for the first time posing problems and finding solutions. They were really motivated to achieve a high standard over a period of time. Thus, as portrayed in Fig. 1.2 and discussed in Chap. 7, the context which is not only the culture and environment but also the project encouraged the teachers to provide a transition between cultural mathematics and school mathematics. The ecocultural context influenced the self-regulating learner which in turn developed their visuospatial reasoning, diagrams, and other representations. In turn, identity as a mathematical thinker was beginning to develop. The analytical, rote-learnt school working out was not always accurate in the examples but in the high majority of cases this was adequate and diagrams were frequent (Owens, 2014).

Furthermore, the ecocultural context was a realistic focus for learning. de Corte, Verschaffel, and Eyende (2000) showed the importance of structuring the problem solving in mathematics by encouraging heuristics such as drawing a diagram, planning, and checking progress with the problem. The ecocultural context for these teachers were real world requiring similar ways of problem solving but not in somewhat contrived questions but in the daily real lives of people in a subsistence culture closely related to neighbours and the land. Teachers spontaneously represented their visuospatial reasoning mostly through diagrams but also through description of dynamic action in story form providing their mental visuospatial imagery and reasoning.

Thus consideration of visuospatial reasoning from an ecocultural perspective is important for extending our understanding of the self-regulating learner. These projects illustrated how self-regulation was more than metacognition but also realistic, involving, requiring guidance in structuring the problem context and intrinsically motivating as cultural connections for mathematics became evident.

Stage of Cultural Activity	Mathematical Ideas/topics Derived from the Cultural Activity	

Clay and water	* Ratio and Rates
	- Ratio
	   Simplifying
	   Using proportion
	- Rates
	   Time and rates
	   Rates graph
		* Circle and Volume
		- Circle
		   Circumference
		   Area of circle
		- Volume (prisms)
		   Volume of cylinder
		*Surface Area

Semi-oval bamboo skeletal shape		* Measurement
		   Length
		* Circle
		   Circumference
		   Area of circle
		* Geometry
		   Shapes

Semi-oval clay head		* Measurement
		- Area
		   Length
		   Width
		* Circle and Volume
		- Circle
		   Circumference
		   Area of circle
		   Surface Area of solids
		- Volume (prisms)
		   Volume of cylinder

Face of the mask		* Symmetry and Construction
		- Symmetry
		   Line symmetry
		   Parallel lines
		* Geometry
		- Construction
		   Bisecting a straight line
		   Angle of 90°
		   Perpendicular from a point on a straight line.

Drying in sun (small dots) or heated house (larger square dots)		*Measurement
		   Time
		*Rates
		   Time and rates
		   Rates graph

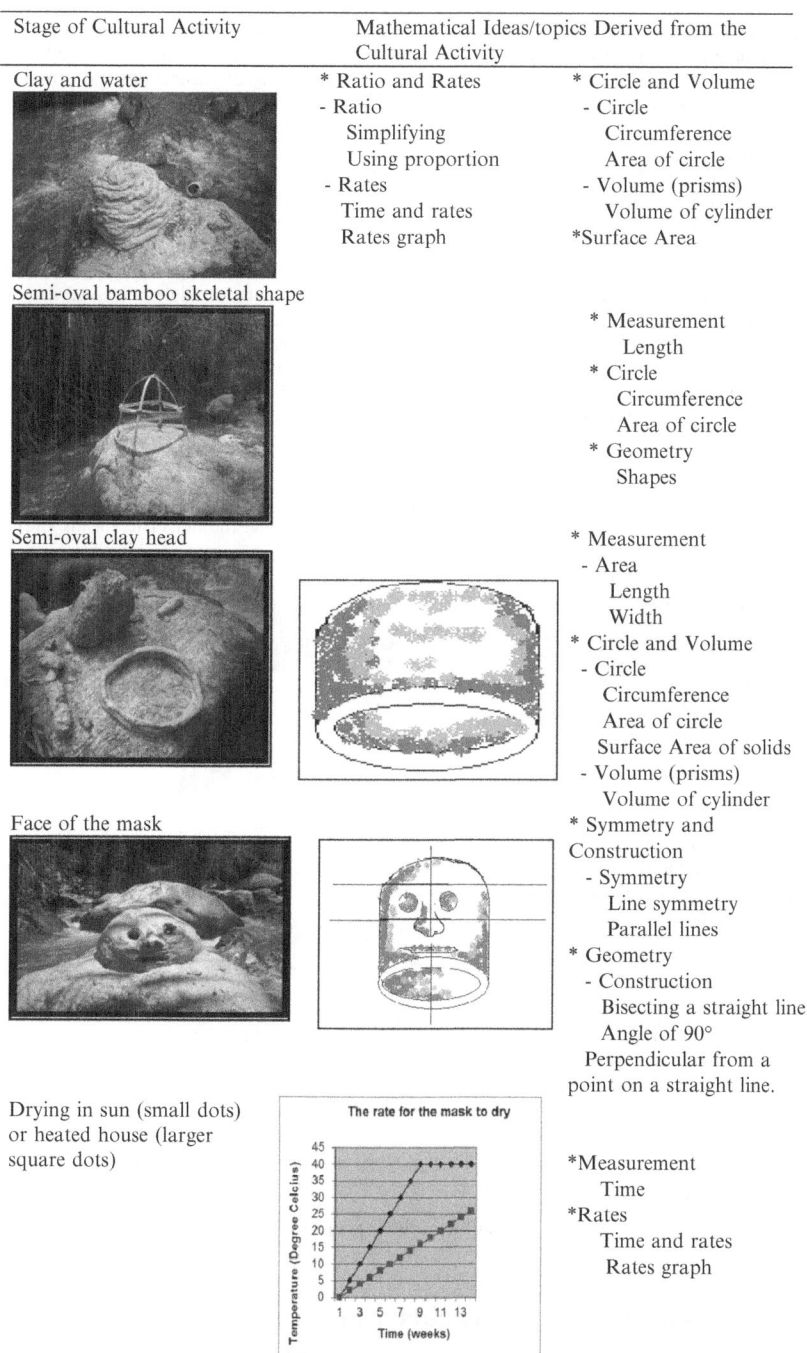

Fig. 8.2 Links between ecocultural mathematics and school mathematics (John, 2007)

This project confronts the idea of Eurocentrism-i.e. the widespread prejudice about mathematics as being predominantly of European origin, which can be used and manipulated as deemed best according to their interpretation. This write-up gives a light by way of illustrations and proofs using cultural designs of pitpit wall, as well as establishing a common understanding of 'cultural conceptualization' in the context of teaching in classroom. ... Within the diversity of rich cultural heritage is embedded the complex and varied sets of mathematical knowledge that, I would say, are not fully exploited to this very day. ... We look at these mystique (sic) geometric forms and patterns of such traditional activities and posed the question; why do these materials or products possess the form they have?' We are all part and puzzle (sic) of these cultural activities and it reflects some sort of mathematical knowledge, experience and wisdom. ... I will use the designs of pitpit wall pattern as illustration to determine the sum of the interior angle of polygons. Special reference will be given to exterior angle of a triangle and sum of interior angle of a quadrilateral. ... Traditionally, these designs are performed by skilled and experienced person(s).There are, in fact, eight different types of designs identified in the Sinasina area namely; X-shape, rending, zigzag, modified, basic pentagonal, hexagonal shape, starry, and octagonal shape. I will use the last four patterns as examples to determine interior and exterior angles of polygons (i.e. triangle & quadrilateral), as well as showing the proof of the formula of the angle sum of polygons. ... traditionally termed; *kewah, egleh, gamlageh, bongeh*, for starry, pentagonal, hexagonal and octagonal shapes, respectively.

a) If we consider the starry shape it seems to look like a quadrilateral. By drawing lines on the edges, as illustrated below, we would come up with the kite

* Starric (sic) pattern from traditional Sinasina *'kewah'*

c) Either looking at the small patterns inside the design or the sides derive the octagonal shape

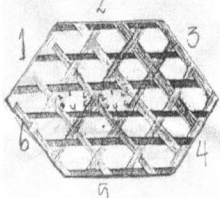

'Gamlageh'
* Hexagonal shape from Sinasina

b) Inscribed in the basic pentagonal weave are three shapes, namely triangle, rhombus (sic, means kite) and pentagon

* Basic pentagonal weave from Sinasina *'egleh'*

d) If we count the sides of the pattern inscribed in this pattern we would identify eight sides.

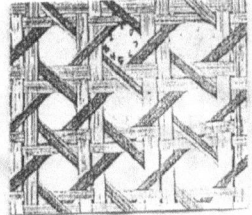

'Bongeh'
* Octagonal shape from traditional Sinasina

Fig. 8.3 Tabare weaving shapes from Julius' (2007) project

PNG Elementary School Teacher Education Study

Having noted the impact on the secondary teachers through this project, it was important to see how elementary school teachers could incorporate an ecocultural perspective to mathematics and how that might impact on their visuospatial reasoning. Elementary schools are built and maintained by the community. Teachers are grade 10 graduates who speak the language of the community. There is a transition from local language to English during elementary schools. Figure 8.4 shows typical rural schools in PNG. In 2013–2014, a design-based study in PNG, has elaborated several principles for teachers that can be applied across PNG's 850 cultures and languages and ecologies, remoteness, and experience. It applies no matter what the use of Tok Pisin or the use of Tok Ples provides. Although Muke (Muke & Clarkson, 2011) illustrated that teachers will use all three languages including English to explain concepts in class (see also Setati & Adler, 2000), there was a much more complex picture emerging in practice. "Cultural Mathematics" required other principles in teachers' knowledge.

Figure 8.5 provides the model of principles used to guide workshops. After the first workshop, it was found the learning experiences appropriate for children to learn mathematics needed to be expounded further so the middle principle was elaborated particularly based on research in mathematics education but also incorporating what we had learnt from our studies of PNG cultural ways of thinking mathematically. The principle on the nature of mathematical thinking deliberately moved away from learning facts and is inclusive of PNG mathematics. We were able to supply analyses of language from previous research. Early childhood principles needed enunciated as many teachers were teaching as they had been taught in

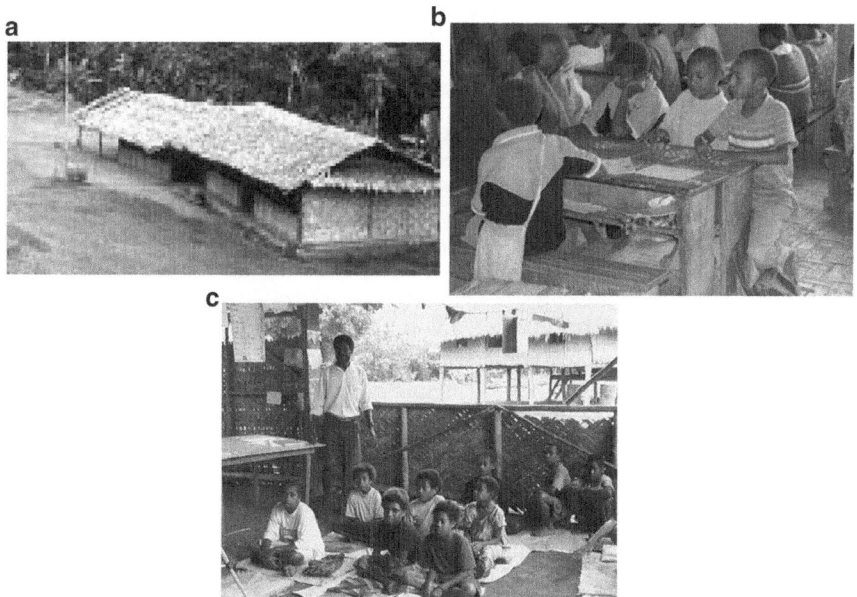

Fig. 8.4 Schools in Papua New Guinea. (**a**) Typical bush material school, Morobe, PNG, ~ 1984. (**b**) Group work in Tsigimil Primary School grade 3. (**c**) Elementary school, Atzera language, Binimap, Morobe, 2006

Mathematical ways of thinking	Activities appropriate for children	Language resources
• Links to school mathematics; • Mathematics is thinking e.g. 　· problem solving 　· Reasoning 　· Fluently applying concepts and procedures 　· Patterning and abstracting relationships	• Learning experiences 'in but outside' school; • Early childhood emphasis: play and inquiry	• Rich diversity • Patterns in counting • Gestures • Decision-making • Grammar and meaning

	Learning experiences to promote children's efficient mathematical thinking	Cultural activities
Early mathematical thinking • Patterning • Sorting, • Ordering • Comparing • One-to-one matching • Symmetry • Recognising equality • Noticing attributes of groups and shapes • Trialling ideas • Posing questions	• Counting • See and know • Visualising • Recognising pairs and groups • Describing and classifying • Arithmetic in language • Group counting • Measuring informally • Locating • Explaining • Investigating • Enjoying challenges **What they learn?** Assess, report, plan	Extend cultural mathematics • patterns, • groupings, • arithmetic inside counting, • representation, • measuring, • ratio, • spatial relations, • traversing the land and sea • language for location, • space and place • constructing and designing • building relationships

Cultural capacity and partnerships
- Values: need to preserve cultures and languages of people
- Elders roles for community cohesion

Fig. 8.5 Design of key principles for teacher professional development in *Cultural Mathematics*

primary school and few assessed their children's learning. Finally, the role of Elders in the school was yet to be realised and applied to *Cultural Mathematics*.

A manual outlined all of these principles together with an inquiry model of planning and examples. A stand-alone website was developed with the manual information; example lessons; video examples of cultural activities, children learning and assessment tasks, and teachers' sharing; and other workshop ideas. This was loaded onto touchscreen computers where possible and these effectively engaged the teachers. The workshop began with a welcome activity setting the scene about mathematics and talking mathematics. Then Elders from a similar ecology were shown explaining their cultural activity and mathematics (see Chap. 5 for examples). This stimulated teachers in small groups to discuss some of their own cultural activities and tease out the mathematics involved. The systematic ways of thinking were supported by activity and discussion but systematic ways tend to be visuospatial reasoning about measurement, space, and geometry.

> **Weekly Learning Plan for Mathematics**
>
> **Purpose:** Children are expected to think and do mathematics through activities linked to cultural practices. Children are expected to have a sense of belonging with the new ideas in culture and school through a good transition that links cultural ways of thinking with school ways of thinking.
> **Key Ideas:** e.g. What is the new pattern and relationship? How does the thinking lead to problem solving?
> **Prior knowledge:** What do they know? How do they think and feel?
> **Resources:** Places to visit; materials for exploring, com paring, measuring, recording, modelling; game cards, spinners, Elders who know the cultural activity
> **Assessment:** Observing ways children try things, what they say, how they problem solve, what they write, what they ask to make clear or to extend their exploring
>
> Day 1
> - **Tuning In**
>   ' Children are motivated, have real world experience e.g. outdoor; listen and participate in a story.
> - **Planning to find out**
> - **Finding Out**
>   ' Children observe, notice, compare, measure, discuss mathematical patterns
>
> Day 2
> - **Sorting Out**
>   ' Children discuss, model, compare, make a table, draw a diagram, find same and different,
>
> Day 3
> - **Going Further**
>   ' Children apply to other numbers or another situation, read and discuss the maths book, use symbols, play a game, solve an open problem,
>
> Day 4
> - **Making Connections**
>   ' Children summarise the mathematics, whole class discussion, or story writing,
> - **Taking Action**
>   ' Share at home, solve a real problem.
>
> Day 5
> - **Sharing, discussing, and reflecting**
>   ' Children explain the mathematics, write a maths story, write their own summary, say what new mathematics they have learnt.
>   ' Teacher reviews and decides what to teach next based on assessment, cultural activities, and syllabus guides.

Fig. 8.6 Design of inquiry method for *Cultural Mathematics* based on Murdoch (1998)

Cultures are quite different across PNG and we deliberately targeted three ecologies for the research: highlands, coastal, and inland in coastal provinces. Starting with culture was a clear way of engaging teachers who are proud of their cultures. The workshop then covered good teaching approaches for young children, what is known from research about how children best learn about arithmetic, measurement, and space and the key concepts in each. A key for bringing together each of these principles in planning learning experiences was the use of an inquiry method. To make it easy, it was suggested that the topic was covered over a week. Although days were allocated to each step, it was made clear that this need not be strictly followed. The inquiry method based on Murdoch (1998) and applied to *Cultural Mathematics* is shown in Fig. 8.6.

Teachers were given example learning plans (see Appendix F for an example) and early readers that related to mathematics topics. Murdoch's steps proved to be a bonus for teachers to bridge the gap between culture and school mathematics. They were able to use it to extend their children's thinking mathematically and to encourage practice of concepts and skills. Since the cultures are rich in visual examples these were linked to research on early arithmetic, measurement, and geometry (Moschkovich, 1996).

Some examples of the teachers preparing learning experiences will help to explain the importance of the visuospatial reasoning within the mathematical thinking of the teachers. The first three examples come from the coastal village of Tubusereia, National Capital District, PNG where Motu is the home language, often with English. In the first example, the teachers of Elementary 2 were referring to the cultural practice of sharing fish. They set up the task of sharing 24 fish between four families (Fig. 8.7e). They knew each family would get fish according to family size. However, their school mathematics at first made them decide it had to be equal shares like division. Through discussion, teachers realised that an open-ended problem that was actually more like reality, was to decide all different ways of sharing the fish. The children responded well to deciding numbers and how they could find other ways of sharing the fish. They began to systematically record. They practiced how to add numbers. All the time, the children were either using the fish they had made from cardboard to explain to each other or they were using the empty number line (which was new to them) to jump to the next number being added on and then decide on the last number to make 24. Note that in practice, the sharing often starts with the larger fish to each family, then the smaller ones (Odobu, 2007). Children used various strategies to make up new sets of numbers such as realising that if they gave 10 and 6 fish, then they needed another 2 numbers that added to 8.

In another class, Elementary 1, the teacher took the tears tattoo, 2, 1, 2 pattern modelled it with steps (large)—2 forward, turn left, 1 forward, turn right, 2 forward; with strips of paper; and with dots on a sheet of paper. The ratio impact was evident to the children for the various activities using the same pattern. The results of visuospatial reasoning of Elementary 2 children who copied a design are shown in Chap. 5, Fig. 5.4. Another teacher asked Elders to make a model garden for yam expecting it to be in rows of two mounds but instead the Elders used the triangular pattern (Fig. 5.6). Nevertheless, the children in Elementary Prep not only asked questions of the Elders (the teacher had prompted children by giving some example questions) but also showed how the centres of the mounds were equidistant using a stick. They also compared the lengths of various yams and put them in order. They showed different ways of comparing and measuring. The open question approach to teaching reduced copying from the board and meant children were using visuospatial reasoning more and more. Questions asked of children after the lessons indicated that they understood well the meaning of half as big again and how to measure informally and to use a smaller unit to measure the remaining length and to give more than one pair of numbers that added to 13, explaining how they reasoned mentally to get another answer. Gestures indicated that they had used the idea, for example, of taking away one visually from one number and adding one to the other number.

In another area where teachers came from a number of different language groups in the mountains and along the Rai Coast of Madang Province, PNG, one group of teachers visually considered how they planted dry land rice and demonstrated using feet lengths (Fig. 8.7c). Then they went further discussing other measuring units for

Fig. 8.7 Bringing ecocultural experiences into the elementary classroom. (**a**) Preparing a dictionary of mathematical terms. (**b**) Trying out the lesson idea of using the body to make shapes following the squares (*diamond*) in the weaving. (**c**) Simulating using steps to measure when planting rice. (**d**) Sharing the making of rope for a bilum and making a bilum relating it to mathematics. (**e**) Sharing fish between families according to need. Recording of children's solutions explaining how they worked from one addition to another with visual supports

length, with both straight lines and curves. Another group first wove patterns including the *diamond* pattern. In going further, while sitting on the ground with crossed legs, they made diamonds with their knees at right angles and used their feet with the man opposite to make another smaller square. Then they discussed the properties of the squares. Other groups of teachers made model houses, models of different kinds of fences, and models of bilums (Fig. 8.7d) but each time discussing the equal lengths of fence posts or different heights of posts for the house for the roof slope, making the right angles, and using the radius to form a circle. One discussion was about the height of the bow and arrows and how to compare them and people's heights, and how to informally measure them. Each time, key ideas about measuring were being emphasised within the activity. Teachers were getting small groups to represent and discuss the key ideas based on their visuospatial representational and mental reasoning. Much of this work was being said in their Tok Ples which was translated into Tok Pisin for me or spoken in Tok Pisin.

The significance of this approach to teaching mathematics is the impact that the design principles have on establishing both teachers and children as mathematical thinkers and identifying not only with cultural mathematics but also school mathematics. The ecocultural mathematical context (see Fig. 1.2) is evident in the mathematical problems, models of cultural practice, cultural tools of measuring and drawing, and valuing cultural practice. The cultural ways of thinking especially about ratio and pattern and size are not being lost. In terms of the model of identity, we see how the visuospatial reasoning within the more open questioning classroom was allowing children to be self-regulated, an important aspect in their mathematical identity but also to use a range of cognitive and affective aspects of learning as indicated also by the model in Fig. 2.17. Teachers and children were goal setting, reasoning (including visuospatial reasoning), planning, self-evaluating by explaining to their peers, reviewing by reflection on the learning, using drawn and mental models, and working out how to present results or their thinking. There was a strong sense of ownership. Furthermore the social interactions and having a go at responding (instead of copying) were relatively new experiences. Thus we can see the beginnings of not only a cultural mathematical identity developing but one related to school. In the case of the teachers they were engaging in mathematical thinking and engaging others while some of the children were doing the same in explaining to peers.

In the workshops, after preparing and where possible trying out their planned learning experiences, the teachers discussed the issues of language and emphasised how language can have real treasures that can be used for explaining the mathematical concepts to children. Teachers considered a number of English words used in mathematics to try to decide on some appropriate Tok Ples words. This was not an easy task and again visuospatial reasoning was used to help make decisions (Fig. 8.6a). Finally we looked at assessment and learning stories. We also introduced a reflective teacher's questionnaire and an interview schedule that teachers could use with a couple of children in their own class. They practiced it on each other and then children. This interview schedule reinforced much of what was being discussed throughout the workshop. Visualisation strategies were assessed as they were in Count Me In Too Schedule for Early Numeracy Assessment (NSW Department of Education and Training, 1998) and Count Me into Space (NSW Department of Education and Training Curriculum Support and Development, 2000). Matang (Matang & Owens, 2014) had modified some of the former in his study and we used simpler and shorter schedules than Matang's but asked some questions on position, shapes, and measurement and questions specifically linked to PNG cultural practices.

Working with Mental and Physical Visuospatial Representations in Africa

For cultural weaving activities to be introduced into school, weaving boards of card with slits and loose cardboard strips were used by Cherinda (2001, 2002). Initially students were asked to copy the pattern and continue it. Children could feel, both tactually and mentally, how twill weaving developed from a cultural experience into visuospatial reasoning related to school mathematics. Questions, such as the following, have been used to stimulate their thinking about weaving designs such as in Fig. 8.8b.

Observe the path of strip L1 (first horizontal *loose* strip). What is the number of the next L-strip with the same path as the path of L1?

Then, the path for L6 is the same as for L2. Consider the set L1–L4 as section A, and the set L5–L8 as section B. What do you say about the appearance of the two sections?

By the already verified repetition of the paths of the L-strips, students can see that the two sections, A and B, are the same. The section B appears as an image of section A when this section is moved down. In such cases the "movement" from A to B (or vice-versa) is called a translation. Students are then encouraged to weave another pattern that shows vertical symmetry. For example, Fig. 8.8b, d makes an attractive design with vertical symmetry or two axial symmetries.

Students are then encouraged to repeat the set which produces a square design (Fig. 8.8b finished) and they can see both axial symmetries through the centre and by turning the board they can note there is a coloured square in each corner. This provides an example of rotational symmetry. For the design in Fig. 8.8d, with a horizontal line of symmetry, students were asked to reflect the design in the weaving. The diagonal pattern in Fig. 8.8c is continued to provide a rotational symmetry design. Each diagonal will have three squares with the zigzag in between. The board can be rotated 180° to show the rotational symmetry.

Then students are asked to find the fundamental block or unit for the weaving and this is then discussed in terms of squares, for example the four columns by five lines in Fig. 8.8d. Then other fundamental blocks are provided and these are used to create intricate repeated designs.

Fig. 8.8 Weaving in Mozambique to encourage visuospatial reasoning in geometry. (**a**) Student motivated to learn in Mozambique. (**b**) Translation experience. (**c**) Repeated to produce rotational symmetry. (**d**) Weaving a reflection pattern

One important finding from Cherinda's work is that weaving boards (WB) provided students with interaction between various representations of concepts. Furthermore, they were able to express their thinking.

> The fact that the subjects have manipulated the WB to certain extent and then continued thinking without it, at least physically (using numbers only, or reproducing woven pattern on squared paper to facilitate the reasoning) reveal the different representational systems that the learners used in attempting the least complex way to acquire and develop mathematical knowledge. (Cherinda, 2012, p. 942)

Cherinda (2002) found that students were motivated to learn (Fig. 8.7a) and the active problem solving created this interaction between representations encouraging mental imagery and reasoning. Both these aspects were found by myself in PNG (Owens, 1999b, 2012a) and Australia (see Chap. 2; Owens, 1993).

Teachers need to be shown how to carry out and incorporate the weaving activities into mathematics. The students did not always copy as expected and needed to be specifically directed to observe the repeated lines. Similarly for students to see the symmetry patterns specific questions were required. Cherinda encouraged the students to draw up number patterns, presenting the line associated with each repeated set of lines. In higher years of school, more complex examples were used. For example, they used two blue vertical colours followed by two yellow vertical strips on the weaving board to make much more complex patterns such as the herringbone pattern.

Mental Mapping of the Navajo, USA

In Chap. 6, we looked at the Navajo's visuospatial reasoning and it is appropriate to note the importance of considering ethnomathematics as the overarching mathematics in which academic mathematics is a part (Owens, 2013a). What works might become just as important as formal proof. There is no reason to doubt what Pinxten and François (2011) suggest could occur for any of the mathematical knowledges in "multimathemacy". Building on Pinxten's earlier work (Pinxten et al., 1983; Pinxten, van Dooren, & Soberon, 1987), their example was that:

> Chee's [Navajo youth] mental mapping and measuring be recognized and generalized to make it a powerful nonwestern geometry which has possibly more potential than we ever guessed, because it would offer problems and solutions about movement through space while starting from an intrinsically dynamic spatial understanding which is typically 'Navajo Indian' and hence beyond the normal scope of the western academic mathematician. (Pinxten & François, 2011)

Quotations and discussions in Chap. 6 justify the need for such comments to be illuminary in mathematics education.

In Chap. 2, mental mapping was raised in discussing its role in attention, intention, and responsiveness. Ecocultural context has a prime role in mental mapping and should not be just regarded as background as other authors have suggested. Language as mentioned in Chap. 4 is a significant part of the ecocultural perspective. For example, as mentioned in Chap. 4, in many PNG cultures (Capell, 1969; Matang & Owens, 2014) and Navajo (Pinxten et al., 1983), the emphasis is on verbs and not nouns so objects, parts, and wholes are not as significant as verbs related to moving rather than being.

> The child pictures the landscape by means of 'significant' rocks (i.e., with particular shapes), the adjacency of dips and water sources, the movements of the sun throughout the day, eventual greenery at different places, the entry of the hogan one finds along the path (i.e., the door of the hogan or dwelling place will always point to the east), the changes of color of the air throughout the day (going from white in the morning, over blue and yellow to black), and so on. All of these elements co-define distance in the Navajo view.
>
> ...we start from the preschool and outside of school knowledge, make the concepts and intuitions explicit and label them in the native language. E.g., a line is conceptualized as the result of the uninterrupted movement of the child through a landscape. It corresponds more

to a path and is closely tied to the moving body. Or it can be understood as the result of two movements in a plane, one rock plateau spreading towards a different moving progression of what we call a river. Where the two movements meet we speak about a line. When looking at the world of experience of the Navajo child, one notes that 'a river' is not a continuous thing … we can then go on and label 'line' in Navajo as a particular movement. Next, we explore the characteristics of the line explicitly and only then move to the lemmas and problems defined in the Euclidean geometry system. In this way, multimathemacy is shaped in the classroom practice. (François, Pinxten, & Mesquita, 2013. p. 30–31)

Thus we find language and visuospatial reasoning are significant in an ecocultural perspective. Curricula that take account of the ecocultural context encourage a sense of ownership but also provide for Indigenous knowledge processes to be used in reasoning especially visuospatially, planning, reviewing, and structuring the learning environment. Thus learner mathematical identity is encouraged (Fig. 1.2).

Yup'ik Mathematics Education

Like in PNG, there are many cultural practices that use proportional reasoning and some common body measurements that are used in a range of activities. However, a first point should be made about realising how Elders who make garments are able to visually consider a person and then cut out a garment that will fit using mental visualisation and reasoning based on experience. In another activity during the fishing period a net of fish is loaded into wheelbarrows and taken to the drying racks providing typical volume units and proportional reasoning. People can estimate the number required and then the amount of fish on the rack becomes a volume unit for filling the smokehouse. Finally another volume unit is used to distribute the fish to families. Berries are picked into specific containers, amalgamated and then placed in plastic bags for the freezer providing further volume units and proportional reasoning between quantities.

In terms of measures of length, the various body parts are used for the kayak, for building the fish racks, and even for houses. A fish rack is five half-fathoms by four elbow to elbow, hands clenched at the centre of body. Children build these to illustrate measurement units, how to measure and how the rack is utilised. Interestingly, the various body measures allow for a range of equivalent fractions as two of the smaller units equals the next larger unit. One unit equals three of the double handspan—little finger knuckle to little finger knuckle, thumbs touching naturally. Measures are also used in determining heights of stars, direction for travel from the star position, and estimated distances by angle of movement across the sky to signify an hour of time passing and distance likely to be travelled. Part of the reasoning relates to the fact that body parts are always available and objects fit the person better if their own body measurements are used. An important point is that children who undertook these cultural mathematics units did significantly better on a test of mathematics compared to the comparable control group (Lipka, Mohatt, & The Ciulistet Group 1998).

Thoughts on an Ecocultural Curriculum

These studies suggest that an alternative approach to curriculum for mathematics education in geometry and measurement might be needed. It is not easy to provide teachers with a curriculum that emphasises processes (see the UK move away from the curriculum *Man: A course of study* in the 1980s). However, early childhood education in mathematics often emphasises processes such as sorting, comparing, matching, one-to-one correspondence, describing, drawing, and ordering. These are processes so it is not unexpected that a curriculum that notes the mathematical processes of an Indigenous group would also have such a list.

Comparing as measure: many of the cultural activities of PNG use comparing as a skill, generally as a visual skill supported by group discussions. Hence the curriculum should contain a development of this skill. Comparing lengths is a beginning first by sight and then by culturally used units such as steps and hand-spans. This would develop for comparing large areas such as gardens or floor space. This might include the number of mounds or people that could fit into the area as an estimate prior to considering in terms of an informal area unit. Volume would also come as a comparison of items such as food like yams and piles of food. Small baskets or wraps of food could be used as informal volume units to supplement the capacity units such as pots or bamboo lengths.

Comparing as ratio: many activities compare two sizes at once. It might be the giving of packets of biscuits or bundle of sago to each child and the total number all together. It might be the floor of one house and the floor of another (half as much again), it might be noting how big a garden will need to be for a bigger family, how large an area of kunai or how large a sago plant to obtain roofing material for two different-sized houses. Comparing how many people can sit on a number of mats compared to another number of mats or how many people are provided for by one basket or bundle of food compared to several baskets or bundles. These can form ready reckoners for the idea of ratio or comparison. In school mathematics these will lead to multiplication and fractions and to ratio including trigonometry ratio but that is much later in school. The important point is to keep these ratio ideas as a strength for learning.

Ordering: many things might be ordered. Size or number may be ordered. Order might represent status. Position in relation to other objects or people in space might also be denoted. Order is often associated with value and this could be in terms of gifts or money or for other relationships. Order might not just be linear in the complex web of family relations. Time can also be ordered.

Sorting for classifying: In geometry, we name shapes with certain properties and we relate shapes according to properties. We analyse problems using these properties and derived information. In schools, these tend to be around Euclidean shapes predominantly 2D shapes. Sorting designs and shapes and objects according to their properties can be around curved designs, related either according to common use or symbolism use in creating a complex design, or by the way they are created. Hence different types of lines, particularly curves, and shapes or designs can be connected to each other. By incorporating shapes other than Euclidean shapes, the curriculum can maintain the richness of traditional understanding of lines and shapes. Classifying is also evident in language such as classifiers for counting. Objects can be grouped and the shapes discussed in both Tok Ples and English.

One-to-one correspondence: There are communities where counting is not carried out, but there may be both one-to-one and one-to-many matches as well as the use of suffixes to indicate number. It is also apparent in body part tally systems. Corresponding matching is associated with distribution, providing an introduction to sharing and division (often by pairs) as well as multiples for groups. It might be a different approach to introduce numbers.

Brazilian Initiatives

Rosa and Orey (2012) suggest that ethnomodelling is an important pedagogy for acknowledging the cultural origins of mathematics for different groups of students. Ethnomodelling encourages students to find out how a particular practice is carried out in the community and then to summarise it in a sequence that provides a mathematical ethnomodel. In Brazil, a number of studies have considered the mathematics of different cultural groups. For the areas involving the Landless Movement, the students can consider the methods for calculating area of irregular quadrilaterals by multiplying the averages of the opposite sides (Knijnik, 2002) or by squaring the result of dividing the sum of the sides by 4 (Flemming, Flemming Luz, & de Mello 2005) cited in Rosa and Orey (2012). In areas where Italian immigrants brought wine making or the German immigrants brought strong European mathematical ways of thinking, alternative ethnomodels could be likely for the students based on their ecocultural context. Rosa and Orey saliently reminded their readers that it is important to recognise the changing nature of cultures in undertaking ethnomodelling.

However, the language and complexity of the classroom illustrates how power dominates the situation in terms of language available to students and the ways in which ethnomodelling is expressed in language (Knijnik, 2002). Nevertheless, by focussing on visuospatial reasoning, teachers like the researchers

> explore the idea of miniature cycles of learning actions to focus on the mathematical learning that is taking place. We describe the dynamics and the complexity of the ongoing activity in the calculation of areas; and, how drawings form a part, and show their influence, in it. We argue that part of this influence was associated with the contradiction between abstract mathematical ideas and their empirical representations, revealed by the tensions perceived in the activities analysed; and, simultaneously, that we could see as an impelling force for the learning of the rules and norms which regulate the use of visual representations in school mathematics. (David & Tomaz, 2012)

The small cycles of learning in the classroom incident are closely linked to the cycle in Chap. 2 that arose from my study in primary schools in Australia (Owens, 1993) but also formed the basis of the establishing of a mathematical learning identity that takes account of the ecocultural context summarised in Chap. 1.

Language and Inquiry for Visuospatial Reasoning in Geometry: Mauritius

This research work was carried out with upper primary level pupils (fourth and fifth graders) from four primary schools in Mauritius. To involve schools of different categories in the study, one high-performing school (School 1 with 213 students), two average-performing schools (Schools 2 and 3 with 367 students), and one low-performing school (School 4 with 165 students) were randomly selected. A quasi-experimental design was selected to test the acquisition of geometric skills (visual, logical, applied, drawing, and verbal) after experimental teaching with inquiry-based teaching with the use of manipulatives and/or local language Creole. The students from each school were classified into four groups: Group1 (traditional teaching with textbook and English), Group 2 (traditional teaching with textbook and Creole), Group 3 (inquiry-based teaching with manipulatives and English), and Group 4 (inquiry-based teaching with manipulatives and Creole). After the experimental teaching, a posttest in the form of a multiple-choice paper-and-pencil questionnaire (MCQ) was collectively administered in all the groups for both fourth and fifth graders. It contained 31 multiple-choice items and there were 4 options for every item and only 1 response was correct. The same posttest was again conducted after 6 or 7 weeks as a retention test. The aim was to measure how well pupils from different groups were able to perform on the MCQ geometry questionnaire and how their performances were affected with the passage of time.

It might be argued that direct drill and practice of the names of the 2D shapes and their properties with visual representation and practice exercises might be more effective at the recognition level and even at the analysis level for each of the geometric skills. However, this research study has shown that the use of inquiry-based methods when combined with manipulative materials encouraged the pupils to become more adept at using the higher order thinking skill of analysis and to acquire the geometric skills with the additional use of Creole as language of instruction. Visual skill at both recognition and analysis levels were significantly influenced by the use of inquiry-based methods with manipulative materials. This shows that acquisition of visual skill at the recognition level was generally within the ability of these primary level pupils. In addition, the use of inquiry-based teaching with manipulatives in Creole had promoted the acquisition of visual skill at the analysis level (basic property analysis of van Hiele level 2). In particular, in four of the eight items in the MCQ questionnaire where there were high success rates (greater than 67.4 %), the pupils from group 4 outperformed pupils from groups 1 and 2. However, the other four items were difficult tasks with low average success rates of 43.3 % in the posttest and 42.9 % in the retention test. In the retention test only, the pupils irrespective of the teaching strategies used, had consistent poor performances in these four items whereas a few significant differences in the posttest were favouring pupils from groups 3 and 4. It must be borne in mind that visual skill is crucial in the acquisition of geometric skills and also helps in the understanding of other abstract mathematical concepts and thinking (Clements, Battista, & Sarama 1998;

Jones, 1998; Lean & Clements, 1981; Wheatley & Cobb, 1990). It is the visuospatial skill that helps pupils to acquire the "geometrical eye" essential in mathematics as stated by Godfrey (1910; Jones & Mooney, 2003). Hence, the acquisition of visuospatial skill is important and the use of inquiry-based teaching methods with manipulatives and Creole is significantly helping in achieving this skill at both recognition and analysis levels.

Logical skill at recognition level is mainly about realising that there are differences and similarities among figures and that figures conserve their shapes in various positions. There were hardly any significant discrepancies in the performances of the pupils from the four groups. Logical skill at analysis level is mainly about using properties to distinguish figures and understand that figures can be classified into different types. Despite the low success rates in both posttest and retention test, pupils from group 4 had performed significantly better.

Applied skill at recognition level is mainly about identifying geometric shapes in physical shapes. There were a few significant discrepancies which mostly favoured pupils from groups 3 and 4. Applied skill at analysis level is mainly concerned with the recognition of geometric properties of physical objects and there was only one item in MCQ questionnaire testing the acquisition of applied skill at analysis level. Despite the low success rates in both posttest and retention test, pupils from group 4 performed significantly better.

Drawing skill is an essential skill required for the acquisition of geometric thinking and it is a crucial part of the geometry curriculum. Its acquisition at the recognition level requires the making of sketches of figures accurately labelling given parts and its acquisition at analysis level requires the translation of given verbal information into picture and the use of given properties of figures to draw or construct the figures. The drawing skill requires a significant amount of visuospatial reasoning. There were more pupils who acquired van Hiele level 2 skill in drawing the equilateral and isosceles triangles than the two quadrilaterals: parallelogram and rhombus in both posttest and retention test. Thus, drawing triangles was easier than the quadrilaterals for the primary-level pupils. Concerning group-wise performances, no significant difference was observed in the performances of the pupils in the drawing of the parallelogram, isosceles triangle, square, and rectangle. However, pupils from group 4 performed significantly better than pupils from the other three groups in drawing the rhombus and equilateral triangle. Consequently, the use of different teaching strategies had little impact on the acquisition of drawing skill. Using isosceles right-angled triangles to construct quadrilaterals seemed to be a relatively difficult task for the pupils (success rate below 52.8 %) and in general, an overall average percentage of 21.5 % had acquired van Hiele level 1 skill and 18.9 % had acquired van Hiele level 2 skill in manipulating the triangles to construct the quadrilaterals. The use of the different teaching strategies had little impact on improving drawing skill. It is observed that the teaching strategies had hardly influenced the acquisition of drawing skill in both the high-scoring and low-scoring items. Perhaps, the applied and practical nature of the items especially the use of identical triangles to construct quadrilaterals had equal impact on the pupils.

It might have been thought that the use of English in the textbook, in class for instruction, and in the test may have assisted students with their verbal skill. However, for the seven items in the questionnaire on verbal skill, it was found that overall the success rate in all groups in both the posttest and retention test were low. The item which required completing a half-drawn rhombus with a line of symmetry needed both verbal and drawing skills and the average success rates were 66.1 % in the posttest and 66.2 % in the retention test. Definitely the given diagram had helped all the pupils in their attempts. Otherwise for the other six items requiring only verbal skill, the success rates were low in both posttest and retention test (success rates 26.1–57.0 %). Generally, the pupils from groups 3 and 4 significantly outperformed their counterparts from groups 1 and 2 in all the seven items in the posttest and in three items in the retention test. Table 8.1 shows results. Writing the properties of some shapes in their own words proved to be a difficult task for the primary level pupils in both posttest and retention test and the low percentages of success confirmed the fact. Concerning group-wise performances, generally pupils from group 4 performed significantly better than the other groups mainly groups 1 and 2 in the tasks in both posttest and retention test. Thus, despite the low success rates, pupils taught using inquiry-based methods with manipulatives had acquired verbal skill at analysis level significantly better than the others. The use of Creole had further improved the acquisition.

The use of the local language can act as a bridge to learning geometry. As mentioned by Khisty (1995), native language is a resource for learning because pupils are more successful when they continue to develop their native language skills

Table 8.1 Performance of pupils in items requiring verbal skill

		Posttest		Retention test	
		Group 4 significantly better than groups listed[a]	Percentage of pupils acquiring the required skill in group 4	Group 4 significantly better than groups listed[a]	Percentage of pupils acquiring the required skill in group 4
1	Properties of sides of an isosceles triangle	1, 2	62.3	1, 2	56.6
2	Identifying a statement which was not true about a rectangle	1, 2	48.7	1, 2	47.5
3	Properties of a square	1, 2	74.7	1, 2	66.9
4	Completing a half-drawn rhombus following a given statement about its symmetry	1, 2	77.4	2	71.9
5	Properties of a rhombus	1, 2	35.5	1, 2	40.1
6	Identifying a statement which was not true about a parallelogram	1, 2	44.2	1, 2	43.8
7	Comparing an angle with a right angle	1, 2	38.5	1	32.7

[a]Based on the partitioning Chi-square test with $p < 0.05$

Fig. 8.9 Children learning from manipulative activities in Mauritius. (**a**) Noting angle size (Group 4). (**b**) Different shapes and sizes (Group 3). (**c**) Different quadrilaterals (Group 4). (**d**) Halving a rectangle (Group 4)

rather than focusing exclusively on learning in English. Teachers and pupils must be encouraged to use their native languages to communicate especially at the basic schooling level. This creates a learning environment in which pupils feel more comfortable and have a greater sense of ownership of mathematics. In addition, pupils are able to acquire the practices of mathematics while at the same time maintaining their cultural and linguistic identities. When pupils present their ideas in their local language, they bring a great deal of thinking resources from their lives and habits which are helpful in the learning endeavour. Thus, the inclusion of local language Creole can be viewed as a step in providing a language-sensitive framework for constructing and reviewing content area assessments.

Figure 8.9a shows two pupils who were collaborating to check the equality of the three angles of the equilateral triangle using a bent pipe cleaner as angle-tester. In the following excerpt, a conversation between the two girls is presented.

Pupil 1: Shall we bend the pipe cleaner and then place on the angles?
Pupil 2: Maybe it is better to place the pipe cleaner on one angle and get a measure (just like an angle-tester) …then check for the other two angles… what do you say?

Pupil 1: It's a good idea…ok let's try it. Here is the angle-tester [she bends the pipe cleaner on one of the angles in the triangle]. Now you check whether it fits the other two angles.

Pupil 2: Yes, see…it is fitting the other two angles…. Let's tell the teacher that we found that all the angles are equal.

This small conversation illustrates how the two girls were able to construct the angle-property of the equilateral triangle while collaborating and using concrete materials. Their collaboration seemed to boost their confidence to explore the angle-property. They used verbal, visual, and logical skills in carrying out the task.

It was very encouraging to see how pupils, especially the low-ability ones, were able to spot the isosceles triangles on their own. Figure 8.9b illustrates students' enthusiasm and achievement. The isosceles triangles were the most commonly constructed triangles and it was observed that pupils from all the four schools were having an instinctive habit of choosing two equal length straws and one different length straw for their construction of triangles. As a result, all the pairs of pupils had at least one isosceles triangle constructed. The kinesthetic actions and semiotic activity had facilitated the learning of different types of triangles.

Figure 8.9c shows the collaboration of pupils to construct parallelograms with elastic bands on geoboards. It was observed that the pair of pupils was constructing their parallelograms separately on the same geoboard then discussing whether the shapes were correct parallelograms. Then they removed or reshaped the ones they doubted as examples of parallelograms. Many pupils were very creative as they first drew squares and rectangles and then stretched one pair of opposite sides to get their parallelograms.

Figure 8.9d shows the pupils' active involvement in drawing the diagonals of a green rectangular paper cut-out shape. Students were physically involved in the activity of drawing diagonals on a range of shapes and then discussing their results. Activity assisted the connection between emotional engagement and learning that was central to the creation of the learning environment in the experimental classes with manipulatives. It not only guided the learning activities in the classroom but also sharpened the creation of the classroom culture and maintained a supportive and emotionally stable classroom environment in which all felt comfortable to take risks and to explore ideas in new ways.

Beside visuospatial reasoning, the classes with manipulatives have led the pupils to achieve experience of the aesthetic. Greene (2001) defined aesthetic learning as an "initiation into new ways of seeing, hearing, feeling, moving, reaching out for meanings and learning to learn integral to the development of persons—to their cognitive, perceptual, emotional, and imaginative development". Thus, it is found that the use of inquiry-based teaching and investigative methods with manipulative materials is promoting cognitive, perceptual, emotional, and imaginative development in the pupils which are in fact the main aims of teaching. Undoubtedly, these strategies will be beneficial to the teaching of other mathematics content area beside geometry.

Moving Forward

This chapter took up the challenge of why an ecocultural perspective on visuospatial reasoning was important for Indigenous communities in transcultural situations. The chapter showed how intuitive or incidental experiences within an ecocultural context strengthen our understanding of mathematics and mathematics education for Indigenous communities. The chapter also provided an example of visuospatial reasoning in classrooms where everyday language differed from the formal school language and the impact that language has on visuospatial learning and reasoning. An ecocultural perspective sheds new light on how gazing, noticing important features, selectively attending, and interpreting visuospatial representations occur in visuospatial reasoning in problem solving in cultural contexts. This extends and in one sense re-interprets recent research on diagrams or other external visual imagery (e.g. Lowrie, Diezmann, & Logan, 2012; Mason, 2003) taking account of the issues for different groups of students especially those undertaking national or state paper-and-pencil tests.

The chapter illustrated how teacher education that takes account of ethnomathematics provides teachers with a means of incorporating the strengths of visuospatial reasoning in an ecocultural context into the learning of students. In particular, the chapter considered geometry and measurement and extends the arguments from earlier chapters on the importance of taking an ecocultural perspective to extend our understand of visuospatial reasoning and the role it plays in problem solving, learning about geometry and measurement, and encouraging a stronger cultural and mathematical identity.

It provided examples of what Castagno and Brayboy claim as important for Indigenous youth, "a more central and explicit focus on sovereignty and self-determination, racism and Indigenous epistemologies in future work on CRS [culturally responsive schooling] for Indigenous youth" (Castagno & Brayboy, 2008, p. 941). The examples tap into at least some of the technologies, worldviews, relativities, place bases, and responsibilities to community, self and the use of power of the learner. However, the holistic perspective must be maintained. It is anticipated that students will have the opportunity to use Indigenous knowledge and language to meet both local and western education goals. In particular, recognition of visuospatial reasoning is in line with a stronger emphasis on holistic learning and seeing the whole picture as a critical aspect of Indigenous ways of learning. An ecocultural perspective of visuospatial reasoning as the examples show is in line with the Alaska Native Knowledge Network principles for CRS:

- A culturally-responsive curriculum reinforces the integrity of the cultural knowledge that students bring with them.
- A culturally-responsive curriculum recognizes cultural knowledge as part of a living and constantly adapting system that is grounded in the past, but continues to grow through the present and into the future.
- A culturally-responsive curriculum uses the local language and cultural knowledge as a foundation for the rest of the curriculum.

- A culturally-responsive curriculum fosters a complementary relationship across knowledge derived from diverse knowledge systems.
- A culturally-responsive curriculum situates local knowledge and actions in a global context. (Alaskan Native Knowledge Network, 1998)

At the same time, can such an ecocultural perspective developed in the last few chapters and considered in this chapter as place-based education with Indigenous or transcultural communities apply to students in a digital age? This is the focus of the next chapter.

Chapter 9
Visuospatial Reasoning in Contexts with Digital Technology

Kay Owens and Kate Highfield

The Challenge

The ecocultural perspective on visuospatial reasoning was established by considering Indigenous communities and their practices and appropriate schooling and other diverse ecocultural practices illustrating visuospatial reasoning. However, today is the digital age so does research on visuospatial reasoning support this ecocultural perspective. Since most of the research on visuospatial reasoning has been focussed on dynamic computer-generated images, it is important to consider digital technological facilities as an ecological context. How can an ecocultural perspective of visuospatial reasoning enhance our understanding and valuing of visuospatial reasoning? In this chapter we consider how a computer-facilitated learning age influences an ecocultural identity and both self-regulation and visuospatial reasoning. It is then important to consider how these personal dispositions impact on mathematical identity.

This chapter focuses on prior-to-school and elementary or primary schooling and the impact of the digital age on visuospatial reasoning. In particular, it will consider how students are reasoning visuospatially in the context of hand-held robots in early childhood (Highfield et al., 2008). Highfield showed that children were capable of reasoning and learning concepts in mathematics through the use of robots. Analogies will be drawn with the use of diagrams for reasoning.

Sections of the chapter cover the importance of dynamic geometry softwares in the way that students reason visuospatially. Considering that there are many research articles in this area, several will be selected, especially those that look at the use of ICTs in different cultural groups and with primary and middle school students. The importance of reasoning in a visuospatial environment (Jones, 2000) and the importance in design of software (Christou, Pittalis, Mousoulides, & Jones, 2006; Jonassen, 1999) as it impacts on visuospatial reasoning, self-regulation and sociocultural identity will be explored.

There is also an increasing interest in computers as tools in modelling (Goos et al., 2003) and in terms of this book in ethnocomputing (Eglash, 2007). However, this modelling approach can be more broadly interpreted in the modelling sense that is developed from a particular cultural group (Rosa & Orey, 2012). This strong support for valuing cultural mathematics joins with ecocultural perspectives in visuospatial reasoning and hence developing mathematical thinking identity that moves beyond the western-dominated perspective.

The argument continues with important reasons in today's society for visuospatial reasoning. A discussion of the importance of ecocultural perspectives for appreciating geographical studies and the mathematical understandings necessary for such studies will illustrate the importance of visuospatial reasoning and the impact that ecology and culture have on this development and thinking.

Ecocultural Perspective of Measurement in Changing Worlds

An historical look at geometry from Fibonacci to the twentieth century shows that diagrams and practical mathematics with measurement was commonplace and proofs such as Euclid's were not always centre stage (Menghini, 2012). Is it possible that the digital era built on this background of measurement and experiment? Even Fibonacci gave a way of calculating the volume of a heap of wheat in a corner by measuring lengths horizontal to the floor on either side, multiplying and dividing by 2. Perhaps the PNG communities who use lengths for assessing volumes are not so different. Measuring is not uncommon in digital dynamic geometry experiences. Nevertheless, the geometric or spatial reasoning needs to link to more theoretical ways of thinking if the technology is to be considered mathematics.

Wassmann (1994) noted that the Yupno of Papua New Guinea employed three different ways of spatial perception and not just the one western way which is egocentric. They used object-centred locations such as relative positioning, absolute positioning (east, west, south, north), and anthropomorphic description to locate themselves depending on the time and context. However, it is possible to explore visuospatial reasoning in a digital environment despite Turkle and Papert's (1990) suggestion that the multiple modes of thinking cannot be known. The ecocultural perspective and the model of mathematical identity within context assist us to understand the process of visuospatial reasoning in the digital context.

We are challenged by Lévi-Strauss's (1968) view that visuospatial reasoning is only of the "primitive mind". While we recognise this as a strength that should not be lost to Indigenous cultures, we can better understand visuospatial reasoning and improve learning in the digital age by realising the importance of visuospatial reasoning in the context of the "human-with-media" (Borba & Villarrea, 2005). The place of visuospatial reasoning within the model of identity assists us to appreciate how

> the computer, with its graphics, its sounds, its text, and its animation, can provide a port of entry for people whose chief ways of relating to the world are through movement, intuition, and visual impression (Turkle & Papert, 1990, p. 131)

and enhance learning for all students through media. Despite

> discrimination in the computer culture (that) takes the form of discrimination against approaches to knowledge, most strikingly against ... an approach we call "bricolage". (Turkle & Papert, 1990, p. 135),

there are various students who, for one reason or another, do not want to do "black box" programming. The creative, visual approach is illustrated by a student Anne

> Instead of thinking of compound objects as a way of getting a picture to be bigger, she thinks of compound objects as a way of getting sprites to exhibit a greater complexity of behavior, an altogether more subtle concept. Thus, Anne's level of technical expertise is as dazzling in its manipulation of ideas as in its visual effects ... ., her path into this technical knowledge is not through structural design, but through the pleasures of letting effects emerge. (Turkle & Papert, 1990, p. 139)

Other students produced unexpected solutions. Thus Turkle and Papert argue that visuospatial reasoning is to be respected as much as formal abstract reasoning, that working with objects is also to be valued. As Sternberg (1987) put it, one of the intelligences is that of practical.

Because young children can form rules and properties that are incomplete, some children may not do as well from the more abstract approach as a child who has "a tendency to see things in terms of relationships rather than properties, access to a style of reasoning that allowed them to imagine themselves 'inside the system'" (Turkle & Papert, 1990, p. 144). They used a relationship to the gears to help them think through a problem but they presented their results in a more formal way.

Furthermore, the characteristics of media and their engagement of students have led to a number of studies connecting visuospatial reasoning to ecocultural contexts. The use of census databases is just one way (e.g. Australian Bureau of Statistics, 2013). Graphing programmes such as Tinkerplot also allow students to move and select and so both physically and visually engage with visuospatial reasoning to support a growing understanding of visuospatial displays of data and statistics. Dynamic, visual software and movies were used by Dalin (2013) to create a powerful means of students becoming mathematicians. He set about to create

> teaching and learning school mathematics in a human environment and through a human learning process. It can be done by translating the mathematical language into graphic, visual-dynamic-quantitative representation and providing the needed tools for active learning through self-experience and exploration. (Dalin, 2013)

Thus learning is understood in terms of the model in Chap. 1 in which ecocultural contexts are significant in the development of the mathematical learner through using visuospatial reasoning as well as other cognitive processes with motivation and self-regulation.

The Role of Digital Media in Developing Self-Regulation for Learning

Self-regulation especially in terms of motivation and self-assessing of actions has generally been assessed by observation and self-reports but the use of computers allows for some monitoring. In young children this is a possible important step forward for self-regulation in problem solving (de la Fuente Arias & Díaz, 2010). Furthermore, we can see the visuospatial reasoning that young children are undertaking.

Highfield's (Highfield, Mulligan, & Hedberg, 2008; Highfield, 2010, 2013) work with young children suggests that simple robotics may provide opportunities for young mathematics learners to engage in self-regulation including metacognitive and problem-solving strategies. This work suggests that the act of planning, programming, and observing the robots movement can act as a catalyst for engagement in a range of mathematical concepts and processes as well as prompt reflection and revision of plans. While this cyclic engagement in problem solving (see Fig. 2.17) highlights the potential of these tools—the context of learning, the child's engagement and responsiveness, multiple representational modes, and the role of a community of learners are key in this process. Figure 1.2 emphasises the context and self-regulation in the cycles.

The role of the teacher, however, may be critical in how well the computer tool and the students' use of it and their collaboration. Laborde, Kynoigos, Hollebrands, and Strässer (2006) note that all papers presented at the PME conferences mentioned that in the dynamic geometry software environment "the notion of dependency is difficult for students and not understood initially" (p. 286). Furthermore, the role of teacher seemed crucial in assisting students to move from the visualisation to another more substantial form of understanding. Laborde et al. emphasised the movement from graphic to mental back to graphic and then to mental activity. This switching, reminiscent of code switching, is an important aspect of both self-regulation and visuospatial reasoning. A strength of dynamic geometry software is the availability of numerical and figural cues and the ability to produce and refine objects to find a solution (Love, 1996). Key to understanding how students find a general solution is a recognition that examples and attention to features is dependent on "a deep, personal, situated structure" (Goldenberg & Mason, 2008, p. 138).

A study by Goos, Galbraith, Renshaw, and Geiger (2003) highlights how the idea of an ecocultural perspective on visuospatial reasoning can occur within the ecocultural context of classrooms with digital media. The tools were becoming extensions of the students' thinking especially under the prompting of the teacher. The teacher intervened on a few occasions encouraging the students to find a solution using an alternative digital means, by seeing what other students were doing, and he also encouraged the group to share their findings. The classroom approach as the teacher portrayed was critical in this ecocultural perspective.

Students' attention and so persistence and self-regulation were the visuospatial representation of the graph that indicated three intersection points. It was then a matter for the group to verify these points. Again the teacher "encouraged the group to use the technology as a *partner* to re-organise their thinking" by using the graphics calculator and spreadsheet on the computer simultaneously. Then when they were still struggling, he encouraged them to see what other groups were doing and the technology became a partner to mediate mathematical discussion between students, resulting in the group coming back together and working out how to set up the spreadsheet to find all solutions. Their hastily prepared presentation was improved by questioning from students and the teacher helping to draw attention to salient aspects of the task and how the different technologies created different representations of the task.

> Mathematical and communications technologies were thus seamlessly integrated to share and support argumentations on behalf of the group of students, suggesting that technology became an extension of *self* for members of this group. ... "we were doing it ourselves, not just listening to the teacher. And seeing something visual helped our understanding." ... The students' recollections of this experience hint at the sense of autonomy and power associated with appropriating technology into one's personal repertoire of mathematical practice, that is, as an extension of self. (Goos et al., 2003)

Reasoning about the nature of the graph, the use of the spreadsheet and the algebraic representation was parallel to the way in which Indigenous communities were reasoning about the visuospatial representations in their ecocultural context. "Tools"—computers and calculators producing graphs, spreadsheets, equations on the one hand; buildings, carvings, paintings, weaving, dancing, navigating, and rituals creating visuospatial representations on the other hand—were used with reasoning and manipulation. Alternative and "hidden" meanings were understood from the mathematical context and ways of thinking and reasoning about the visuospatial representations. Both required technical expertise. Both required knowledge of the mathematical visuospatial ways of reasoning.

Furthermore, both achieved self-regulation, goal setting, cognitive processing with visualisation. Both involved communication with others within an ecocultural context and both resulted in a sense of achievement and belonging. Both resulted in being a member of the community of practice. Both connected the members to a sense of autonomy and power associated with appropriating technology. One was seen as mathematical but was also ecocultural; the other was mathematical but seen as ecocultural. Both were visuospatial reasoning from an ecocultural perspective.

Dynamic geometry software has been shown to encourage internalisation of motion from the visual screen that students manipulate. The dragging and trace tools in the dynamic geometry software are seen and manipulated being transformed into psychological tools supporting students' reasoning.

> From the combination of observation and action students grasped variability as motion, while the idea of covariation, incorporated in the coordinated movement of points on the screen, was experienced through the coordination between eyes and hands. In most of the cases, students' formulations reflect the asymmetrical nature of the independent and dependent variables and the twofold meaning of trajectory. (Falcade et al., 2007, p. 331)

These researchers showed how the classroom conversations encouraged the abstraction and recognition of the meaning of trajectory at a point and as a "journey" illustrating how the ecocultural context (a communicating classroom with DGS facility) was significant in students' visuospatial reasoning and sense of creating the mathematical notions associated with functions.

Rivera (2011) established the importance of the role of technology for visuospatial reasoning. Computers as "servants" in Goos et al.'s terminology, not only "produce such static displays (i.e. the concrete objects) quickly and easily, but in addition it then becomes straightforward to create rotation and morphing animations that can bring the known mathematical landscape to life in unprecedented ways" but they also allow users to "obtain fresh insights concerning complex and poorly understood mathematical objects" (Palais, 1999, pp. 647–648) as illustrated by Goos et al. above. Rivera particularly notes the evolving processes that occur with animation and how interaction with the computer and others assists the development of relations and theory implied by diagrams or codes that model a structure and display the relationship or concept. The tool becomes an extension of thinking.

In another example, the Singaporean use of representational rectangles that are manipulated, for example in fraction work, provide a strong visual analogy giving meaning through the classroom and curriculum culture of labelling the components and defining the spatial relations among the components, and thus becoming visuospatial reasoning. Similarly, diagrams that represent geometric relationships, often as a theorem, involve culturally accepted ways of marking vertices and segments, and a classroom accepted way of understanding the diagrams. Rivera (2011) established the existence and importance of visuospatial reasoning associated with the world of computer technology in education. In each case, the technology is "an extension of self", a position established as part of visuospatial reasoning in an ecocultural context in the earlier chapters and again by Goos et al. (2003).

If we turn to the younger age group, there are benefits of virtual manipulatives for visuospatial reasoning. For example, virtual Pattern Blocks have colours that can be changed, they can be "snapped" into position, unlike concrete material and they "stay where they're put" (Clements, 1999, p. 51). The development of simple repetition, and transformation skills such as reflection, rotation, and scaling are enhanced through on-screen manipulations. Virtual Pattern Blocks and dynamic interactive software can provide representations of concrete manipulatives that allow children to experiment with a broader range of patterns with ease and flexibility. Moyer, Niezgoda, and Stanley (2005) found that children's patterns were more creative, complex, and prolific using virtual manipulatives compared with patterns formed with concrete materials. Highfield and Mulligan (2007) found technological tools allowed ease of representation, with children using virtual manipulatives consistently engaged in increased experimental patterning producing a broader range of patterns, and edited or deleted them before completion. In part, this could be attributed to the "delete tools" that held "novelty value", with the children enjoying "rubbing out" and "chucking" things in the "bin". However, from observations, this can mean children fail to stop and reflect on their pattern making thus requiring teacher intervention to encourage greater self-regulation in the problem solving. Interestingly, children not only use colour but can also orient blocks to form their pattern as illustrated in Fig. 9.1. Transformations are also explored for fun as captured in the following conversation associated with Fig. 9.1b, c:

> Nicholas: Oh he's really big now. He's really, really big. Wee ... Oh ... Big ... Fat (*scaling the lion, enlarging it*)
> Yvette: Make him long (*pointing to the seals*).
> Nicholas: Flat (*after shearing the seal*).
> Yvette: They're both flat (*pointing to the seals*).

The fact that virtual objects can be cloned and repeated also allows for measurement units. However, with young children, using the mouse and accidentally

Fig. 9.1 Children's visuospatial reasoning when playing by manipulating objects on computer screens. (**a**) Block patterning. (**b**) Enlarged lion. (**c**) Lengthened seals

clicking on the wrong icon can prevent good construction requiring some guidance. The use of touch screens overcomes some of the difficulties of using the mouse. In some instances, the screen manipulation can prevent good conversations and interactions between the children, again requiring some guidance. In this respect, adult support can be important in the context of play with technology. Further, it is found that children are more likely to click to be entertained and to choose books read to them rather than interactive programmes with non-routine problems. Hence the immediate class context as well as the digital age cultural context influences learning.

Particularly effective for problem solving with technology is the flexibility of multiple strategies (Sarama & Clements, 2009; Siegler, 1999), including: identification of mathematical relationships, inference, generalisation, representation, analogy, recursive cycles of trial and error, and verification (Greenes et al., 2003). However, as stated by Sophian (1999), successful problem solving is "more than the aggregate of the strategies they use; they also know something about the goal" and structure of problems and responses (p.18). Thus self-regulation assists with ensuring the cognitive approaches are effective in students' responsiveness (Fig. 1.2).

Visuospatial Reasoning in Geometry and Measurement Learning Through Digital Technology

When we consider visuospatial reasoning in geometry and measurement, research predominantly focuses on older children and on screen based tools. The focus on older children is likely due in part to curriculum based expectations—with older children encountering formalised geometry and thus this content area being given increased prevalence in research. However, it could be argued that this focus on older children is misplaced with essential measurement and geometry understandings developing at a much younger age (Clements & Sarama, 2007a). The focus on screen based tools is also key here and while this research (as outlined above) provides insight into a range of tools for use in developing geometry and measurement learning there are a range of alternate digital tools that also have potential for learning.

Besides the studies above and studies on digital technology for area and three dimensional stacking for young children, research has mainly focused on the use of Logo programming but there have been differences in the success and usefulness of Logo. Some research has not always found Logo to be effective for young children. The dynamic representation of angle was found to cause confusion for some children. The pathway that the turtle moved through and the angle of turn were not always easily interpreted even when laser beams illustrated the direction of the turtle (Kieran, 1986). Cope and Simmons (1994) also suggested that the immediate feedback obtained from Logo programming may inhibit the development of angle concepts. Their research with students aged 9–11 years, indicated that some learners utilised trial-and-error strategies rather than moving to more advanced, higher level understandings of angle and rotation. Nevertheless, Clements and Battista (1989) and Noss (1987) describe children's increase in understanding of angle concepts when using Logo. Misconceptions may in part be mitigated by appropriate pedagogic structures (Sarama & Clements, 2004). Lehrer and Littlefield (1991) proposed mediated instruction, including structured teaching of Logo skills, as beneficial for children in mastering Logo. Clements and Battista (1991) also recommended tasks that are carefully planned to encourage comparison and avoid misconceptions. To this end, Lehrer, Jacobson, Thoyre, Kemeny, Strom, Horvath et al. (1998) espoused potential benefits of sequenced tasks and inquiry-based learning with Logo.

As a child plans and programmes the turtle's movement in Logo their actions are inherently linked with spatial and geometric concepts, including shapes and angles, directionality, linear measurement, location and position and pathways (Clements, Battista, Sarama, Swaminathan, & McMillen, 1997). Clements and colleagues found that children's engagement with shape construction in Logo enables children to progress quickly in geometric understandings (Battista & Clements, 1991; Clements, 1998). Children's active construction of shapes in Logo facilitated the noticing of properties, verbal descriptions, and integration of geometric understandings. Butler and Close (1989) also found that work in Logo enabled children to develop understanding of two-dimensional shapes. The construction of shapes in a dynamic environment pushed children beyond the static representations they would normally view in traditional representations of geometric concepts. Similar findings are supported by research with children in primary school (Battista & Clements, 1991; Hoyles & Noss, 2003; Lehrer & Littlefield, 1991) and high schools (Khasawneh, 2009).

Simple easily programmable robotics engages students and avoids some of the issues of Logo on screen. Lack of interest partly results from other aspects of the digital age, namely fast moving, noise-producing manipulative screens. Spatial issues are also reduced with the floor turtle. Children are in the same three dimensional space and can face in the same direction as the turtle whereas a vertical screen made some tasks, especially on angles, difficult for students. In Highfield's (2012) study, evidence of visuospatial reasoning is demonstrated not only by the children's activities and conversations captured on video but also by their drawings. Highfield classified the drawings made by 30 children (4 three-year olds, 6 four-year olds, and 20 year 1, around age 6) as idiosyncratic/non mathematical, emergent spatial struc-

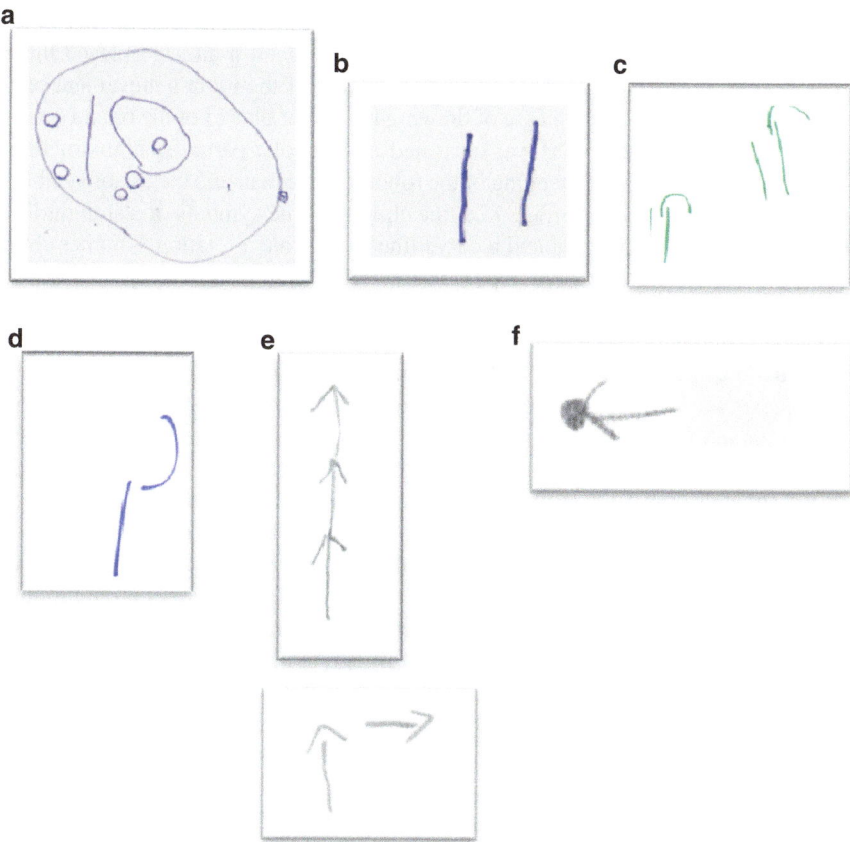

Fig. 9.2 Drawings by children of the movement of their robot. (**a**) Pictorial idiosyncratic. (**b**) Emergent spatial. (**c**) Symbolic emergent spatial. (**d**) Symbolic partial-spatial. (**e**) Symbolic spatial. (**f**) Integrated spatial

ture, symbolic non-spatial structure, symbolic partial-spatial structure, symbolic spatial structure, and integrated symbolic spatial structure. Although only the four youngest children (3 year olds) produced the first two of these categories, nevertheless, one child appeared to be drawing the Beebot (commercial robotic toy) (Fig. 9.2a) with reasonable spatial arrangement and some movement by a line and two additional circles. The drawings of two lines labelled emergent spatial structure (see an example in Fig. 9.2b) may have shown the child's dynamic visuospatial reasoning by the physical order in which they were drawn. It should be remembered that the child's ability to draw to represent their thoughts might lag behind their spatial thinking as found in the Count Me into Space project where a child could verbalise the structure of four squares covering a larger square but could not draw it and knew it did not represent the image in his mind (see also other examples in Chap. 2).

One 4-year-old child's drawing was classified as symbolic emergent spatial structure. His representation indicated the number of steps and a turn but placed these symbols side-by-side so that there was no indication of the robot's movement path (Fig. 9.2c). The most common type of drawing for 4 year olds (5 of the 6) and 6 year olds (11 of 19 actually drawn) was classified as symbolic partial-spatial structure. These demonstrate an understanding of the robot's movement and the use of symbols to demonstrate movement. In Fig. 9.2d, the child uses the symbols of a straight line to indicate movement forward and a curve line to indicate a rotation. Another child has used arrows to indicate three steps forward, with the number of arrows indicating the number of steps taken. The children's use of spatial structure is classified as partial as the step length and the angle of rotation is indicative of the movement rather than structured with a measured or good estimate of the angle of rotation.

The next two drawings classified as symbolic spatial structure represent sufficient information in themselves to convey the movement and direction that the robot took with evidence of the programme steps (Fig. 9.2e). These were produced only by 5 year 1 students (around 6 years old). The last category of representation is integrated spatial structure produced by three children of this older group. They show evidence of integrating a representation of programming (for example correct number of equally sized steps and correct direction for turns) and incorporating programming elements in a coherent manner with the use of both symbols and spatial structure (see Fig. 9.2g). Interestingly these children integrated their symbol for turn (like the dot) into other representations during the project.

Like any study attempting to classify diagrams into a fixed set of structures, some drawings were not easily classified. This indicates the diversity of visuospatial reasoning in their robot action, their thinking, and their drawing. Nevertheless, the study indicates that young children are reasoning visuospatially. More interesting are the videotapes of children problem solving in specific ways from easy movements such as make the robot move backwards to make the robot move forward three steps and to move forward and rotate. The children could respond to these tasks confidently and without need for multiple attempts or using metacognitive strategies.

The more complex tasks of programming the robot to move in a square and to move through "house" tasks which required the child to move from the home position in specified ways (see Highfield & Mulligan, 2009 for further information) presented an opportunity to observe children's use of problem-solving strategies and tools. Frequently children required multiple attempts and they used embodied action and gesture to problem solve such as using hands to indicate the steps they were considering, using the toy to model the planned movement, using whole body action to act out the steps for the toy to move or using symbols such as arrows to plan steps or movement. For example, there were 20 examples of children pointing to a position on the mat, 29 instances of them using their hand to iterate steps and plan length, and 40 instances of children moving their hand in an arc or sliding motion to indicate general movement or rotation without indicating distinct steps. These were embodiments of visuospatial reasoning. In 56 examples, either the eyes or head were used to "point" or indicate steps of movement also indicating visualising reminiscent of the way children learn to count and count on in their heads for early addition. In eight cases children used the toy to act out and plan movement.

One interesting occurrence in the classroom play with robots happened when 4-year-old children communicated in their pairs by observing and recording on their diagram a hook shape similar to that used by another pair. This was reminiscent of the problem solving that occurred in Owens study (Chap. 2) in which groups would use what other children were doing to assist their heuristic of assessing their own work or as Goos recorded above when students were working with dynamic algebra systems or the architecture students (Chap. 5) said that they asked opinions of their friends or looked at what others were trying to do to assist their visuospatial reasoning to get started with problem solving or consider the aesthetics of their paper sculpture. In the last case, they noted that others used cultural practices, and they subsequently, used cultural practices. In a similar way, children observing others in a digital technological classroom illustrate a cultural affinity to that kind of classroom, an ecocultural context.

Yelland and Masters (2007) articulate three types of collaborative scaffolding: cognitive, technical, and affective, and demonstrate that children who were scaffolded using these techniques demonstrated more sophisticated strategies in solving problems. Effective teacher cognitive scaffolding includes ensuring that the children have understood the task and are utilising and articulating the specific strategies, intervening at appropriate times to assist students with a difficulty for which they need a little piece of information, and in larger more formal classrooms encouraging children to share at different steps in the inquiry (McCosker & Diezmann, 2009; Williams, 2008). Thus the role of the teacher in the ecocultural classroom in the digital age has an important role just as Elders in an Indigenous community.

Highfield and Mulligan (2009) provided a list of ways in which robots could be used to establish processes and concepts in geometry and measurement in particular. These are shown in Table 9.1. The use of the robot can facilitate children's visuospatial reasoning and learning.

Highfield and Mulligan note:

> In this project it is significant that the children engaged in multiple mathematical processes concurrently and sequentially; and they demonstrated perseverance, motivation and responsiveness to these tasks that would not usually be evident in their regular programs. (Highfield & Mulligan, 2009, p. 27)

Educational robotic application (ERA) principles (Catlin & Balmires, 2010) for effective learning are grouped into three areas pertaining to technology, student, and teacher.

- Technology should demonstrate a range of intelligent behaviours, interact through a range of semiotic systems and use embodiment, enable the student to learn through meaningful interactions situated in space and time.
- Students should have engagement fostered, be able to engage in sustainable and long-term learning and be able to personalise the robotic learning experience.
- Teachers should be able to access and demonstrate effective pedagogy, present tasks that intersect with curriculum and assessment opportunities, ensure equitable access to the technology, meeting the practical needs of organising and delivering educational opportunities.

Table 9.1 Uses of robots to develop visuospatial reasoning for concepts

Spatial concepts	Capacity: Creating and measuring space that is large enough for the toy to move through (such as a tunnel) or fit inside (such as a garage)
	Angle of rotation: Exploring the rotation of the toy as a pre-set 90° angle, creating pathways that utilise a 90° angle
	Directionality: Examining concepts such as forward, backward, rotate, left, right, and positional language
	Position on a plane: Using increasingly complex language, "over there" becomes "in the far left corner". Using terms such as over, under, beside, through, near, and far
	Transformational geometry: Exploring concepts such as rotation and linear motion
Measurement	Informal and formal units: Using informal units, such as hands, counters, blocks, or the toy's length, and formal units such as measuring tapes to ascertain distances and assist in creating programmes
	Identification and iteration of a unit of measure: Using the toy's pre-set step as a unit of measure, when moving the toy; using hand and eye gestures as place holders in measuring distance
	Direct comparison: Using the toy's length to compare directly the distances needed to complete a pathway
Structure	Grid: Developing and using grids showing the toy's step length to assist in planning and developing programmes
	Gesture and movement: Using gestures and body movement to indicate and imagine the structure of regular steps, For example, when asked how she knew what the programme required, a child responded "I imagined where the steps would be"
Number	Perceptual and figurative counting: Engaging in both perceptual and figurative counting to ascertain the number of steps required to complete a given pathway
	Comparison of number: When comparing programmes or movement pathways the children frequently compared number; for example: "I went eight forward and you only went six forward and so mine went further"
Problem solving	Estimation: Predicting and estimating the number of steps required to complete a pathway; examining the estimation to assess reasonableness before programming
	Reflecting: Observing a programme, reflecting on attempts, and making the changes required
	Trial and error: developing confidence to trial a programme, even if incorrect and identifying errors
	Recall of prior knowledge: recalling prior knowledge and skills to apply in programmes
	Investigating multiple solutions: Predicting and developing multiple solutions to tasks; for example, travelling clockwise, or anti-clockwise
	Evaluating solutions: Examining the efficiency of a programme to decide if it was most effective
Representation	Semiotic understanding of symbols: In order to programme the robot to move the children needed to develop an understanding of what each symbol meant. The forward arrow meaning one step forward, arrows to the left or right meaning rotation (not movement to the right)
	Constructing and recording programmes using symbols: After completing a programme the children represented what they had done in the "robot diaries". This required learners to develop a symbol system representing their programme

Note: Source—Highfield and Mulligan (2009), p. 26

Although not yet widely adopted Catlin and colleague's ERA principles are most relevant to the design and pedagogic affordances of a broad range of robotics and robotic toys.

Thus we find that children's ecocultural context in the digital age influences not only their ecocultural identity but their self-regulation in terms of both affective and cognitive processes and hence responsiveness. This responsiveness is assisting in establishing their ecocultural (digital) mathematical identity. In turn they were influencing each other in the classroom. The role of the teacher who has also established an ecocultural (digital) mathematical identity in this digital context is also critical.

Visuospatial Reasoning in the Digital Age Taking Account of Ecocultural Contexts

It is no wonder then that a number of researchers have used computer technology to engage Indigenous and disenfranchised students. Eglash (2007) has prepared a number of different programmes to engage students with pleasing results. Brown (2008) has also carried out a study in Australia emphasising

> Mathematics programs that accentuate Aboriginal students' life experiences and contexts bring relevance to their learning, thus providing purpose and in turn increased levels of motivation and engagement. Mathematical modelling and problem solving can inject curiosity into what is sometimes considered by students to be a boring subject: when the two are properly combined, they can improve students' attitudes towards mathematics (Falsetti & Rodríguez, 2005; Brown, 2008, p. 95)

Brown's study involved urban Indigenous Grade 4–7 students (primary school) in Queensland where cyclones are becoming more prevalent and have always been a concern. She utilised visual and written texts including graphs about cyclones and chocolate. Students in groups participated well saying they had a job to do and the mathematics was genuinely useful, and some shared the work, but the mathematics they were utilising, they did not necessarily recognise as themselves doing mathematics at the time.

> Students are offered a variety of modes to deliver their findings and indeed some students have requested to formulate their own. It is this level of student interest that indicates that mathematical modelling can be perceived by students to be a productive and worthy enterprise. (Brown, 2008, p. 97)

Thus we see a sense of self-regulation, ownership, and identity with the requirements of the task, not necessarily seeing it as mathematical. Interpreting the visuospatial representations was given a context of relevance to the students in their ecocultural environment.

Simulations are a digital age tool that can encourage visuospatial reasoning but also empathy for the tools and for the sources of content (Holton, 2010).

Holton reminds us that learners' views about learning from the computer and beliefs about control are critical in self-regulation and working with digital media. Jonassen (Jonassen, 1999; Jonassen et al., 1999) whose work was critical for establishing the model in Chap. 1 (Fig. 1.2) noted the importance of information and computer technologies facilitating meaningful learning experiences that were active, constructive, collaborative, intentional, complex, contextual, conversational, and reflective. These aspects all interact with each other. Thus students will create visuospatial representations but also discuss these so they have shared meanings. Contextual experiences take account of the ecocultural contexts.

Eglash's work on culturally situated design tools (Eglash & Rensselaer Polytechnic Institute, 2003) including the VBL (virtual bead loom) is established on a careful discussion of the ecocultural background from which the digitized designs are linked. So, for example, he discussed the extensive use of four-fold symmetry in Native American cultures for the VBL included on the webpage.

> Before reading the text, teachers can ask students to look at the designs and describe them; such discussions offer opportunities to introduce symmetry as a term and concept. The text describes, (as he does in the paper), how four-fold symmetry is a deep design theme in many Native American cultures, and is evident not only in a wide variety of native arts, but also indigenous knowledge systems such as base four counting, four-quadrant architecture, the "four directions" healing practice, etc. A second web page shows how such structures are analogous to the Cartesian coordinate system. Finally, the webpage introduces the Native American bead loom as another example in which we find an analogue to the Cartesian grid. (Eglash, 2009)

Not only can students create given designs but from their creative design on the virtual tool, then can recreate a real example on a bead loom. Eglash noted

> There are three pedagogical frameworks that can be used with VBL. In application/reinforcement we start students with the task of simulating one of the original beadwork designs. Teachers have reported success in using this software for teaching Cartesian coordinates, reflection symmetry and its relation to Cartesian values, numeric aspects of translation, and other subjects. In structured inquiry specific math challenges can be proposed by teachers: developing rules for the reflection of polygons about the axis, numeric descriptions for color sequences, etc. For example, teacher Kristine Hansen at the Shoshone-Bannock reservation school had students create a rectangle in quadrant I (the positive-positive quadrant), and then apply the following:
>
> 1. Reflect your rectangle into quadrant II with the following transformation $(x,y) \rightarrow (-x,-y)$
>   Students then created transformation rules to place the rectangle in other quadrants. Doing this with asymmetric triangles might be even more effective since it would help visualize the reflections. Another exercise carried out by Hansen:
> 2. Program a green isosceles triangle at the bottom of the screen. Use the transformation $(x,y) \rightarrow (x,y+5)$ to translate your triangle up 5 units. Continue to iterate this translation by translating your last triangle up 5 units until you reach the top of the grid.
>   This was assigned in early December; she reports that she had intended that the students create a Christmas tree, but to her surprise the students modified the assignment and closely overlaid the triangles using a multitude of colors, creating what she describes as "the feathered bead pattern we see in a lot of the beadwork here on the reservation." This indicates that one advantage to this more open-ended approach to ethnomath is that it lends itself better to "appropriation" (Eglash et al., 2004), thus offering a more constructivist-based learning environment in which students' cultural sensibilities can be used as a bridge to math education.

Finally there is <u>guided inquiry</u>, in which students chose their own challenges. For example, one student of Puerto Rican heritage decided to create a beadwork image of the Puerto Rican flag, which includes an equilateral triangle. At first he tried to create an equilateral triangle by having the same number of beads on each side, but that did not work because the beads along the diagonal are spaced farther apart than the beads along the vertical or horizontal. He finally arrived at a solution by using the ratios of a 30-60-90 triangle to arrive at a discrete approximation (Fig. 8); a challenge that he might have balked at had it simply been assigned to him. (Eglash, 2009)

Adam (2010) went back and discussed possible food covers with the weavers. Eglash (2009) went back to the Shoshone-Bannock to find the algorithm they used and built that into his programme. "Using iterative rules—e.g. "subtract three beads from the left each time you move up one row." It worked better than the standard computer algorithm.

Moving Forward

This chapter has outlined some research that has considered the value of digital technology in encouraging visuospatial reasoning in problem solving. The digital age provides digital tools that can be engaged to enhance visuospatial reasoning as students learn mathematical processes and concepts. The ecocultural background is significant for the students of today and influences the self-regulating student in terms of affective and cognitive strategies. The impact of the classroom context is evident. Furthermore, Eglash and others have shown how there can be a synergy between ecocultural Indigenous contexts and ecocultural digital-aged contexts. There is evidence to show that both engage the students' self-regulation and visuospatial reasoning.

The last chapter encapsulates the arguments presented throughout the book providing a synthesis of research from across the world, across time, and across paradigms of psychology, anthropology, and psychological education and critical philosophical approaches to education.

Chapter 10
An Ecocultural Perspective on Visuospatial Reasoning in Geometry and Measurement Education

> *Education that consists in learning things and not the meaning of them is feeding upon the husks and not the corn.*
>
> (Mark Twain)
>
> *Only when a result fits into a wider context do you really begin to see its significance.*
>
> (Mason, Burton, & Stacey, 1985)
>
> *One way of capitalizing on the strength of social studies of science, and avoiding the reflexive dilemma is to devise ways in which alternative knowledge systems can be made to interrogate each other*
>
> (Watson-Verran & Turnbull, 1995, p. 138)

The Challenge

The introduction to this book gave a preliminary description that visuospatial reasoning incorporates a wide range of spatial abilities and skills together with visual and spatial imagery, representations, processes, and related concepts. Visuospatial reasoning is the mental process of forming images and concepts and mentally modifying and analysing these visual images. Visuospatial imagery involves the relationship, position, and movement of parts of an image or sequence of images. The spatial component of imagery may result from bodily movement as well as visual perception. Frames of reference may be bodily rather than in terms of visual or cardinal frames. Mental often dynamic or patterned imagery that embodies relationships is part of visuospatial reasoning. Imagery signifies a schema of the relationships, one in which the cognitive processes, some innate, some intuitive are not separated from the social context when reasoning. However, visuospatial reasoning was only one aspect of the development of an identity as a mathematical thinker. The role of visuospatial reasoning among other cognitive and affective aspects was positioned as important for self-regulation but more importantly the overall perspective showed how an ecocultural context influenced not only the ecocultural identity of the learner but its impact on the self-regulated mathematical learner.

Chapter 2 discussed a number of studies from educational psychology on spatial abilities and visual imagery and their relevance to our understanding of how children learn mathematics. It presented the strong legacy of this research to our understanding of visuospatial reasoning. In particular the chapter discussed linked studies that drew together the literature on visuospatial reasoning and illustrated its occurrence in classrooms. Aspects of these classroom studies illustrated how context—the particular problems, the materials, classroom expectations, and interactions with people influenced the cognitive and affective learning of children. The key study that showed the impact of a series of geometry problem-solving experiences on visuospatial reasoning emphasised that intention and attention played key roles in the way that the context influenced thinking. Nevertheless, without responsiveness there was no forward movement in problem solving and in turn the context was affected. However, mental schema, including visuospatial imagery and processes together with beliefs and values impacted on visuospatial reasoning. The effect of sociocultural and ecological situations was particularly relevant to what are often considered intuitive ways of visuospatially reasoning.

The third chapter established the importance of a critical perspective to understand visuospatial reasoning in terms of place. Place has both spatial and cultural aspects. Each impacts on visuospatial reasoning. Education, it was argued, should take account of place, ecology, and culture. Thus an ecocultural perspective on visuospatial reasoning was established. The fourth chapter showed how relevant this was from a linguistic point of view in terms of locating and comparing, two key aspects of geometry and measurement. The ecocultural perspective was expounded in terms of the rich cultures of Papua New Guinea in Chap. 5. The influence of culture on representations and creating representations was particularly evident. Visuospatial reasoning was illustrated in many ecocultural activities such as designing and constructing houses, boats, artefacts and in village activities such as food production and reciprocity exchanges. The importance of cultural identity and cultural ways of thinking was particularly evident in the areas of mathematical reasoning associated with space. Thus visuospatial reasoning was not only defined in terms of the complexities of psychological literature but it was also established in terms of an ecocultural perspective. This was supported by studies from other cultural groups in Chap. 6.

Importantly, in Chap. 7, it was argued that an emphasis on visuospatial reasoning addressed perception of mathematics itself. Reasoning was logical and systematic within the visuospatial realm. It was evident in Indigenous cultures that visuospatial reasoning is a mathematical strength that is currently not recognised and nurtured in schools. The ecocultural perspective challenged perspectives of mathematics. Support for this argument was granted by a number of school education programmes reported in Chap. 8 that took an ecocultural perspective. Visuospatial reasoning was valued and valuable in transitions between home and school that took account of ecology and culture.

Chapter 9 indicated that the ecocultural perspective also applied to digital worlds in which visuospatial reasoning was evident and where self-regulation was encouraged by the very nature of the inbuilt investigative system requiring reflection to try another approach. The context and tools encouraged children to use visuospatial reasoning, keeping their interest in investigating, and evaluating their progress resulting in their responsiveness and action. As a result, they established a sense of identity in working with the tools in a mathematical way.

The argument was established for an ecocultural perspective of visuospatial reasoning especially in geometry and measurement from the studies of Indigenous groups in various places especially in PNG. This ecocultural perspective of education applied to the digital age. This final chapter draws these arguments into establishing the importance of visuospatial reasoning from an ecocultural perspective in enriching the discipline of mathematics and mathematics education. This perspective provides a synergy for education to be for both local and global contexts.

Mathematics

A recent report for the Programme for International Student Assessment (PISA) stated:

> Mathematical literacy is an individual's capacity to formulate, employ, and interpret mathematics in a variety of contexts. It includes reasoning mathematically and using mathematical concepts, procedures, facts, and tools to describe, explain, and predict phenomena. It assists individuals to recognise the role that mathematics plays in the world and to make the well-founded judgments and decisions needed by constructive, engaged, and reflective citizens. (Organisation for Economic Co-operation & Development, 2013, p. 25)

This definition describes people with a mathematical identity as those who can engage with mathematics at the contextual level. In other words, people are able to think mathematically to make decisions related to their place, albeit local and global. Whatever processes and concepts are used, both involvement of context together with critical and consequential application are present in a developing ecocultural mathematical identity (Gresalfi & Barab, 2011). It involves the disposition to think mathematically to understand situations and to solve problems.

Furthermore, the quoted definition applies across cultural and ecological contexts. Thus a person creating a *kapkap* decorated disc in PNG will be mathematical in the way they create and make the disc with an arrangement of shapes representing human and cultural relationships. A person who is considering school trigonometry will apply trigonometry to the making of a bridge but could also use cultural visuospatial reasoning to select and join wood for the bridge. Both support each other when the synergy of ecocultural visuospatial reasoning is combined with western school mathematics. Experiences in the local and global world will provide for individual differences and excellences.

This book is not just probing ethnomathematics of Indigenous cultural groups in terms of visuospatial reasoning. It is arguing that these Indigenous ways of visuospatial reasoning are indeed strengths for mathematics, the growing field of geographic representations (National Research Council Committee on Geography, 2006), and other mathematical sciences.

> Visualization and visual thinking are far more than access, preparation, and motivation: They are worthy content themselves. Visual representations help people gain insights into calculations in arithmetic and algebra. Properties of mathematical processes are often discovered by studying the geometric properties of their visual representations. ... By ignoring visualization, curricula not only fail to engage a powerful part of students' minds in service of their mathematical thinking, but also fail to develop students' skills at visual exploration and argument. (Goldenberg et al., 1998, p. 6)

Today, mathematics curricula appear to be placing less emphasis on visuospatial reasoning under national and state testing regimes that find it hard to test except in terms of diagrammatic representations. However, visuospatial reasoning by comparison, ratio, and alternative relationships that link different dimensional knowledges are important for geographic and other sciences. The importance of recognising differences in non-numeric representations and collaborative decision making is a way forward in mathematics and hence needs to be addressed in mathematics education. It is crucial that Indigenous ways of knowing are investigated and become springboards for school education especially in the transition from ecocultural contexts to school.

Theories of Mathematics Education

This book has illustrated some of the newer synergistic ways of researching in mathematics education. It has gone further than what Presmeg (1998) has suggested as ways forward with theory building. Presmeg noted that some research had already used linguistic and philosophical disciplines to integrate cognitive and imaginative rationality through prototypes, metaphors, and metonymies in mathematics since these processes underlie human reasoning itself. The study of language about size and position for PNG languages reflects the ways of thinking of Indigenous PNG communities and their use of prototypes, metaphors, and metonymies in their representations, reasoning, and connectivity in designs and cultural activities. This book demonstrates how these theoretical approaches enlighten our understanding of imaginative visuospatial reasoning. Visuospatial reasoning is embedded in language, gestures, activities, and visual representations of objects and mathematical systems related to roles and relationships, beliefs and systems. General principles of education that take account of Indigenous knowledges associated with the environment and place (Gruenewald, 2008) are explicated in the chapters of this book.

Other research studies, as Presmeg (1998) noted, have brought the sociological and anthropological together with the psychological perspectives in discussing social interactionism, mathematical development, mathematical meaning, and the nature of mathematics itself. The psychological studies in Chap. 2 led to a recognition of context and responsiveness to context in terms of mathematics. Many of the studies presented in later chapters showed that context, specifically ecocultural context, was significant in learner and mathematical identity. This book in fact illuminates the important role of considering visuospatial reasoning from an ecocultural perspective to appreciate how the ecocultural context really impacts on reasoning that is important for self-regulation, responsiveness and hence mathematical identity.

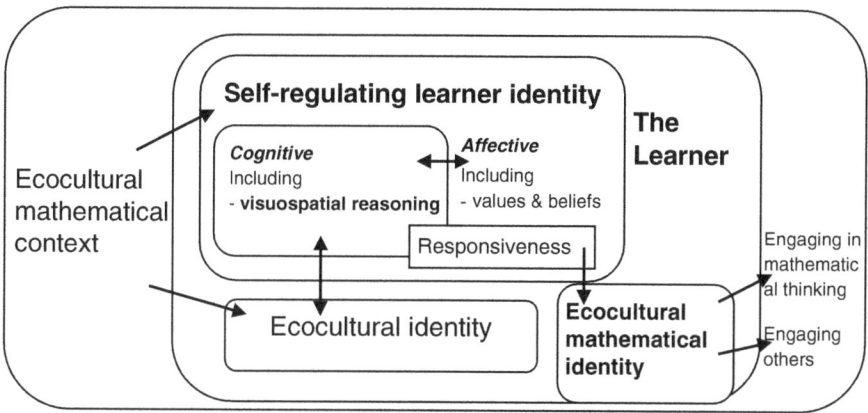

Fig. 10.1 Developing identity as a mathematical thinker (full details given in Fig. 1.2)

Anthropological studies of the making of *kapkap* (Chap. 5) and other objects (Chaps. 5, 6, and 8), the qualitative studies of house and other constructions (Chap. 5), and ways of discussing location (Chap. 4) underline how visuospatial reasoning is enmeshed in cultural connectivity. The study of artefacts and activities summarised in Chap. 5 and by teachers (Chap. 8) illustrated the hybridity (González et al., 2005) and in some cases the synergy of traditional and school ways of thinking mathematically. Underlying the many commentaries in Chaps. 5 and 6 of mathematical visuospatial reasoning and the significant role of culture in these ways of thinking is a valuing of the mathematical systems inherent in the cultures.

The arguments put forward in this book have crossed the thirdspace of educational spatiality (Luitel, 2009; Soja, 2009) in providing a rich understanding of the ecocultural perspective. In other words, the area of mental transition, of creating an understanding of space from different perspectives permits a local and global perspective and various degree of each for the individual if not for the sociocultural group. The ecocultural context is represented in Fig. 1.2 from Chap. 1, presented in this chapter as Fig. 10.1. This figure gives the theoretical picture for discussing visuospatial reasoning from an ecocultural perspective. Significantly, the figure represents the way in which ecocultural context directly and through an ecocultural identity impact on the developing learner's cognitive and affective self-regulating and hence being and becoming an ecocultural mathematical thinker.

Figure 10.1 provides an example of what Rosa and Orey (2012) suggest as a way of modelling mathematics and mathematics education from an ethnomathematics perspective that takes account of non-western approaches to mathematics education. This is particularly evident in the PNG elementary school project for *Cultural Mathematics* in which key principles highlight a strong emphasis on culture, home language treasures, and partnerships. Nevertheless, the proposal became practical through an inquiry approach that encourages the teacher and students to explore the mathematics of cultural activity in order to recognise, develop, and use unique ecocultural visuospatial ways of thinking (Fig. 10.2). In doing so, strong school mathematical thinking is established. The synergy between cultural ways of reasoning

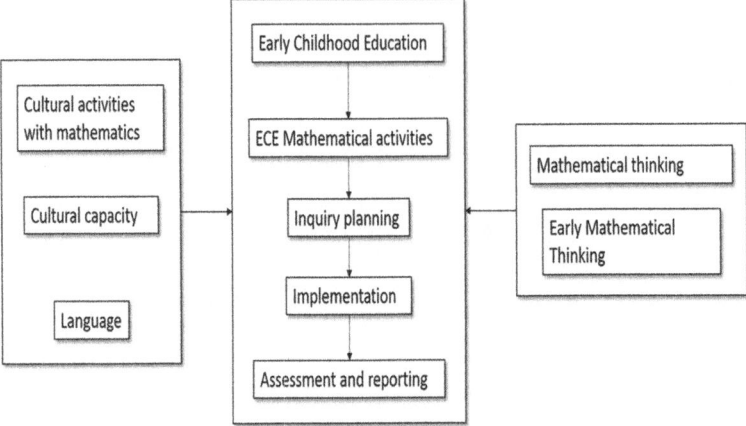

Fig. 10.2 Simplified design of principles for teaching ecocultural mathematics (Owens, Edmonsd-Wathen, & Bino, 2014)

and school mathematics ways of reasoning is illustrated in terms of multiplicative and ratio thinking but more importantly in recognising, maintaining, and strengthening cultural visuospatial reasoning.

Within theories of education, the emphasis on ecocultural contexts is significant in that it not only recognises the natural and physical environment of children at home and at school but it recognises the cultural groups' responses and systematic ways of thinking about geometry and measurement. The approach of this book leads to recognising not only a way of incorporating culturally responsive education but also an important approach for all mathematics education to recognise visuospatial reasoning and ecocultural perspectives more in order to strengthen mathematical learning for all students. The study of learning in a digital age justified this position. Thus local and global interactions and emphases are achieved in education.

The Synergy of Research Studies on Visuospatial Reasoning

I used to watch a clown fish take food and feed the anemone before he took food for himself. Of course, the anemone also provided the clown fish with protection and he ducked quickly into the tentacles if a large fish came near. This is a synergistic relationship in which two things come together for a greater outcome than the sum of the two independently. Combining reasoning, especially visuospatial reasoning, with geometry content will result in strengthening students' visuospatial reasoning needed in a wide range of experiences and geometry, another synergy. Visuospatial reasoning is an essential part of mathematical literacy applied in geometry and other areas of mathematics (Lehrer & Chazan, 1998). However, visuospatial reasoning requires development in the mind through many hands-on experiences. It is important to use activities that develop pattern and dynamic imagery. Actions of the learner—predicting, classifying, translating, and scaling—are important in visuospatial reasoning whether abstract or contextual (Kim et al., 2011; Rivera, 2011; Trninic & Kim, 2012). Students' active aesthetic and technical judgements connect to geometry, architecture, and other cultural activities through reflection and embodied activities of inhabiting, drawing, and theorising (Rawes, 2007). If the task itself is contextualised and knowledge shared among those of the community, the attention of the student is focused accordingly but still individually (Leinhardt et al., 1990; Owens & Clements, 1998).

The perspective developed from bringing together a broad range of research studies is represented in Fig. 10.1. A diagram is used to capture the complexity of the theoretical perspective emerging from this book. Within this perspective is the role of visuospatial reasoning.

Ecocultural mathematical context includes:

- Mathematical problems
- Ecocultural assistance
- Ecocultural tools
- Ecocultural supports
- Valuing the ecocultural identity and ecocultural mathematical identity

Mathematical problems occur in and outside school. In the Indigenous communities of PNG, these include construction of buildings and artefacts, building relationships, providing food and finances, and other aspects of living. Developed over time are ways of thinking and interacting that are mathematical, often involving visuospatial reasoning. Tools are often readily available such as ropes and sticks and body parts but there are also the shared visuospatial reasoning tools that are like mental visual ready reckoners incorporating relationships between spaces such as house size and areas of kunai for roofing. In addition, relationships between people, sharers of knowledge and cultural activities, and ways of sharing reasoning are all supports for developing both ecocultural identity and learner identity. Within the cultures, respect is held for those who have knowledge for various cultural activities but also those with school knowledge. The synergy of these two aspects ensures that the school learner investigates, supports, and extends the ecocultural ways of thinking. This knowledge in turn forms a strong foundation for school mathematical learner identity. The projects written by the teachers, some of which were incorporated into Chaps. 5, 7, and 8 illustrated the valuing of ecocultural knowledge and those who were using this knowledge to solve mathematical problems in cultural activities.

Ecocultural identity involves:

- Responsive social interaction
- Clear access to social meaning and relations
- Responsive self-regulation
- Co-participation
- Alignment

These aspects of ecocultural identity, being within a person are also mentioned in the self-regulating learner.

The cognitive aspects of the self-regulating learner are:

- goal setting
- reasoning
- planning
- self-evaluating
- reviewing
- information seeking
- using a tool box of strategies
- structuring the learning environment

Ecocultural identity was shown in the projects as a driving force in the teachers' projects resulting in responsive products in terms of the reports, the use of language words, and stages of the activities or representations of cultural artefacts, and the identification of people involved in the cultural activities. Furthermore, the students were responsively self-regulating both in the teachers' ethnomathematics projects (Chaps. 5 and 8) and the architecture projects referred to in Chap. 5. They set themselves goals, reasoned about how to achieve these goals, made plans but also self-evaluated their progress. In many cases, the teachers found out further cultural knowledge from their Elders and relatives. I trust my PNG ethnomathematics research incorporated into this book have been an adequate conduit to share some of this knowledge to illustrate its richness.

Affective aspects of the self-regulating learner include:

- engagement
- imagination & creativity
- resilience in problem solving
- confidence
- sense of ownership
- values and beliefs

The teachers and architects used other strategies including asking others for opinions but they also tried things out, applied ideas from other contexts, applied information, and made decisions about their work and learning environment. The architecture students particularly noted their feelings and engagement with the sculpture project while the teachers expressed their amazement and gratitude and cultural identity with their Elders' knowledges. All were responsive in problem solving moving forward in solving their problems that involved thinking mathematically. Responsiveness is the mental disposition that results in actions such as speaking, doing, and writing which in turn influences the problem-solving situation (see Chap. 2, Owens, 1993; Owens & Clements, 1998). There is both an affective and cognitive depth to responsiveness. It implies a sense of ownership and comfort with the mathematical situation. The teachers' generally positive reflections indicated their developing mathematical identity which specifically incorporated the fact that they knew their Elders, ancestors, and families were thinking in culturally appropriate mathematical ways. Interestingly part of that recognition often noted that it was "by eye", in their imagination, only the person knew and other terms indicating that this was not easily verbalised knowledge suggesting that the Elder or actor was using visuospatial reasoning.

The framework on mathematical identity presented in Fig. 10.1 meets the six criteria suggested by Dasen and Ripaupierre (1987), for good theories of cross-cultural and differential psychology. Such studies included those of Berry (Berry, 1966, 1969, 2003) that showed different and similar developmental growth in visuospatial reasoning in different groups. I contend the criteria apply to the model represented in Fig. 10.1 although it is a synthesis of studies from educational psychology and sociocultural perspectives. First, cultural difference is built in the model in Fig. 10.1 by emphasising ecocultural contexts and encouraging self-regulation and responsiveness to lead conceptual development. Stage theories of learning often focus only on the conceptual development and reduce the role of visuospatial reasoning and affect in learning. Not only is metacognitive or evaluative thinking of self-regulation occurring but this is embedded in affective and meta-affective thinking. In all this, there are individual differences; so the model permits "both commonality and individual, situational and cultural differences" (Dasen & de Ribaupierre, 1987, p. 805).

Second, the model permits differences at different levels of thinking. If engagement in mathematics is at a superficial procedural level, there would be no evident collation of self-regulating ways of thinking such as goal setting and checking processes. When a deep level of engagement is reached self-regulating is evident in a number of ways and will act with affective processes and dispositions to create a responsiveness inherent in a developing mathematical identity. Third, the ecocultural context is influencing and being influenced by the learner in specified ways. This raises the fourth criteria: through valorisation or as the model suggests ecocultural identity, individual performance will be affected by the ecocultural context. Evidence of cultural practices involving visuospatial reasoning in PNG and other countries is described in Chaps. 4–8. Fifth, validation has been enacted describing and analysing several studies on visuospatial reasoning in different ecocultural communities. Discussion of developmental differences in cultures and the differences in beliefs and values and ecologies that influence visuospatial reasoning have been highlighted in the studies. Finally, the model allows for individual differences and outlines a range of ways such as cognitive processing in which self-regulation develops.

Visuospatial Reasoning

Having established the importance of self-regulation, responsiveness, and ecocultural identity in the development of an ecocultural mathematical identity, I return to developing the importance of visuospatial reasoning. Imagination and intuition for concepts and known processes (see studies on adults discussed above and Rivera, 2011) play a part in establishing visuospatial reasoning. There is a two-way link in the diagram between visuospatial reasoning and imagination and creativity. Affect and cognitive processing are not discrete but are experienced together. Discussions in Chap. 5 in particular illustrate just how much the ecocultural perspective provides for this affective and cognitive linking. Visuospatial reasoning has strong cultural, emotional, and visual aspects. However, visuospatial reasoning also develops from problem solving as shown in the studies discussed in this book. Chapter 2 outlined not only heuristics (closely linked to self-regulating) but numerous cognitive processes such as conceptual, visual imagining, selective attention, and perceptual factors that are important in learning. Visuospatial reasoning incorporates these cognitive areas. Nevertheless, contexts impact on these cognitive processes.

Cultural practices that are taken-as-shared rules and habits-of-seeing impact on visuospatial reasoning (Hutchins, 1983). Nevertheless, structures can be instantaneous, result from practice (observed in ecocultural studies described above), or be recognised through the assistance of others. The whole may be held, details may be discerned, variation recognised, relationships recognised, properties declared as common to many instances, and reasoning undertaken on the basis of these properties and relationships (Mason, Stephens, & Watson, 2009). Importantly, the structure permits problem solving, generating change, and accepting variability consistently. Change in the object representing the geometry (figure or diagram, virtual or concrete, words or symbols) are representative of the change in all examples. The relationships and structures are part of visuospatial reasoning as shown in the studies with young children. However, there is a development both within the individual and within society as these change and develop (Saxe, 2012). Shared intention motivates for sharing visuospatial reasoning within the cultural situation (Tomasello, Carpenter, Call, Behne, & Moll, 2005).

Visuospatial reasoning incorporates the many spatial abilities required to investigate. For example, re-seeing a shape assists the problem solver to see and assess alternative relationships between parts or between shapes as illustrated by responses to the assessment tasks on visuospatial reasoning discussed in Chap. 2. Interpreting visual information also becomes significant in a cultural context depending on the degree of exposure to similar images and to their position within other areas of education and culture. The links between spatial abilities, visualising, and investigating were explored in Chap. 2. Particularly important in visuospatial reasoning is pattern imagery and dynamic imagery. The former is seen as a step towards generalisation while the latter is increasingly recognised as a way of students developing their concepts, justifying their reasoning about concepts, and imagining and developing imagery that began as static or prototypical images.

Visualising and visuospatial reasoning will impact on geometric concept development but at the same time concept development impacts on visuospatial reasoning as shown in the angle study with adults. Critical to concept develop is language whether this is through deliberate describing and classifying or as is often the case in Indigenous communities embedded in language structures and ways of classifying to identify certain objects, shapes, or other things that impact on visuospatial reasoning and concept development. Appropriate activities focusing on visuospatial reasoning with manipulatives such as those used in studies by Owens et al. (2003), Lehrer, Jenkins, and Osana (1998), Ness and Farenga (2007), Cherinda (2012), and Jawahir

(2013), and digital technology such as used by Highfield (2012), Highfield, Mulligan, & Hedberg, (2008) and Sarama and Clements (2004) extend students' space and geometry knowledge through visuospatial reasoning (discussed in this book).

Cognition involves conceptual and imagistic processing as well as heuristic processing during problem solving while affect often empowers or limits cognitive processing (Goldin, 1992, 2000; Owens & Clements, 1998) (see Chap. 2 for discussion and Fig. 2.17). For example, acceptance of open-ended questions may impact on students' willingness to attempt this kind of question as well as their knowledge of how to attempt such questions. Positive beliefs about being a problem solver develop as problems are successfully undertaken with cognitive strategy assistance or "aha" moments drawing the problem solver back to positive attitudes (Goldin, 2000). Transition of cultural identity to incorporate cultural mathematics can positively impact on establishing a mathematical identity incorporating culture and school mathematics. In particular, cognitive and imagistic strategies, self-regulation, and goal setting that are culturally centred develop this sense of identity (Owens, 2014, see Chap. 8).

Imagination creates new images of the mathematics and the learner as a mathematical thinker in a social context (Owens, 1997b; Wenger, 1998). Imagery is a dynamic tool for giving concepts meaning (B. Davis, 1999). For example, the person can imagine a triangle changing shape or turning around. Imagery also allows one to visualise oneself as an actor solving the problem (Owens & Clements, 1998). Imagination precedes action and provides the opportunity to perceive different possibilities. Imagination partners risk taking (Owens, 1998a; Wenger, 1998) as learners attempt activities beyond their current repertoire of procedures and feelings of familiarity and comfort. Imagination also partners resilience (Goldin, 1992; Zimmerman, 1990) whereby the person will persist to solve a problem despite discomfort. The idea of seeing oneself as an actor in solving the problem is more transient than Sfard and Prusak's (2005) view of learning as moving a person from their actual identity to their designated identity. It is a part of each problem-solving experience that builds up to a belief in oneself as a problem solver (Goldin, 1992, 2000). This learning is part of the dynamic formation of identity and is not solely directed by the cultural knowledge and expectation of being a mathematical thinker. The identity is a state of being and becoming; it is constantly evolving (in B. Davis, 1999 terms); it is current rather than a goal or a narrative as suggested by Sfard and Prusak (2005).

The Importance of Visuospatial Reasoning

An important aspect of learning is self-regulation and the different types of imagery play their role in visuospatial reasoning at different times and in different situations permitting self-regulation. Furthermore different types of imagery are enhanced by different ecocultural experiences. For example, cultures that encourage careful watching and attending allow for procedural imagery to develop early. Ways of making artefacts that entail patterns or dynamic imagery encourage these kinds of imagery and need fostering throughout schooling. Nevertheless, visuospatial reasoning plays a crucial role in self-regulating during problem solving and learning.

The review of studies of ethnomathematics in Chaps. 5–8 shows that visuospatial reasoning is important in practice in Indigenous communities. Taking an ecocultural approach to mathematics in schools and considering the links or alternatives to western mathematics requires making these links and alternatives explicit. This approach enhances mathematics and mathematics education. It is not so unlike the development of western mathematics, that Indigenous communities use a generate and test process (Van Moer, 2007). While some western mathematics is deductive, much is comparative and inductive. Furthermore, the use of visuospatial reasoning may be more important than proof or generalisation (Rivera, 2011).

Culturally, visuospatial reasoning plays a crucial role in self-regulation and it in turn develops identity as a mathematical thinker. Culture and cultural identity are not only directly impacting on identity as a mathematical thinker through appreciating that culture has rich ways of thinking mathematically but indirectly in developing self-regulation through visuospatial reasoning influenced by ecocultural contexts. The projects presented in Chaps. 5, 7 and 8 illustrate this influence. The secondary teachers' projects showed the motivation that cultural practices had. The vision to link cultural practice to school mathematics drove the teachers to produce good projects. Their self-regulation came from setting their own goals, reasoning about the mathematics and particularly visuospatially as they considered artefacts and activities, they planned appropriately for teaching and for presenting their reports, they evaluated whether their cultural practice did involve mathematics by comparing with the syllabus, they sought information from Elders, craftsmen, and books, and they used technology, observing, questioning, and visuospatial representations to develop their knowledge and to present it. They showed a sense of pride and ownership in their projects and resilience to complete them. They acted or responded to develop their ecocultural mathematical identity by presenting their cultural mathematics and linked school mathematics.

The elementary teachers likewise developed learning through discussion and reasoning as they came to grips with the mathematics of their cultural practices. Often they acted out their plans illustrating their visuospatial realisation of mathematical knowledge in their cultures. Their pride in culture was evident in their keen participation and their grasp of their cultural mathematics. They valued their subject but they also valued school mathematics, especially the way they learnt it. The challenge was difficult as they developed their mathematical identity but they valued their new learning through the workshops.

The architecture students also developed their identity building on their cultural identity to show self-regulation in terms of goal setting, problem solving, reviewing their work, and visuospatially reasoning to create their sculptures. They responded and showed how pleased they were with their imagination and creativity and their final products. They could articulate when their ecocultural context influenced their thinking.

Without an ecocultural perspective, it would be hard to express the importance or nature of visuospatial reasoning. It is for this reason that many mathematics education programmes have paid little attention to its value. The importance of self-regulation and cultural identity on learning and mathematical identity is now well established. Within this perspective, visuospatial reasoning is a key to problem solving, self-regulating, and identifying with mathematics.

Contexts for Visuospatial Reasoning

Early visuospatial experiences (such as Japanese and Chinese writing, and socio-economically provided experiences with "writing", drawing, and viewing) have shown positive differences in developing spatial manipulation and later visuospatial reasoning (Clements & Sarama, 2007a). However, the PNG studies presented in Chap. 5 show how objects that are made in a culture have particular significance in learning about what to attend to and what is valued in different relationships in society and how it is represented in the artefact. When there is western education, then a hybridity or confusion may develop in terms of what is seen as significant shapes. Language either facilitates or hinders a hybridity of understanding which is apparent with the ubiquitous use of Tok Pisin in PNG and an English-language, western dominated curriculum. Tok Pisin may not express the cultural depth as Tok Ples does and it does not necessarily express the depth of the school mathematics. Pinxten (Pinxten, 1997; Pinxten et al., 1983) noted a similar issue for parts and wholes for the Navajo. Nevertheless, in many environments and cultures, visuospatial representations do show wholes (e.g. the dominance of circles), sections (e.g. of line combinations), connections (considering where and how parts are combined), and later boundaries (see the lack of boundary and orientation in PNG children's mapping, Chap. 4).

Cross-cultural studies showed that not all children develop along the same lines that Piaget had suggested in terms of egocentric to cardinal descriptions of location. Psychological studies also suggested that the diversity of spatial abilities and visual imagery classifications were so extensive and used in such diverse ways that hierarchical structural theories restricted opportunities if they dominated the curriculum. Children's development in visuospatial reasoning follows possible trajectories from emerging strategies resulting from informal experiences, perceptual strategies encouraged by movement with and around objects, initial pictorial imagery strategies in which re-seeing and viewing in alternate ways including those influenced by culture were important, and pattern and dynamic imagery encouraged by problem solving in cultural and novel situations. Visuospatial efficiencies develop through experiences that are culturally relevant, motivating, and practiced but also associated mentally with verbalised patterns and relationships. However, these visuospatial strategies are not necessarily ordered or age or stage related but rather alternative strategies used as required. These strategies remain important and occur at different times in development of concepts and visuospatial reasoning throughout life.

Describing and classifying are dependent on language and the sociocultural background of the student as illustrated in Chap. 2. Interestingly, both Lehrer, Jacobson, Thoyre, Kemeny, Strom, Horvath et al.'s (1998) and Owens and consultants' (Owens & Reddacliff, 2002; Owens, McPhail, & Reddacliff, 2003) research with children showed the role of context, and the richness and interaction of cognitive processes

(Owens, 1993), and the diverse ways in which students learn. Both studies considered that stages such as those purported by van Hiele's theory or later variants, usually based on SOLO taxonomy (e.g. Pegg & Davey, 1998), did not adequately explain what was happening. Lehrer et al.'s main argument was that learning was more incremental and that variance was so great that imposing stages on children's development was not so helpful. Part of the reason for this was the complexity of mental processes that impact on learning, not to mention the impact of experiences. For example, Clements (1998) pointed out one 3-year-old was able to achieve higher scores on his test than all the 6-year-olds he tested.

Mathematics Teacher Education for Visuospatial Reasoning

The right in Sweden to education in your home language for the first 3 years of schooling means that schools employ or involve speakers of the child's language, specific Sámi schools are established, and the Sámi Education Board considers the need for Sámi mathematics in schools. One project involving the University in Luleå aims to increase capacity building in schools and provide teacher education about colonising processes. Teacher education focuses on cognitive, cultural, communicative, creative, critical, social, and didactic competences (Johansson, 2008). At the same time, teachers develop projects with parents and the local community around language training, dance, music, joik (a Sámi way of singing), natural sciences and spirituality, storytelling pedagogy, the use of Elders in the daily work, independency and responsibility, and transferring knowledge e.g. branding reindeer calves and reading the signs of nature. Education begins outside with an Elder. Other challenges for teachers occur around bilingual education and revitalising of the Sámi language (Jannok Nutti, 2013).

Such an approach was developed in Dubbo, NSW with a forum organised by the local Wiradjuri community and the local University (Owens et al., 2011, 2012). This forum established the importance of sharing, of valuing Indigenous knowledge, and specifically noting the past as it impacts on the present and future. It becomes important for Indigenous people to have empowerment through inclusion, at various levels including that of their local community. Group interests are still an important component of the political machine but sameness has been a focus of politics and liberal democracy rather than recognition of religious or cultural differences. Project funding rather than long-term funding has reduced the all important impact of personnel in this regard. The teacher and community members who participate long-term will have a greater impact on the child's continuous education. Yolgnu in the north of Australia talk of *yothu yindi*, literally mother child, "a metaphor for the balance and negotiation which runs through the natural world and should govern the social world" (Thornton & Watson-Verran, 1996, p. 6). Furthermore, teachers are expected to plan to engage students in their own classrooms in mathematical thinking and thus fulfill the position taken by Wenger (1998) regarding the importance of identity for creative teaching and learning.

Given that Indigenous cultures have much to offer in terms of visuospatial reasoning, then the values and strategies outlined in the previous two paragraphs provide for ways of incorporating visuospatial reasoning of Indigenous cultures in schools. The descriptions of Sámi cultural activities in Chap. 7 are indicative of ways of implementing in the school. Similarly the various studies referred to in Chap. 8 indicate the synergy between culture and school ways of visuospatial reasoning results in an emphasis on visuospatial reasoning and a strengthening of this aspect of mathematical thinking and mathematical identity.

Visuospatial Reasoning in Geometry and Measurement

From the beginning of this book, I have argued that problem solving has been one activity in which visuospatial reasoning occurs. A decision to place geometry and measurement into other areas of mathematics and the curriculum is one way of generating problems to solve. Examples of learning occurred in "Cultural Mathematics" as in PNG and in Alaskan mathematics units based on cultural practices like collecting eggs or building fish racks. Courses involving probability and data management can also have a focus on cultural practices especially measurement practices linked closely to ethnoscience and ethnotechnologies and the mental ready reckoners of practitioners as found in PNG. Links with geography that recognises the importance of place also encourage a valuing of cultural mathematics of people.

Ecocultural activities such as finding the middle of the side of a wall with a fixed length stick (a Kopnung village example) or using two equal ropes joined at the centre to draw house floor plans (an African practice) generate problems that can be extended and investigated systematically. Such an activity could preempt students finding the shape of the join of the midpoint of quadrilaterals which might be of all kinds (concave and convex, some sides the same, all sides different), and even find the areas of the whole shape and the smaller shape. Links to designs that might use this feature are possible. Estimating, tabulating, predicting, generating geometric figures with specific properties, modelling situation to explore mathematical relationships, and making sensible measurement are all involved in such a problem. It can even be tackled at various school grades in various ways. Such problems provide for a transcultural learning experience that values cultural practices and it encourages visuospatial reasoning (Enderson, 2003). The teachers' ethnomathematics projects and the elementary school project both aimed to show teachers how they could take cultural practice and extend and go further to develop the mathematical concepts and reasoning.

A key argument in this book has been the centrality of visuospatial reasoning in mathematical thinking especially in geometry and measurement. Strengthening, and at least not losing, the wealth of skills learned in community in Indigenous groups is argued as a key to improve mathematics education for these students. These are not only strong logical mathematical skills but they are also imbued with culture that has been shown to have a strong role in establishing mathematical identity through self-regulation and motivation. When cultural mathematics is left unconnected to school mathematics dissonance develops (Esmonde & Saxe, 2004; Presmeg, 2002). Furthermore, strengthening visuospatial reasoning through problem solving for all students was shown by studies described in Chap. 2, in Cherinda's and Jawahir's studies (Chap. 8), and Highfield's study (Chap. 9) as effective for learning geometry.

Developing Views of Spatiality

Having perused a range of Indigenous mathematical ways of reasoning, let me quickly review the way in which western views of space developed. The ancient Greek philosophical positions gave western mathematics the view of infinity of space, of unity or homogeneity of space, of immaterial but real space. Objects were in space but were perceived and experienced in terms of learning. In similar ways, Indigenous views of space incorporate a timeless perspective of place and a cohesiveness of place and objects with people and ideas. The immaterial is as strong as the real. Visuospatial reasoning mentally with or without objects or drawings or equipment is linked to the past and future through relationships to people and ecocultural activities.

Structuring space and place is dependent on culture and environment as evident from the diversity of ways of giving position in space and relating to that space (Chap. 4). This was evident from PNG and Pacific cultures but also Navajo, Wiradjuri, and Yolngu. Interpreting, making, and using paintings and other objects were structured visuospatially as evident in Chap. 5 on PNG cultures. Furthermore, the early childhood research studies illustrate children's visuospatial reasoning with Logo and other non-Logo navigational programmes that move small robots on the floor. Spatial concepts such as right and left, movement and order of movement are established through visuospatial reasoning in exploratory activities.

Place-based education fosters connection and attachment to local places and provides interrelationships between one place and another. Place-based education contextualises opportunities to explore the ecological, social, and political dimensions of those places and recognises Indigenous conceptions of place as an inseparable link between person and country (Cameron, 2003). Critical pedagogy of place involves educating within a local and ecological context, identifying and challenging oppressions of race, class, and gender (and nature), decolonizing, and reinhabiting (Gruenewald, 2008). Mathematics plays a critical role in this. First with the involvement of the community, the curriculum can be decolonized and challenged. Second, students' skills in understanding and creating maps of position including those with geographical information are critical for their decision making. Understanding graphs and systems helps in recognising that ecological systems are mathematically non-linear and are self-regulating but that global warming is moving in a trajectory that will take it outside the recovery orbit. Thus visuospatial reasoning is critical in a global world.

The Issue of Equity

An ecocultural perspective in mathematics education is a critical pedagogy of place that impacts on learning about location, space, and place. Teaching and curriculum that take account of the complexity of classrooms (François, 2010) take the philosophical perspective that each child has an equal right in the classroom. This goes beyond saying that each child should be presented with the western cultural view of mathematics. It suggests that diversity of ecocultural perspectives offers a better perception of one's own mathematical practices and provides opportunity for all students to appreciate, learn from, learn with others having alternative visuospatial ways of reasoning, and not to be limited especially by Euclidean ways of thinking or structuralists' restrictions on curriculum (e.g. from zealous application of Piagetian and van Hiele stages).

François (2010) noted that it is often said that one learns better about language including one's own if one learns more than one language. People's outlook is broadened in this globalised world.

> This comparison could even be extended to the mathematics education where knowledge of mathematical practices of several cultural contexts and throughout time proves to be advantageous. (François, 2010, p. 199)

Tuinamuana (2007) in Fiji and González et al. (2005) in the USA also show that critical pedagogy with an inquiry-oriented paradigm results in a hybrid way of teacher thinking that serves their students better than a curriculum based on the western view of mathematics and psychological theories. Not only did students become engaged but teachers noted how their thinking was modified and their views of mathematics developed. Furthermore, when school mathematics is imposed without regard to cultural background, it encourages a belief that mathematics is rote practice of procedures rather than about understanding and relevance (Lave, Smith, & Butler, 1989).

Chapter 8 illustrated the inquiry method (Murdoch, 1998) with examples that elaborated:

- Tuning in with cultural activities
- Finding out more about the mathematical ways of thinking culturally
- Sorting out to find patterns and relationships
- Going further to make additional connections and investigate the problem further
- Drawing conclusions that often involved visuospatial representations
- Taking actions that applied the mathematical thinking and concepts or involved sharing with others or created cultural objects
- Reflecting on the mathematics and taking ownership of the mathematics

The challenges for teachers were in creating the connections between tacit cultural knowledge and language structures, to develop the mathematical ideas and extend them to incorporate school mathematics.

An interactive and collaborative classroom approach in multicultural classrooms benefits all and assists in establishing the equal valuing of mathematical ways of thinking (César, 2009; Verlot & Pinxten, 2000). If the human, cultural aspects of mathematical reasoning are promoted in the classroom, then engagement of students occurs and mathematical thinking is extended by the multiculturality of a collaborative setting. Part of this involves acknowledging one's own cultural approach and ways of engaging with mathematics (Goldin, Epstein, Schorr, & Warner, 2011; Verner, Massarwe, & Bshouty, 2013). Reflecting on learning this ethnomathematics is critical for the individual and for the group to achieve a strong ecocultural approach to visuospatial reasoning.

Challenges Addressed

Rivera and Rossi Becker (2007) explored the issues of whether ethnomathematics could be incorporated into the schooling of children (Adam, Alangui, & Barton, 2003). While social justice and identity are addressed by incorporating ethnomathematics into classrooms, there are several issues. As these authors pointed out, sometimes the ethnomathematics of groups was limited and far from a full mathematics curriculum, other times the language was so different to the school language such as English that it was not possible to express the mathematical thinking of the group in English while the worst situations were just using ethnomathematics as an introduction or a side issue to teaching the school curriculum mathematics. This last approach may have motivated and linked children to school mathematics but it could also have led to an identity of a discarded past that was not valued. Truly globalising and internationalising mathematics require yet another perspective in which the curriculum and planning could incorporate not only similarities between mathematics but also alternatives and synergies of the mathematical ways of thinking. In fact school mathematics should be seen as one mathematics among many ethnomathematics.

This book has explored in depth one of the strengths of many of the Indigenous and non-mainstream mathematical ways of thinking, namely visuospatial reasoning. By emphasising this aspect of learning and reasoning it is possible to develop a new approach to mathematical curriculum and to find a common ground between different types of mathematics. In Chap. 8, we considered several examples of ways forward for schools:

- The use of language and manipulatives in investigating the mathematical concepts albeit from a standard school curriculum.
- Using the ecocultural context for embedding concepts in material culture, language, and problems that were relevant to students.
- Using an inquiry approach in which the key focus of the learning plan was the cultural mathematics which was investigated and how that investigation could be developed in terms of good practices for learning concepts and procedures appropriate for western mathematics.

Thus the western mathematics was regarded as an aspect of ecocultural mathematics. The strengthening of visuospatial reasoning in a digital age as illustrated in Chap. 9 supports the argument. In each example there was visuospatial reasoning about the cultural context or activity that was to lead to an expansion that might have been generated by other cultural practices or western mathematics. The argument expounded in this book emphasised the role of an ecocultural perspective in the framework of the development of a mathematical identity. Ecocultural context influences ecocultural identity and impacts on the developing learner's cognitive and affective processing including visuospatial reasoning and self-regulating. Through being responsive, the learner is being and becoming an ecocultural mathematical thinker.

Abbreviations

DEEWR	Department of Education Employment and Workplace Relations, split to Department of Education (Australian National Government)
GIS	Geographic information systems
NCTM	National Council of Teachers of Mathematics (North America)
NDOE	National Department of Education, Papua New Guinea
NRCCG	National Research Council Committee on Geography (UK)
NSWDET	New South Wales Department of Education and Training, became Department of Education and Community (DEC)
PNG	Papua New Guinea
Unitech	PNG University of Technology
UoG	University of Goroka

Appendix A
A Synthesis of Problem-Solving Processes

Table A.1 Problem-solving processes from some research literature

	Heuristic	Conceptual	Imagistic	Affective
Goldin (1987)	• Heuristic • "aha"	• Verbal syntactic • Formal notational	• Imagistic	• Affective
Lester (1983)	• Control	• Knowledge		• Affects • Beliefs • Sociocultural
Clarke (1989)	• Structural strategies	• Mathematical principles, procedures, facts		• Personal
Schoenfeld (1985)	• Control strategies	• Resources		• Beliefs
Polya (1957)	• Plan • Implement	• Understand • Check		
Krutetskii (1976)	• Analysis, synthesis	• Understand the kind of problem		• Perception of problem
Yee (1990)	• Problem orientation heuristics • Problem solution heuristics • Metacognition	• Domain-specific knowledge		• Affective behaviours
Collis et al. (1992)	• Structure recognition • "aha"	• Symbols	• Diagrams • Images	• Beliefs • Reality • Common sense

(continued)

Table A.1 (continued)

	Heuristic	Conceptual	Imagistic	Affective
Pirie/ Kieren (1991)	• Folding back	• Property noticing............................		• Primitive knowing
			• Image making	
			• Image having	
		• Formalising............................		
		• Structuring............................		
		............................• Observing		
		• Inventising............................		

Note: Several of the terms used are best placed in more than one category as indicated by the *dotted line* stretching across to the other category

The conceptual processing category refers to the verbal and symbolic aspects of conceptual processing

Appendix B
Test of Visuospatial Reasoning for Young Children

This test has instructions which may be obtained from the author. Each section is introduced by examples with cardboard cutouts. It is not timed so children are requested to have a book to read or picture to draw if they finish before their classmates as it is done by section. To administer each child requires stickers coloured appropriately. For scoring in my study I selected items that had good validity according to a Rasch analysis (Owens, 1992a). (Items were initially chosen from a trial set using discriminant analysis.) A simplified black-and-white version with few items is also available.

Appendix B: Test of Visuospatial Reasoning for Young Children

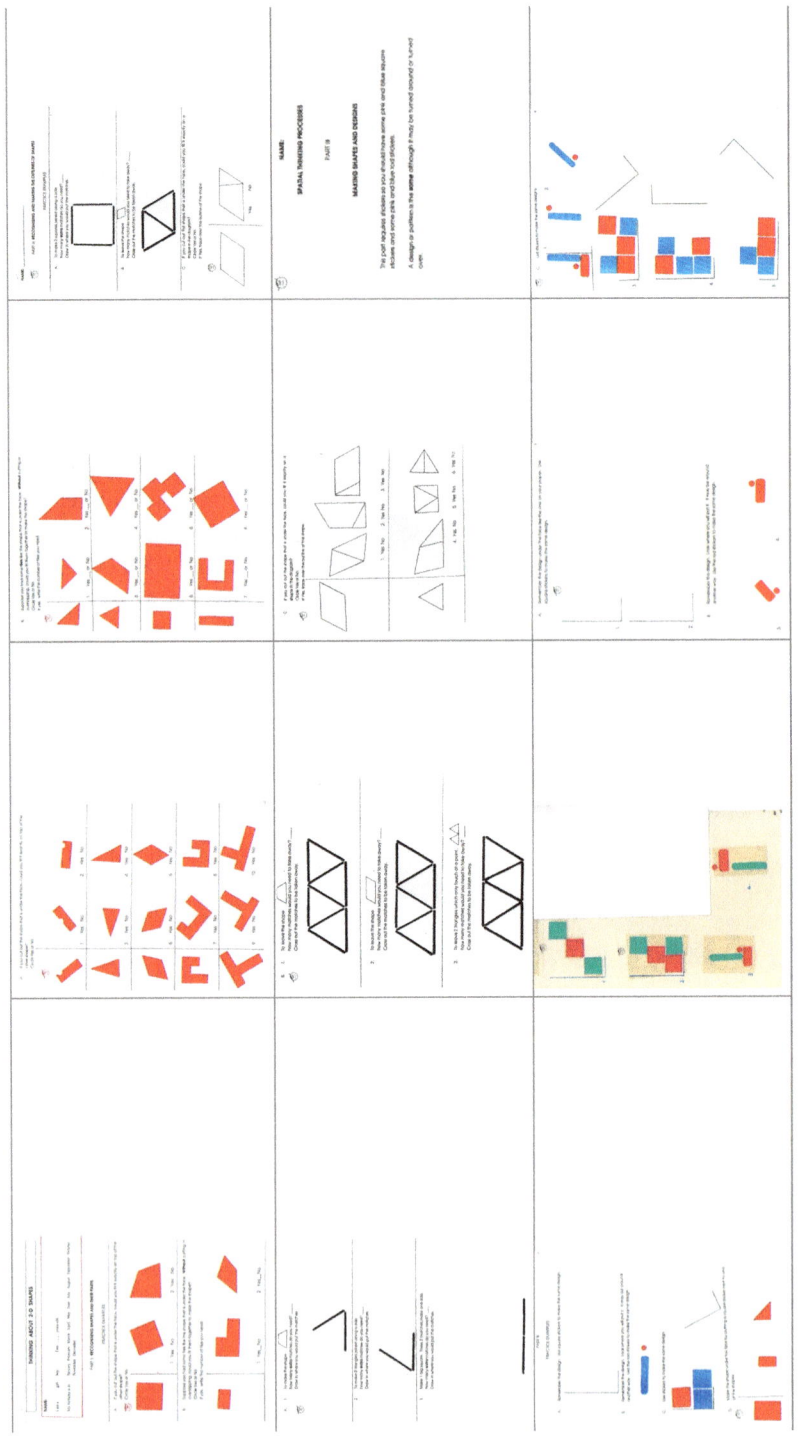

Appendix B: Test of Visuospatial Reasoning for Young Children

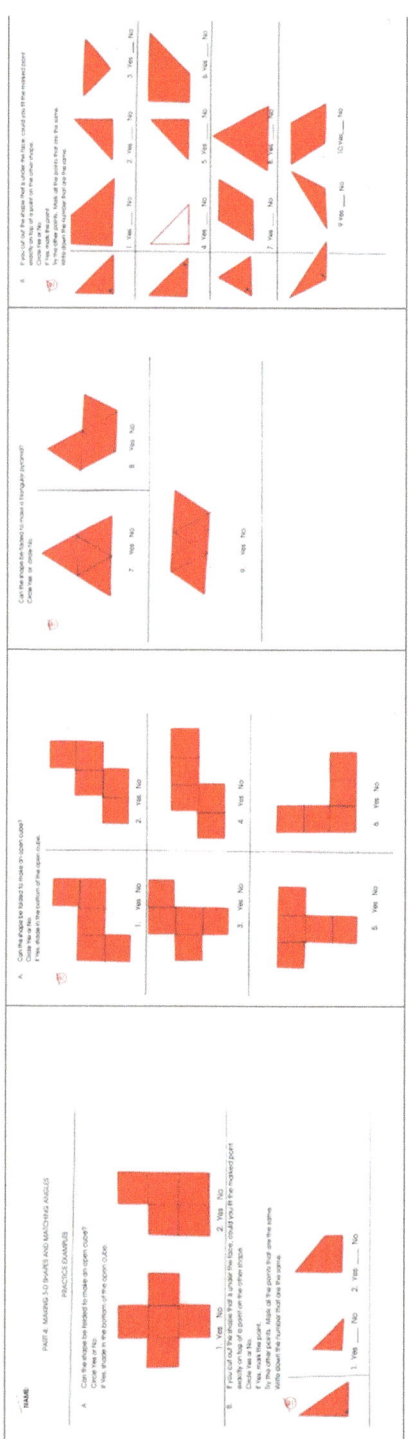

Appendix C
Map of PNG and Some Counting System Details

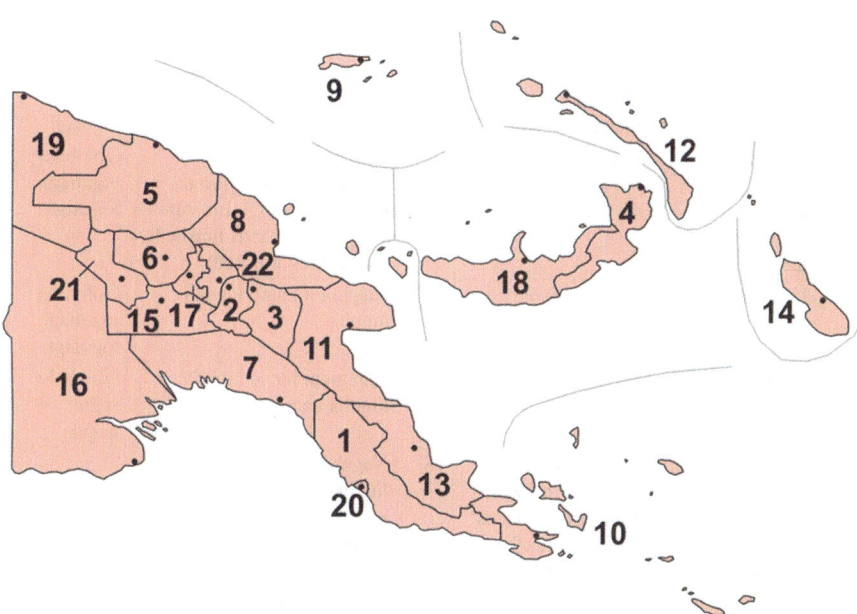

Map of Papua New Guinea with provincial boundaries. Source: Wikipedia. 1 Central, 2 Simbu, 3 Eastern Highlands, 4 East New Britain, 5 East Sepik, 6 Enga, 7 Gulf, 8 Madang, 9 Manus, 10 Milne Bay, 11 Morobe, 12 New Ireland, 13 Oro, 14 Autonomous Region of Bougainville, 15 Southern Highlands, 16 Western, 17 Western Highlands, 18 West New Britain, 19 Sandaun, 20 National Capital, 21 Hela, 22 Jiwaka

Papua New Guinea lies to the north of Australia with considerable history linked to Australia. The western half of the main island is under Indonesia now but often called West Papua.

Details of Counting Systems

Table C.1 Alternative numbers in Gahuku provided by two men from Kaveve, Eastern Highlands, PNG (Table 3.1 from this table)

	First man	Second man		First man	Second man
1	Hamo	Hamako	11	Nigizani logosi asu oko ligisaloka hamo oli'o malago (two hands finished and one on the leg)	Golohaki hamakoki
2	Logosita	Logosi	12	Nigizani logosi asu oko ligisa loka logosi oli'omalago (two hands finished and two on the leg) or nagahuni makoki logosita	Golohaki logosigi
3	Logidigi hamoki	Luguha (logosigi moka)	13	Nagahuni makoki logosigi makoki	Golohaki luguhagi
4	Logosivi logosive	Logosigi[2] "Logosigi squared meaning 2 plus 2"	14	Nagahuni makoki logosi logosi	Golohaki logosigi logosigi
5	Logosigi logosi hamo or nigizani hamo asu igo (one hand finished)	Logosigi luguhagi	15	Nagahuni makoki logosi logosi hamo or nigizani logosi asu 'olo ligisa hamo asuigo (two hands finished and one leg finished)	Golohaki luguhagi logosigi
6	Luguha luguha	Luguha logosi	16	Nagahuni hamo luguha luguha	Golohaki luguhagi logosigi
7	Luguha luguha hamoki	Luguha logosigi makoki (segininaga)	17	Nagahuni hamo luguha luguha hamoko	Golohaki segini nagaki
8	Nigizani hamo asu o'oko makotoka logsive hamo ol omalago	Logosi[4] means logosi is repeated four times "means 2 plus 2 plus 2 plus 2 plus"	18	Nagahuni hamo, luguha luguhagi logosi or nigizani logosi asu'oko ligisahamo asuiko mako toka luguha oli'o'mallago (two hands finished, one leg finished, and three on the other side)	
9	Nigizani hamo a su o'ko logosive logosive oli'o malago	Luguhagi luguhagi luguhagi	19	Nagahuni logosi, hamo hakene igo (20-1) or nigizani logosi asu okake nigisa mako asu oko makotoka logosigi logosi oli'o'malago	
10	Nigizani logosi asu igo (nagahuni hamo)	golaha	20	Nagahuni logosi or nigizani logosi aso'oko nigisa logosi asu igo (two hands finished and two legs finished)	

Appendix C: Map of PNG and Some Counting System Details 319

Table C.2 Names for classificatory groups in Kilivila[1] (Table 5.1 is a selection from this table)

Participle	Category description
Group 1	
Tay (Te)	Human beings; males (men, boys)
Na	Persons of female sex; animals (pigs)
Day (ke)	Trees and plants; wooden things; long objects (canoes, sticks, poles)
Dway (Kwe)	Round, bulky objects; stones; abstract nouns (betel nut, houses, yams)
Ya	Leaves; fibres; objects made of leaf or fibre; flat and thin objects (coconuts, spherical containers, clothes, string)
Sisi	Boughs (branches)
Li	Forked branches; forked sticks
Kavi	Stone blades
Kwoya (mweya)	Human and animal extremities (legs, arms); fingers of a hand
Luva	Wooden dishes
Kwoyla (kwela)	Clay pots (cups, containers)
Kada	Roads
Kaduyo	Rivers, creeks, sea passages
Vilo	Villages
Group 2	
Kila	Clusters ("hands") of bananas
Sa	Bunches of betel nut
Bukwa	Bunches of coconut
Group 3	
Pila	Parts of a whole; divisions; directions (books)
Vili	Parts twisted off
Bubwa	Parts cut off by transversal cutting
Utu	Parts cut off; small particles
Si (Sisili)	Small bits (slices of meat, bread)
Kabila	Parts of meat cut off from animals
Group 4	
Kabulo	Protuberances; ends of an object
Nutu	Corners of a garden
Niku	Compartments of a canoe
Kabisi	Compartments of a yam house
Nina	Parts of a song, of a magical formula
Mayla	Parts of a song, of a magical formula
Kubila	Large land-plots-ownership divisions
Siwa	Sea portions-ownership divisions with reference to fishing rights
Kala	Days
Siva	Times (number of occurrences)
Group 5	
Kapwa	Bundles—wrapped up (packages)
Oyla	Batches of fish

(continued)

---

[1]Malinowski (1917) with additions from Counting System Questionnaire data given by tertiary students provided in parentheses and noted at the end.

Table C.2 (continued)

Participle	Category description
Um'mwa	Bundles of taro
Kudu	Bundles of lashing creeper (dried reeds)
Yuray	Bundles of four coconuts, four eggs, four water bottles
(Kupwa)	(Fish counted in twos)
(Kayo)	(Crabs counted in twos)
Group 6	
Kasa	Rows
Gili	Rows of spondylus shell disks on a belt
Gula	Heaps
Group 7	
	Numerals without a prefix are used to count baskets of yams and numbers can be very large
Group 8	
Uwa	Lengths, the span of two extended arms, from tip to tip (fathoms)
Respondents to the Counting System Questionnaire also gave	
(Ta)	(Baskets)
(Tam)	(Vines)
(Yata)	(Watermelons, pumpkins)
(Bwa)	(Short or thick solids)
(Kaula)	(Groups of 20)

Appendix D
Making the *Bilum* Pattern 50 Toea or Soccer Ball

An example of this design can be found in Fig. 5.16g. Stana Amos (2007) provided the following instructions relating it to school mathematics.

Requirement: Four (4) different colours of choice: orange, white, green, and black: twist the yarn	– 1 loop (green)
Start by making:	– 1 loop (black)
– 5 loops (orange)	– 5 loops (orange), it goes up to where orange is and it stays there
– 4 loops (white), leaving one loop of orange	– 1 loop (white)
– 3 loops (green), leaving one loop of white	– 1 loop (green)
– 3 loops (black), it goes up to the mouth of the bilum and make two loops and it stays there	– 3 loops (black), it goes up to where black is and it stays there
– 1 loop (orange)	– 4 loops (orange), it goes up to where orange is and it stays there
– 1 loop (white)	– 1 loop (black)
– 3 loops (green), it goes up to the mouth of the bilum and make ten loops and it stays there	– 7 loops (white), it goes up to where white is and it stays there
– 1 loop (black)	– 1 loop (orange)
– 1 loop (green)	– 10 loops (green), it goes up to where green is and it stays there
– 3 loops (white), it goes up and it makes seven loops, leaving one loop of green and it is below green	– 1 loop (white)
– 1 loop (black)	– 1 loop (orange)
– 3 loops (orange), it goes up and makes four loops and it is below white, it leaves one loop of white	– 3 loops (black), it goes up to where black is and it stays there and the process continues, so every time when these ladies make bilum they have in mind how many loops to make and where to leave and start the next half of the design

(continued)

(continued)

- 3 loops (black) and it is below orange. It takes one loop of orange, white, and green and it goes to the mouth of the bilum making two (2) loops and it stays there
- 1 loop (green)
- 1 loop (white)
- 3 loops (orange), and it goes up to the mouth of the bilum making ten (10) loops and it stays there
- 1 loop (black)
- 1 loop (green)
- 3 loops (white) and it goes up and makes seven (7) loops leaving one loop of orange and it stays there
- 1 loop (black)
- 3 loops (green), it goes up, makes four (4) loops, leaving one (1) loop of white
- 4 loops (black), it goes up taking one (1) loop of green, white, and orange, and it makes two (2) loops at the mouth of the bilum. The same process continues until you have half of the design done. When you come to the other end, tie the yarn together, making sure you make five (5) loops of orange so that you will have even numbers and then you tie the yarn with the same colour that you have started with that is orange with orange, white with white, green with green, and black with black

Second half of the pattern

In order to complete the second half of the pattern, the four (4) colours are reversed, you start with black, green, white, and orange

- 3 loops (black), it stays there
- 3 loops (green), it goes up to where green is and it stays there
- 1 loop (black)
- 4 loops (white), it goes up to where white is and it stays there

The mathematics derived from bilum-making pattern

(A) Ratio (Grade 10A ratio)
(B) Geometry (Grade 7A shapes)
(A) Ratio (using bilum pattern in teaching)
One round—half of the pattern is done
Two rounds—one complete pattern
Three rounds—one and a half patterns

Four rounds—two complete patterns
Five rounds—two and half-complete patterns
Six rounds—three complete patterns

Number of Patterns	Number of Rounds
1	$1/2$
2	1
3	$3/2$
4	2
5	$5/2$
6	3

A woman makes a bilum, she makes the second round, and one complete pattern is formed. Now if she makes ten (10) rounds, how many complete patterns will she make

(B) Geometry
A teacher can ask questions such as:

(1) How many different shapes can be identified from the bilum design?
(2) Congruent shapes have same angle measure and when cut out it can fit exactly into the other same shape. Can you identify congruent shapes and their angle measure? [figure of shapes supplied]
(3) How many lines of symmetry can you draw? [figure of shapes supplied]

Appendix E
Example of Learning Plan for Cultural Mathematics

Highlands Round House: Area

Purpose: Children will see what area is, how to look and compare, and how to put one area on top of another to help compare, and how to start measuring areas.

Key ideas: Spaces where people live are familiar areas. Areas have a size that can be compared and measured. Measuring area is about covering with an area unit without any gaps or overlaps. Rectangular areas are easier to measure with rectangular or square area units represented by an object.

Prior knowledge: Children cover areas by sleeping on spaces, laying out cloth and food on spaces, wrapping themselves up, noticing how many people can lie or sit in a space. (Earlier parts for Grade 1.)

Resources: A round house, a rectangular house, and another round house of a different size or use.

Assessment: Observing ways children suggest to compare areas and do compare areas to assess their sense of area as a 2D space inside a boundary. Children select area units for measuring (not length units but cultural ways are a good discussion point—note relative accuracy for specific areas).

Day 1

- *Tuning in*
  - Visit the (round) house. What areas are there in the house? What are they used for? What shapes are they? Why? Are some areas covered up? Which areas are bigger than others?
  - How will we compare areas?

- *Finding out*
  - What are the area places in the house (small laplap with karuka, bigger laplap with karuka, space for saucepans, space for fire, space for firewood, space for clothes, space for food, water).

- Some of the small areas where the karuka are drying, are also used for sleeping, how many people would sleep on this platform?
- Look at the rectangular house and its divided floor spaces, are they the same size, double each other.
- Look at another round house and compare the areas to the first house. Are they bigger? Are they used for the same things?

Day 2

- *Sorting out*
  - Take areas in the classroom. Which are bigger spaces, what takes up the bigger space, estimate, directly compare or use a unit
  - How can we compare the spaces? (Cover with laplap, big leaves, coffee bag, rice bag, banana leaf, breadfruit leaf, canvas—depending on house, whether rectangular or round)
  - Children cover the spaces with a "standard" area measure. For example, coffee bags
  - No spaces between the units (notice mathematical language of unit), estimating

Day 3

- *Going further*
  - Garden areas. Visit a garden that is divided up into areas of a similar shape and size. Discuss the pattern of smaller areas tessellating without gaps to cover the area.
  - Read the book, *Areas in the Garden*.
  - Discuss each page and do activities to reinforce the concepts and story of the book.
  - Bigger areas, e.g. joined classrooms in the building

Day 4

- *Making connections*
  - Area is the 2D space inside a boundary, area can have different shapes, area can be compared and measured.
  - Area units should take up all space inside and there can be parts of units. Different shapes can be the area unit.
  - Read the book, *Seli Measures Area*[2], discuss, and do the activities.
  - Try a rectangle or square paper and fold and tear into four equal triangles and rearrange into different shapes.
  - Use clay or playdough areas to represent the areas of a classroom; model or draw the plan and areas of the house in the sand (links to shapes, curved and straight lines).

---

[2] Books for early readers prepared for the project (Owens, 2013d).

Appendix E: Example of Learning Plan for Cultural Mathematics

- *Taking action*
  - Find out how many A4 papers or large leaf would cover the area where you sleep.

Day 5

- *Sharing, discussing, reflecting*
- What is area?

  How do we compare area?
  Can we have round areas and areas with straight sides?

- What are some ways of comparing and measuring areas? Children select an area, and a measuring unit and measure and record their findings. (Teacher can use for learning story of children's learning, noting aspects of visuospatial reasoning used by children and knowledge of measurement.)

  (Owens, 2013c)

Appendix F
Selection of Curriculum Statements and Issues

Agenda for Action of National Council of Teachers of Mathematics

use visualisation and spatial reasoning to solve problems both within and outside of mathematics

Reasoning about Congruence and Similarity
Reasoning in Two Dimensions
Reasoning about Surface Area and Volume
Reasoning in Geometric Modelling

Key issues:

- Geometric ideas are useful in representing and solving problems. Students should gain experience in using a variety of visual and coordinate representations to analyse problems and study mathematics.
- Instructional programs should not repeat the same measurement programs year after year and students should learn to choose appropriate units for measurement.
- Many students have difficulty with understanding perimeter and area.
- Understanding that all measurements are approximations is a difficult but important concept for students.
- Opportunities to use problem-solving strategies must be embedded naturally in the curriculum across the content areas.
- Reasoning and proof begin in the early years. A mathematical proof is a formal way of expressing particular kinds of reasoning and justification.
- Conjecture is a major pathway to discovery. (Owens & Perry, 1998)

Singapore Curriculum

Key issues:
The curriculum is centred around a mathematical problem-solving framework that emphasises concepts, skills, processes, attitudes, and metacognition. Succinctly the concepts are numerical, geometrical, algebraic, and statistical. The skills begin with estimation and approximation; mental calculation; communication; use of mathematical tools; and then manipulations. The processes are deductive reasoning (including logical thinking, deducing new information from existing; and drawing conclusion), inductive reasoning (including recognising patterns and structures and forming generalisation), and heuristics for problem solving (including using diagrams, tables, making suppositions, and so on). Words such as *estimate, calculate, visualise, use* are common among the objectives (Owens & Perry, 1998).

Republic of South Africa

Key issues:

Implementation of the intention:

Mathematics enhances and helps to formalise the ability to be able to grasp, visualise, and represent the space in which we live. In the real world, space and shape do not exist in isolation from motion and time. Learners should be able to display an understanding of spatial sense and motion in time.

The outcome requires that learners:

> ...describe and represent experiences with shape, space, time and motion, using all available senses. (Department of Education, 1997)

The MALATI program noted the importance of visualisation under vision in stating:

Descriptions of the orientation of an object is an indicator of Position, Vision
Demonstrate an understanding of the interconnectedness between shape, space, and
  time is an indicator of Position, Vision, Motion, Shape

Australian Curriculum

The Australian Curriculum: Mathematics aims to ensure that students:

- Are confident, creative users and communicators of mathematics, able to investigate, represent, and interpret situations in their personal and work lives and as active citizens.
- Develop an increasingly sophisticated understanding of mathematical concepts and fluency with processes, and are able to pose and solve problems and reason in Number and Algebra, Measurement and Geometry, and Statistics and Probability.

Appendix F: Selection of Curriculum Statements and Issues   329

- Recognise connections between the areas of mathematics and other disciplines and appreciate mathematics as an accessible and enjoyable discipline to study.

Mathematics has its own value and beauty and the Australian Curriculum: Mathematics aims to instil in students an appreciation of the elegance and power of mathematical reasoning. Mathematical ideas have evolved across all cultures over thousands of years and are constantly developing. Digital technologies are facilitating this expansion of ideas and providing access to new tools for continuing mathematical exploration and invention. The curriculum focuses on developing increasingly sophisticated and refined mathematical understanding, fluency, logical reasoning, analytical thought, and problem-solving skills. These capabilities enable students to respond to familiar and unfamiliar situations by employing mathematical strategies to make informed decisions and solve problems efficiently.

Measurement and Geometry are presented together to emphasise their relationship to each other, enhancing their practical relevance. Students develop an increasingly sophisticated understanding of size, shape, relative position, and movement of two-dimensional figures in the plane and three-dimensional objects in space. They investigate properties and apply their understanding of them to define, compare, and construct figures and objects. They learn to develop geometric arguments. They make meaningful measurements of quantities, choosing appropriate metric units of measurement. They build an understanding of the connections between units and calculate derived measures such as area, speed, and density.

Proficiency strands are understanding, fluency, problem solving, and reasoning. Using units of measure, shape, location, and transformation begin in the foundation year of school while geometric reasoning begins in grade 3 and relates to identifying angles in the environment in various situations. Pythagoras and trigonometry begin in grade 9. Expected knowledge, concepts, skills, and processes are given but approaches to teaching are not prescribed (Australian Curriculum Assessment and Reporting Authority, 2010).

Key issues:
Culture is supposed to be considered but it is not mentioned in the content strands or any examples.

There is no mention of visuospatial reasoning. Representation and explaining are mentioned.

Papua New Guinea Elementary Syllabus

The learning outcomes and indicators will:

- Give teachers individually or in groups the flexibility to write programs and units of work, which should be developed to suit local conditions and individual student needs.

In Cultural Mathematics, a manageable number of outcomes are identified for each grade in Elementary Prep, 1 and 2. They are: … created using an active verb to ensure students actively participate in the learning.

The language of instruction at Elementary is the students' vernacular, which will enable teachers to enhance the students' understanding of mathematical concepts. Students need to use mathematics in different contexts. Teachers teaching this course must be competent in the language the children speak, so that they can explain the mathematical concepts clearly to their students. Students at Elementary will be able to link new mathematical concepts from the five strands in this syllabus to their existing cultural knowledge. The students will integrate this knowledge so that they can confidently use mathematics in their everyday lives. The Elementary Cultural Mathematics course provides many opportunities for relevant and purposeful learning in an environment that is built on the principles of home life.

Aims:

Students develop:

- A sound foundation for further mathematical learning
- Confidence in applying mathematical skills
- Curiosity leading to the understanding of concepts
- Determination to persist with difficult problems
- Critical judgement in selecting approaches to problems
- An appreciation of the cultural diversity in numeracy

Space

This strand deals with giving information and directions to be followed to move from location to location. It also deals with the concept of shape and the language required to describe various shapes.

Measurement

This strand concentrates on the units used to describe length, weight, capacity, area, and time and how they are measured. The concepts in this strand focus on ways of measuring using local measurements as well as common formal measurements. Students will also estimate and calculate time using traditional ways.

Space outcomes for each grade	P.1.1 Follow and give simple directions for moving in a space	1.1.1 Follow and give directions to move from place to place	2.1.1 Follow directions from simple maps
Indicator examples	• Give simple directions such as walk three steps forward, stop, take three steps left, turn right, take three steps backward • Demonstrate using directions given • Make a list of local names for directions such as east (sun rises) • Use direction words to play games	• Give directions to a person to find a place in the community	• Talk about direction names in vernacular such as east (sun rises), west (sun sets), north, and south • Draw simple maps to show directions to find the school, villages, gardens, church • Make a list of directions to guide people to certain places

Appendix F: Selection of Curriculum Statements and Issues 331

Space outcomes for each grade	P.1.2 Identify locally known shapes by their visual appearance	1.1.2 Compare and group shapes in the community	2.1.2 Investigate and describe the features of geometric shapes
Indicator examples	• Collect and display locally known shapes in everyday life	• Sort shapes according to their differences and similarities	• Talk about the features of shapes such as edges, angles, curves, faces, sides, and corners
	• Name and label locally known shapes in vernacular • Make a mobile of locally known shapes • Make a collage of locally known shapes	• Make new shapes by putting simple shapes together	• Group shapes according to the number of sides, shapes, and angles
		• Display groups of different shapes from the community such as baskets, pots, kundus, and shells	• Count faces, corners, and edges of geometrical figures
	• Sort and match shapes	• Name features of shapes in vernacular such as edges, angles, curves, and corners	• Build new shapes using three-dimensional shapes
	• Make models of locally known shapes using sand, clay, or mud	• Group local shapes according to their features such as *all* objects with curved edges	• Label using vernacular and say English words for features such as corners, edges, and angles
	• Identify lines found in the local environment such as *r*oads, rivers, and drawings • Draw line drawings of things such as gardens, houses, and playing fields	• Build objects using local materials and label the shapes used • Make line drawings of regular and irregular shapes found in the community	• Identify and discuss types of lines that make up objects such as long, straight, sharp, curved, wavy, thick, and thin • Group lines according to their differences and similarities
	• Find things in the local community where one half looks exactly the same as the other half	• Draw objects from different places that are symmetrical	• Make symmetrical shapes such as butterfly wings, leaves, and flower petals

Measurement outcomes for each grade	P.2.1 Measure the length, weight, and capacity of things using their own informal measuring units	1.2.1 Measure and compare the length, weight, and capacity of things using local informal units	2.2.1 Compare the accuracy of local measures of length, weight, and capacity
Indicator examples	• Match and compare lengths and heights using their own units	• Use things such as hand span, arm lengths, pacing, sticks, or other items to measure length	• Measure lengths in different ways and compare their accuracy
	• Measure length, width, and height using their own units such as bottle tops, seeds, and leaves	• Use comparison words for measuring such as light, lighter, lightest, heavy, heavier, heaviest, long, short, tall, full, empty, and not much	• Talk about and practise some traditional measures from their local community
	• Collect items of different weight and arrange them in order using their own units	• Use different containers and other items to measure capacity	• Compare local ways of measuring weight
	• Match and compare light and heavy objects		• Measure capacity using various common containers
	• Arrange light and heavy objects in a simple order		
	• Collect containers of different sizes and arrange them in order		
	• Use traditional ways to guess and order the amount of liquid such as water and juice		(National Department of Education PNG, 2003)

Key issues:
Curriculum appropriate for many cultures but requires good teacher preparation.

Appendix G
Types of Geometry

334 Appendix G: Types of Geometry

Invariant under mapping	Position	Length	Angle and ratio	Parallelism	Cross-ratio	Neighbour-liness
Identity	•	•	•	•	•	•
Isometry		•	•	•	•	•
Similarity			•	•	•	•
Affinity				•	•	•
Perspectivity					•	•
Topology						•

References

Abe, K., & Del Grande, J. (1983). Geometric activities in the elementary school. In M. Zweng, T. Green, J. Kilpatrick, H. Pollak, & M. Suydam (Eds.), *Proceedings of the Fourth International Congress on Mathematical Education* (pp. 161–164). Boston: Birkhauser.

Adam, N. A. (2010). Mutual interrogation: A methodological process in ethnomathematical research. *Procedia—Social and Behavioral Sciences, 8*, 700–707.

Adam, S., Alangui, W., & Barton, B. (2003). A comment on: Rowlands and Carson "Where would formal, academic mathematics stand in a curriculum informed by Ethnomathematics? A critical review". *Educational Studies in Mathematics, 52*, 327–335.

Adams, T., & Harrell, G. (2003). Estimation at work. In D. Clements & G. Bright (Eds.), *Learning and teaching measurement*. Reston, VA: National Council of Teachers of Mathematics.

Adendorff, R. (1993). Code-switching among Zulu-speaking teachers and the pupils. *Language and Education, 7*, 141–162.

Adler, J. (2002). *Teaching mathematics in multilingual classrooms*. New York: Kluwer.

Aikenhead, G. (2010). Academic science, cultural intransigence, and devious educo-politics. *Cultural Studies of Science Education, 5*(3), 613–619.

Akerblom, K. (1968). *Astronomy and navigation in Polynesia and Micronesia*. Stockholm: Ethnogratiska Museet.

Alaskan Universities Council. (2012). *National best practice for Indigenous education*. Anchorage, Alaska: Department of Education Alaska.

Allport, D. (1987). Selection for action: Some behavioral and neurophysiological considerations of attention and action. In H. Heuer & A. Sanders (Eds.), *Perspectives on perception and action* (pp. 395–420). Hillsdale, NJ: Lawrence Erlbaum.

Amos, S. (2007). Geometry and ratio in bilum (string bag) making in PNG. In W. Kaleva (Ed.), *Ethnomathematics Projects 2007* (pp. 126–131). Goroka, Papua New Guinea: University of Goroka. Retrieved from http://www.uog.ac.pg/glec/teaching.

Anderson, R. (1978). Arguments concerning representations for mental imagery. *Psychological Review, 85*(4), 249–277.

Architectural Heritage Centre. (1996). *Architecture and designs in the Sepik (CD)*. Lae, PNG: PNG University of Technology.

Ascher, M. (1994). *Ethnomathematics: A multicultural view of mathematical ideas*. New York: Chapman & Hall.

Ascher, M. (2002). *Mathematics elsewhere: An exploration of ideas across cultures*. Princeton, NJ: Princeton University Press.

Atkin, T., & Krinsky, C. (1996). Cultural identity in modern Native American architecture: A case study. *Journal of Architectural Education, 49*(4), 237–245.

Atran, S. (1990). *Cognitive foundations of natural history: Towards an anthropology of science*. Cambridge: Cambridge University Press.

Atweh, B., Barton, A. C., & Borba, M. (Eds.). (2007). *Internationalisation and globalisation in mathematics and science education*. Dotrecht, The Netherlands: Springer.

Aust, R. (1989). *Constructing mental representations of complex three-dimensional objects*. Lawrence, KS: University of Kansas.

Australian Bureau of Statistics. (2013). *Census at school*. Retrieved from http://www.abs.gov.au/censusatschool

Australian Council for Educational Research. (1989–1991). *Aspects of numeracy. Basic skills tests for years 3 and 6*. Sydney: NSW Department of School Education.

Australian Curriculum Assessment and Reporting Authority. (2010). *Australian curriculum mathematics*. Retrieved from http://www.australiancurriculum.edu.au/mathematics/Rationale

Averill, R., Anderson, D., Easton, H., Te Maro, P. N., Smith, D., & Hynds, A. (2009). Culturally responsive teaching of mathematics: Three models from linked studies. *Journal for Research in Mathematics Education, 40*(2), 157–186.

Baddeley, A. (1992). Working memory. *Science, 255*, 395–406.

Baenninger, M., & Newcombe, N. S. (1989). The role of experience in spatial test performance: A meta-analysis. *Sex Roles, 20*(5/6), 327–344.

Baillargeon, R. (2004). Infants' reasoning about hidden objects: Evidence for event-general and event-specificexpectations.*DevelopmentalScience,7*(4),391–414.doi:10.1111/j.1467-7687.2004.00357.x.

Barnett, R. (2010). *The coming of the ecological curriculum*. Paper presented at the Educating for 2020 and Beyond, Charles Sturt University, Bathurst, Australia.

Barnhardt, R. (2007). Creating a place for Indigenous knowledge in education: The Alaska Native Knowledge Network. In D. Gruenewald & G. Smith (Eds.), *Place-based education in the global age: Local diversity* (pp. 113–133). New York: Lawrence Erlbaum.

Barratt, E. S. (1953). An analysis of verbal reports of solving spatial problems as an aid in defining spatial factors. *Journal of Psychology, 36*, 17–25.

Barton, B. (2008). *The language of mathematics: Telling mathematical tales*. New York: Springer.

Barton, B., Poisard, C., Do, C. S., & Domite, M. (2006). Cultural connections and mathematical manipulations. *For the Learning of Mathematics, 26*(2), 21–24.

Battista, M. (2007a). The development of geometric and spatial thinking. In F. Lester (Ed.), *Second handbook of research on mathematics teaching and learning* (pp. 843–903). Reston, VA: National Council of Teachers of Mathematics.

Battista, M. (2007b). Learning with understanding: Principles and processes in the construction of meaning for geometric ideas. In W. G. Martin, M. Strutchens, & P. Elliott (Eds.), *The learning of mathematics (NCTM 69th yearbook)* (pp. 65–79). Reston, VA: National Council of Teachers of Mathematics.

Battista, M. T., & Clements, D. H. (1991). *Logo geometry*. Morristown, NJ: Silver Burdett & Ginn.

Battista, M., & Clements, D. (1991). Using spatial imagery in geometric reasoning. *Arithmetic Teacher, 39*(3), 18–21.

Bauersfeld, H. (1991). The structuring of the structures: Development and function of mathematizing as a social practice. In L. Steffe (Ed.), *Constructivism and education* (pp. 1–26). Hillsdale, NJ: Lawrence Erlbaum Associates.

Bellugi, U., Sabo, H., & Vaid, J. (1988). Spatial deficits in children with Williams Syndrome. In J. Stiles-Davis, M. Kritchevsky, & U. Bellugi (Eds.), *Spatial cognition: Brain bases and development* (pp. 273–298). Hillsdale, NJ: Lawrence Erlbaum.

Bellwood, P. (1979). Settlement patterns. In J. Jennings (Ed.), *The prehistory of Polynesia* (pp. 309–322). Cambridge, MA: Harvard University Press.

Ben-Chaim, D., Lappan, G., & Houang, R. (1988). The effect of instruction on spatial visualization skills by middle school boys and girls. *American Educational Research Journal, 25*, 51–71.

Berry, J. (1966). Temne and Eskimo perceptual skills. *International Journal of Psychology, 1*, 207–229.

References

Berry, J. (1969). On cross-cultural comparability. *International Journal of Psychology, 4*, 119–128.

Berry, J. (2003). Ecocultural perspective on human psychological development. In T. Saraswathi (Ed.), *Cross-cultural perspectives on human development: Theory, research and applications* (pp. 51–69). New Delhi: Sage.

Berry, J. (2011). Intercultural relations and acculturation in the Pacific region. *Journal of Pacific Rim Psychology, 4*(2), 95–102.

Bhabha, H. K. (1995). *The location of culture*. London: Routledge.

Bino, V., Owens, K., Tau, K., Avosa, M., & Kull, M. (2013). Chapter Eight: Improving the teaching of mathematics in elementary schools in Papua New Guinea: A first phase of implementing a design. In J. Pumwa (Ed.), *Mathematics digest: Contemporary discussions in various fields with some mathematics—ICPAM-Lae* (pp. 84-95). Lae: Papua New Guinea University of Technology.

Bishop, A. (1973). Structural apparatus and spatial ability. *Research in Education, 9*, 43–49.

Bishop, A. (1979). Visualising and mathematics in a pre-technological culture. *Educational Studies in Mathematics, 10*(2), 135–146.

Bishop, A. (1983). Space and geometry. In R. Lesh & M. Landau (Eds.), *Acquisition of mathematics concepts and processes* (pp. 176–204). New York: Academic Press.

Bishop, A. (1988). *Mathematical enculturation: A cultural perspective on mathematics education*. Dordrecht, The Netherlands: Kluwer.

Booth, D. (1994). Art and geometry learning through spontaneous pattern making. *Journal of Institute of Art Education, 9*(2), 28–42.

Borba, M., & Villarrea, M. (2005). *Humans-with-media and the reorganization of mathematical thinking: Information and communication technologies, modelling, experimentation and visualization*. Dordrecht, The Netherlands: Springer.

Bowden, R. (1990). Kwoma ceremonial houses. In N. Lukehuas, W. Kaufmann, D. Michell, L. Newton, L. Osmundsen, & M. Schuster (Eds.), *Sepik heritage: Tradition and change in Papua New Guinea. Proceedings of 1987 Conference in Switzerland*. Durham, NC: Carolina Academic Press.

Bowers, C. A. (2001). *Educating for eco-justice and community*. Athens, Georgia: The University of Georgia Press.

Brofenbrenner, U., & Ceci, S. (1994). Nature-nurture reconceptualized in developmental perspective: A bioecological model. *Psychological Review, 101*(4), 568–586.

Brown, K. (2008). Employing mathematical modelling to respond to Indigenous students' needs for contextualised mathematics experiences. In M. Goos, R. Brown, & K. Makar (Eds.), *Proceedings of the 31st Annual Conference of the Mathematics Education Research Group of Australasia, Brisbane* (pp. 93–99). Adelaide: MERGA.

Brown, A., & Clark, L. (Eds.). (2006). *Learning from NAEP: Professional development materials for teachers of mathematics (with CD)*. Reston, VA: National Council of Teachers of Mathematics.

Bruner, J. (1964). The course of cognitive growth. *American Psychologist: Anthropology and Education, 19*, 1–15.

Bryan, E. (1938). Marshall Islands stick chart. *Paradise of the Pacific, 50*(7), 12–13.

Burden, L., & Coulson, S. (1981). *Processing of spatial tasks*. Masters thesis, Monash University, Melbourne, Australia.

Bush, W. (2005). Improving research on mathematics learning and teaching in rural contexts. *Journal of Research in Rural Education, 20*(8), 1–11.

Butler, D., & Close, S. (1989). Assessing the benefits of a Logo problem-solving course. *Irish Educational Studies, 8*, 168–190.

Butterworth, B. (1999). *The mathematical brain*. London: Macmillan.

Callingham, R. (2004). Primary students' understanding of tessellation: An initial exploration. In M. Hoines & A. Fuglestad (Eds.), *28th Conference of the International Group for the Psychology of Mathematics Education* (Vol. 2, pp. 183–190). Bergen, Norway: PME.

Cameron, J. (2003). Educating for place responsiveness: An Australian perspective on ethical practice. *Ethics, Place and Environment, 6*, 99–115.

Campbell, K., Collis, K., & Watson, J. (1995). Visual processing during mathematical problem solving. *Educational Studies in Mathematics, 28*(2), 177–194.

Campbell-Jones, S. (1996). *Horizon: Twice five plus the wings of a bird*. London, UK: BBC.

Capell, A. (1943). *The linguistic position of southeastern Papua*. Ph.D. thesis, University of London, London, UK.

Capell, A. (1969). *A survey of New Guinea languages*. Sydney: Sydney University Press.

Carpenter, P., & Just, M. (1986). Spatial ability: An information processing approach to psychometrics. In R. J. Sternberg (Ed.), *Advances in the psychology of human intelligence* (Vol. 3, pp. 221–253). Hillsdale, NJ: Lawrence Erlbaum.

Carraher, T. (1988). Street mathematics and school mathematics. In A. Borbas (Ed.), *12th Conference of the International Group for the Psychology of Mathematics Education* (Vol. 1, pp. 1–23). Veszprem, Hungary: PME.

Casey, B., Dearing, E., Vasilyeva, M., Ganley, C., & Tine, M. (2011). Spatial and numerical predictors of measurement performance: The moderating effects of community income and gender. *Journal of Educational Psychology, 103*(2), 296–311.

Castagno, A., & Brayboy, B. (2008). Culturally responsive schooling for Indigenous youth: A review of the literature. *Review of Educational Research, 78*(4), 941–943. doi:10.3102/0034654308323036.

Catlin, D., & Balmires, M. (2010). The principles of Educational Robotic Applications (ERA): A framework for understanding and developing educational robots and their activities. Paper presented at Proceedings for Constructionism 2010: The 12th EuroLogo conference, Paris.

César, M. (2009). Listening to different voices: Collaborative work in multicultural maths classes. In M. César & K. Kumpulainen (Eds.), *Social interactions in multicultural settings* (pp. 203–233). Rotterdam: Sense.

Cheng, K., Huttenlocher, J., & Newcombe, N. (2013). 25 years of research on the use of geometry in spatial reorientation: A current theoretical perspective. *Psychonomic Bulletin & Review, 20*(6), 1033–1054.

Cherinda, M. (2001). *Weaving board activities: A way to introduce and develop mathematical ideas in the classroom*. Retrieved from http://sunsite.witsac.za.math/weaving.htm

Cherinda, M. (2002). *The use of a cultural activity in the teaching and learning of mathematics: Exploring twill weaving with a weaving board in Mozambican classrooms*. Ph.D. thesis, University of the Witwatersrand, Johannesburg, South Africa.

Cherinda, M. (2012). Weaving exploration in the process of acquisition and development of mathematical knowledge. *12th International Congress on Mathematics Education* (Vol. Regular Lectures, pp. 5–8). Seoul, Korea: ICME.

Christou, C., Pittalis, M., Mousoulides, N., & Jones, K. (2006). Developing the 3DMath dynamic geometry software: Theoretical perspectives on design. *International Journal for Technology in Mathematics Education, 13*(4), 168–174.

Civil, M., & Andrade, R. (2002). Transitions between home and school mathematics: Rays of hope amidst the passing clouds. In G. De Abreu, A. Bishop, & N. Presmeg (Eds.), *Transitions between contexts of mathematical practices* (pp. 148–168). Dordrecht, The Netherlands: Kluwer.

Clarkson, P. (2009). Mathematics quality teaching in Australian multilingual classrooms: Developing a relevant approach to the use of classroom languages. In R. Barwell (Ed.), *Multilingualism in mathematics classrooms*. Clevedon, UK: Multilingual Matters.

Clarkson, P., & Kaleva, W. (1993). Mathematics education in Papua New Guinea. In G. Bell (Ed.), *Asian perspectives on mathematics education* (p. 11). Lismore: Northern Rivers Mathematical Association.

Clarkson, P., Owens, K., Toomey, R., Kaleva, W., & Hamadi, T. (2001). The development of a process for the evaluation of teacher education. In P. Jeffery (Ed.), *Annual Conference of Australian Association for Research in Education*. Fremantle: AARE.

Clarkson, P., & Presmeg, N. (Eds.). (2008). *Critical issues in mathematics education: Major contributions of Alan Bishop*. New York: Springer.

Clements, D. H., & Battista, M. (1989). Learning of geometric concepts in a Logo environment. *Journal for Research in Mathematics Education, 20*, 450–467.

Clements, D. (1998). *Geometric and spatial thinking in young children* (pp. 1–40). Arlington, VA: National Science Foundation.

Clements, D. (1999). 'Concrete' manipulatives, concrete ideas. *Contemporary Issues in Early Childhood, 1*(3), 45–60.

Clements, D. (2004). Geometric and spatial thinking in early childhood education. In D. Clements, J. Sarama, & A.-M. DiBiase (Eds.), *Engaging young children in mathematics: Standards for early childhood mathematics education* (pp. 267–298). Mahwah, NJ: Lawrence Erlbaum Associates.

Clements, D., & Battista, M. (1991). Van Hiele levels of learning geometry. In F. Fulvinghetti (Ed.), *15th Conference of the International Group for the Psychology of Mathematics Education* (pp. 223–230). Haifa, Israel: PME.

Clements, D., Battista, M., Sarama, J., Swaminathan, S., & McMillen, S. (1997). Students' development of length concepts in a logo-based unit on geometric paths. *Journal for Research in Mathematics Education, 28*(1), 70–95.

Clements, D., Battista, M., & Sarama, J. (1998). Development of geometric and measurement ideas. In R. Lehrer & D. Chazan (Eds.), *Designing learning environments for developing understanding of geometry and space* (pp. 201–226). Mahwah, NJ: Lawrence Erlbaum.

Clements, D., & Sarama, J. (2007a). Early childhood mathematics learning. In F. Lester (Ed.), *Second handbook of research on mathematics teaching and learning* (pp. 479–530, especially 488–530). Reston, VA: National Council of Teachers of Mathematics.

Clements, D., & Sarama, J. (2007b). Effects of a preschool mathematics curriculum: Summative research on the Building Blocks project. *Journal for Research in Mathematics Education, 38*, 136–163.

Clements, D., Swaminathan, S., Hannibal, M., & Sarama, J. (1998). Young children's concepts of shape. *Journal for Research in Mathematics Education, 30*(2), 192–212.

Clements, D., Wilson, D., & Sarama, J. (2004). Young children's composition of geometric figures: A learning trajectory. *Mathematical Thinking and Learning, 6*, 163–184.

Codrington, R. H. (1885). *The Melanesian languages*. Oxford: Clarendon Press.

Coiffier, C. (1990). Sepik river architecture: Changes in cultural traditions. In N. Lukehuas, W. Kaufmann, D. Michell, L. Newton, & L. Osmundsen (Eds.), *Sepik heritage: Tradition and change in Papua New Guinea. Proceedings of 1987 Conference in Switzerland* (pp. 491–497). Durham, NC: Carolina Academic Press.

Cole, M. (1998). Can cultural psychology help us to think about diversity? *Mind. Culture and Activity, 5*(4), 291–304.

Cook, E. (1967). A preliminary statement of Narak spatial diexis. *Anthropological Linguistics, 9*(1), 1–9.

Cooper, L., & Shepard, R. (1973). Chronometric studies of the rotation of mental images. In W. Chase (Ed.), *Visual information processing* (pp. 76–176). New York: Academic Press.

Cope, P., & Simmons, M. (1994). Some effects of limited feedback on performance and problem-solving strategy in a Logo Microworld. *Journal of Educational Psychology, 86*(3), 368–379.

Corte, E. d., Verschaffel, L., & Eynde, P. O. t. (2000). Self-regulation: A characteristic and a goal of mathematics education. *Handbook on self-regulation* (pp. 687–726). Philadelphia, PA: Elsevier.

Costigan, K. (1995). *The patterns of structure in the Trobriand Islands*. M.S. Architecture Master of Science, University of California, Berkeley, CA.

Cox, M. (1978). Perspective ability: A training programme. *Journal of Educational Research, 71*, 127–133.

Critchlow, K. (1992). *Islamic patterns: An analytical and cosmological approach*. London: Thames and Hudson.

Cross, C. T., Woods, T. A., Schweingruber, H. A., & National Research Council, Committee on Early Childhood Mathematics. (2009). *Mathematics learning in early childhood: Paths toward excellence and equity*. Washington, DC: National Academies.

Currie, P., & Pegg, J. (1998). Three sides equal means it is not isosceles. In A. Olivier & K. Newstead (Eds.), *22nd Conference of the International Group for the Psychology of Mathematics Education* (Vol. 2, pp. 216–223). Stellenbosch, South Africa: PME.

D'Ambrosio, U. (1990). The history of mathematics and ethnomathematics: How a native culture intervenes in the process of learning science. *Impact of Science on Society, 40*(4), 369–378.

Dalin, J. (2013). *A pedagogical revolution in school mathematics education*. Retrieved from http://www.MathematiX.com; www.NovoMath.com

D'Ambrosio, U. (2006). *Ethnomathematics: Link between traditions and modernity (M. Borba, Trans.)*. Rotterdam, The Netherlands: Sense Publishers.

Dasen, P., & de Ribaupierre, A. (1987). Neo-Piagetian theories: Cross-cultural and differential perspectives. *International Journal of Psychology, 22*(5/6), 793–832.

Davenport, W. H. (1960). Marshall Island navigational charts. *Imago Mundi, 15*, 19–26.

David, D. (2007). Mathematics from the "five stone" game played by Finschafen people of Morobe Province. In W. Kaleva (Ed.), *Ethnomathematics Reports 2007* (pp. 28–33). Goroka, PNG: PNG University of Goroka.

David, M., & Tomaz, V. (2012). The role of visual representations for structuring classroom mathematical activity. *Educational Studies in Mathematics, 80*(3), 413–431.

Davis, B. (1999). Thinking otherwise and hearing differently: Enactivism and school mathematics. In W. F. Pinar (Ed.), *Contemporary curriculum discourses: Twenty years of JCT* (pp. 325–345). New York: Lang.

Davis, W. (2009). *The Wayfinders—Why ancient wisdom matters in the modern world, Lecture 1, Season of the Brown Hyena, 2009 Massey Lectures*. Toronto, Canada: The House of Anansi Press.

Davis, R., Maher, C., & Noddings, N. (Eds.). (1990). *Constructivist views on the teaching and learning of mathematics. Journal for Research in Mathematics Education, Monograph 4*. Reston, VA: National Council of Teachers of Mathematics.

de Abreu, G. (2002). Towards a cultural psychology perspective on transitions between contexts of mathematical practices. In G. de Abreu, A. Bishop, & N. Presmeg (Eds.), *Transitions between contexts of mathematical practices* (pp. 170–189). Dordrecht, The Netherlands: Kluwer.

de Castello Branco Fantinato, M. C. (2006). Quantitative and spatial representations among working class adults from Rio de Janeiro. In F. Favilli (Ed.), *Ethnomathematics and mathematics education* (pp. 29–38). Pisa, Italy: Tipografia Editrice Pisana.

de la Fuente Arias, J., & Díaz, A. L. (2010). Assessing self-regulated learning in early childhood education: Difficulties, needs, and prospects. *Psicothema, 22*(2), 278–283.

De Villiers, M. (1998). To teach definitions in geometry or teach to define? In A. Olivier & K. Newstead (Eds.), *22nd Conference of the International Group for the Psychology of Mathematics Education* (Vol. 2, pp. 248–255). Stellenbosch, South Africa: PME.

Deacon, A. B., & Wedgwood, C. H. (1934). Geometrical drawing from Malekula and other islands of the New Hebrides. *Journal of the Royal Anthropological Institute of Great Britain and Ireland, 64*, 129–175.

Dehaene, S., Izard, V., Pica, P., & Spelke, E. (2006). Core knowledge of geometry in an Amazonian indigene group. *Science, 311*, 381–384.

Del Campo, G., & Clements, M. (1990). Expanding the modes of communication in mathematics classrooms. *Journal fur Mathematik-Didaktik, 11*, 45–79.

Del Grande, J. (1990). Spatial sense. *Arithmetic Teacher, 27*, 14–20.

Del Grande, J. (1992). *Geometry and spatial abilities*. Paper presented at the Subgroup 11.1: Geometry as a Part of Education in Early Childhood in Working Group 11: The Role of Geometry in General Education. International Congress on Mathematical Education, ICME 7, Quebec.

Dennett, H. (1975). *Mak bilong Sepik: A selection of designs and paintings from the Sepik River*. Wewak: Wirua Press.

References

Department of Education Employment Workplace Relations. (2009). *National Aboriginal and Torres Strait Islander Education Policy (AEP)*. Retrieved December, 2011, from http://www.deewr.gov.au/Indigenous/Schooling/PolicyGuidelines/Pages/aep.aspx. Replaced by http://education.gov.au/aboriginal-and-torres-strait-islander-education-action-plan-2010-2014-0

Department of Education formerly DEEWR. (2009). *Belonging, being and becoming—Early years learning framework*. Retrieved from http://education.gov.au/early-years-learning-framework

Deregowski, J. B. (1980). *Illusions, patterns and pictures: A cross-cultural perspective*. London: Academic Press.

Derewianka, B., & Jones, P. (2012). *Teaching language in context*. Melbourne: Oxford.

Dewey, J. (1938). *Experience and education*. New York: Collier.

Diezmann, C., & Lowrie, T. (2012). Learning to think spatially: What do students "see" in numeracy test items? *International Journal of Science and Mathematics Education, 10*(6), 1469–1490. doi:10.1007/s10763-012-9350-3.

Doig, B., Cheeseman, J., & Lindsay, J. (1995). The medium is the message: Measuring area with different media. In B. Atweh & S. Flavel (Eds.), *Galtha: Proceedings of the 18th Annual Conference of the Mathematics Education Group of Australasia* (pp. 229–240). Darwin, NT: MERGA.

Donald Coxeter (1907–2003). Quote came from Hagen, P. (2003). *Mathematics and imaginative education*. Paper presented at the IERG Conference, Vancouver, BC, Canada.

Dörfler, W. (2004). *Mathematical reasoning: Mental activity or practice with diagrams*. Paper presented at the International Congress on Mathematics Education ICME11, Denmark. Retrieved from http://www.icme10.dk/proceedings/pages/regular_pdf/RL_Willi_Doerfler.pdf

Dreyfus, T. (1991). On the status of visual reasoning in mathematics and mathematics education. In F. Furinghetti (Ed.), *15th PME Conference* (Vol. 1, pp. 33–48). Italy: Program Committee for the International Group for the Psychology of Mathematics Education.

Dreyfus, T., & Eisenberg, T. (1990). On difficulties with diagrams: Theoretical issues. In G. Booker, P. Cobb, & T. Mendicuti (Eds.), *14th Conference of the International Group for the Psychology of Mathematics Education* (Vol. 1, pp. 27–36). Mexico: PME.

Edmonds-Wathen, C. (2011). What comes before? Understanding spatial reference in Iwaidja. In M. Setati, T. Nkambule & L. Goosen (Eds.) *Proceedings of the 21st ICMI Study Conference, Mathematics Education and Language Diversity* (pp. 89-97). São Paolo, Brazil: ICMI.

Edmonds-Wathen, C. (2012a). *Frames of reference in Iwaidja: Towards a culturally responsive early years mathematics program*. Ph.D. thesis, RMIT, Melbourne, Australia.

Edmonds-Wathen, C. (2012b). Spatial metaphors of the number line. In D. Jaguthsing, L. P. Cheng, & S. F. Ng (Eds.), *Mathematics education: Expanding horizons. Proceedings of 35th Annual Conference of Mathematics Education Research Group of Australasia* (pp. 250–257). Singapore: MERGA.

Edmonds-Wathen, C., Owens, K., Sakopa, P., & Bino, V. (2014, July). *Improving the teaching of mathematics in elementary schools by using local languages and cultural practices*. Paper presented at the International Group for Psychology of Mathematics Education, Vancouver, Canada.

Egan, D. E. (1979). Testing based on understanding: Implications from studies of spatial ability. *Intelligence, 3*, 1–15.

Egan, K. (1992). *Imagination in teaching and learning: Ages 8 to 15*. London: Routledge.

Eglash, R. (1997). When math worlds collide: Intention and invention in ethnomathematics. *Science, Technology, & Human Values, 22*(1), 79–97.

Eglash, R. (1999). *African fractals*. New Brunswick, NJ: Rutgers University Press.

Eglash, R., Croissant, J., Di Chiro, G., & Fouché, R. (2004). *Appropriating technology: vernacular science and social power*. Minneapolis: University of Minnesota.

Eglash, R. (2007). *Culturally situated design tools: Teaching math through culture*. Retrieved from http://www.rpi.edu/~eglash/csdt.html

Eglash, R. (2009). Native American analogues to the Cartesian coordinate system. In B. Greer, S. Mukhopadhyay, A. Powell, & S. Nelson-Barber (Eds.), *Culturally responsive mathematics education*. New York: Routledge.

Eglash, R., & Rensselaer Polytechnic Institute. (2003). *Culturally situated design tools: Teaching math and computing through culture*. Retrieved from http://csdt.rpi.edu/

Ehrich, V. (1991). *Hier und Jetzt: Studien zur positionalen und temporalen Deixis im Deutschen* (Mimeoth ed.). Germany: Nijmegen/Cologne.

Eisenman, P. (1988). Architecture as a second language. In M. Diani & C. Ingraham (Eds.), *Restructuring architectural theory* (pp. 69–73). Evanston, IL: Northwestern University.

Eliot, J. (1987). *Models of psychological space: Psychometric, developmental, and experimental approaches*. New York: Springer.

Eliot, J., & McFarlane-Smith, I. (1983). *International directory of spatial tests*. Windsor: NRER-Nelson.

Ellerton, N., & Clements, M. (1994). *The national curriculum debacle*. Perth, WA, Australia: Meridian Press.

Enderson, M. (2003). Using measurement to develop mathematical reasoning at the middle and high school levels. In D. Clement & G. Bright (Eds.), *Learning and teaching measurement (NCTM 2003 yearbook)* (pp. 271–281). Reston, VA: National Council of Teachers of Mathematics.

English, L. (1994). Reasoning by analogy in constructing mathematical ideas. In G. Bell, B. Wright, N. Leeson, & J. Geake (Eds.), *Challenges in mathematics education: Constraints on Construction* (pp. 213–222). Lismore: Mathematics Education Group of Australasia.

Esmonde, I. (2009). Ideas and identities: Supporting equity in cooperative mathematics learning. *Review of Educational Research, 79*(2), 1008–1043.

Esmonde, I., & Saxe, G. (2004). *'Cultural mathematics' in the Oksapmin curriculum: Continuities and discontinuities*. Paper presented at the 6th International Conference on Learning Sciences, Santa Monica, CA.

Falcade, R., Laborde, C., & Mariotti, M. (2007). Approaching functions: Cabri tools as instruments of semiotic mediation. *Educational Studies in Mathematics, 66*(3), 317–333.

Falsetti, M., & Rodríguez, M. (2005). A proposal for improving students' mathematical attitude based on mathematical modelling. *Mathematics and its Applications, 24*(1), 14–28.

Farnham-Diggory, S. (1967). Symbol and synthesis in experimental "reading". *Child Development, 38*(1), 221–231.

Fassler, R. (2003). *Room for talk: Teaching and learning in a multilingual kindergarten*. New York: Teachers College Press.

Fennema, E. (1984). Girls, women, and mathematics: An overview. In E. Fennema & J. Ayer (Eds.), *Women and education: Equity or equality* (pp. 137–164). Berkeley, CA: McCutchan.

Ferrare, J., & Apple, M. (2012). Spatializing critical education: Progress and cautions. *Critical Studies in Education, 51*(2), 209–221. doi:http://dx.doi.org/10.1080/17508481003731075.

Finau, K., & Stillman, G. (1995). *Geometrical skill behind the Tongan tapa designs*. Paper presented at the International History and Pedagogy of Mathematics Conference on Ethnomathematics and the Australasian Region, Cairns.

Fiorentino, G., & Favilli, F. (2006). The electronic yupana: A didactic resource from an ancient mathematical tool. In F. Favilli (Ed.), *Ethnomathematics and mathematics education* (pp. 39–48). Pisa, Italy: Tipografia Editrice Pisana.

Fischer, K., & Silvern, L. (1985). Stages and individual differences in cognitive development. *Annual Review of Psychology, 36*, 613–648.

Fisher, J. (2010). *Enriching students' learning through ethnomathematics in Kuruti elementary schools in Papua New Guinea*. Lae, Papua New Guinea: ResearchGate.

Flavell, J. (1977). *Cognitive development*. Englewood Cliffs, NJ: Prentice-Hall.

Flavell, J. (1987). Speculations about the nature and development of metacognition. In F. Weinert & R. Kluwe (Eds.), *Metacognition, motivation, and understanding* (pp. 21–29). Hillsdale, NJ: Lawrence Erlbaum.

Flemming, D., Flemming Luz, E., & de Mello, C. (2005). *Tendências em educação matemática [Tendencies in mathematics education]*. Pallhoça, SC, Brazil: UNISUL.

Flores, A. (1995). Bilingual lessons in the early-grades. *Teaching Children Mathematics, 1*(7), 420–424.

References

Frade, C., & Falcão, J. (2008). Exploring connections between tacit knowing and situated learning: Perspectives in the context of mathematics education. In A. Watson & P. Winbourne (Eds.), *New directions in situated cognition* (pp. 205–232). New York: Springer.

François, K. (2010). Ethnomathematics in a European context: Towards an enriched meaning of ethnomathematics. *Journal of Mathematics and Culture, 6*(1), 191–208.

François, K., & Pinxten, R. (2012). *Multimathemacy: Mathematics in situated learning*. Paper presented at the International Congress on Mathematics Education ICME12, Korea.

François, K., Pinxten, R., & Mesquita, M. (2013). How anthropology can contribute to mathematics education. *Revista Latinoamericana de Etnomatematica: Perspectivas socioculturales de la education matematica, 6*(1), 20–39.

Freire, P. (1992). *Pedagogy of the oppressed*. New York: Seabury Press.

Frostig, M., & Horne, D. (1964). *The Frostig program for the development of visual perception*. Chicago: Follett.

Furinghetti, F., & Paola, D. (2002). *Defining within a dynamic geometry environment: Notes from the classroom*. Paper presented at the 26th Conference of the International Group for the Psychology of Mathematics Education, Norwich, UK.

Fuson, K., & Murray, C. (Eds.). (1978). *The haptic-visual perception, construction, and drawing of geometric shapes by children aged two to five: A Piagetian extension*. In R. Lesh (Ed.), *Recent research concerning the development of spatial and geometric concepts* (pp. 49–84). Columbus, Ohio: ERIC.

Gagné, R., & White, R. (1978). Memory structures and learning outcomes. *Review of Educational Research, 48*(2), 187–222.

Gammage, B. (1998). *The sky travellers: Journeys in New Guinea 1938–1939*. Melbourne, Australia: Miegunyah Press, Melbourne University Press.

Garbutt, R. (2011). *The locals: Identity, place and belonging in Australia and beyond*. Oxford: Peter Lang.

Gee, J. (1992). *The social mind: Language, ideology and social practice*. New York: Bergin and Garvey.

Gell, A. (1998). *Art and agency: An anthropological theory*. Oxford: Clarendon Press.

Genkins, E. (1975). The concept of bilateral symmetry in young children. In M. Rosskopf (Ed.), *Children's mathematical concepts: Six Piagetian studies in mathematics education* (pp. 5–43). Columbia, NY: Teachers College Press.

Gerdes, P. (1998). *Women, art and geometry in southern Africa*. Trenton, NJ: African World Press.

Gerdes, P. (1999). *Geometry from Africa: Mathematical and educational explorations*. Washington, DC: The Mathematical Association of America.

Gervasoni, A. (2005). *Australian Catholic University—Pursuing its mission to engage with communities: A case study of the Melbourne and Ballarat campuses* (A research report for the Australian Consortium of Higher Education). Melbourne: Australian Catholic University.

Giaquinto, M. (2011). *Visual thinking in mathematics: An epistemological study*. Oxford: Oxford University Press.

Gibson, J. (1979). *The ecological approach to visual perception*. Hillsdale, NJ: Lawrence Erlbaum.

Gilsdorf, T. E. (2009). Mathematics of the Hñähñu: The Otomies. *Journal of Mathematics and Culture, 4*(1). Retrieved from http://nasgem.rpi.edu/pl/journal-mathematics-culture-volume-3-number-2.

Ginsburg, H. P., Lin, C.-l., Ness, D., & Seo, K.-H. (2003). Young American and Chinese children's everyday mathematical activity. *Mathematical Thinking and Learning, 5*(4), 235–258.

Giroux, H. (1997). *Pedagogy and the politics of hope: Theory, culture, and schooling: A critical reader*. Boulder, CO: The Edge.

Godfrey, C. (1910). The Board of Education circular on the teaching of geometry. *Mathematical Gazette, 5*, 195–2000.

Goetzfridt, N. (2012). Pacific ethnomathematics: The richness of environment and practice. *Journal of Mathematics and Culture, International Conference on Ethnomathematics, 4*(Focus Issue), 223–252.

Goldberg, M. (1997). *The art of the question.* New York: Wiley.

Goldenberg, P., Cuoco, A., & Mark, J. (1998). A role for geometry in general education. In R. Lehrer & D. Chazan (Eds.), *Designing learning environments for developing understanding of geometry and space* (pp. 3–44). Mahwah, NJ: Lawrence Erlbaum.

Goldenberg, P., & Mason, J. (2008). Shedding light on and with example spaces. *Educational Studies in Mathematics, 69*(2), 183–194. doi:10.1007/s10649-008-9143-3.

Goldin, G. (Ed.). (1987). *(a) Levels of language in mathematical problem solving; (b) Cognitive representational systems for mathematical problem solving.* Hillsdale, NJ: Lawrence Erlbaum.

Goldin, G. (1992). On the developing of a unified model for the psychology of mathematics learning and problem solving. In W. Geeslin & K. Graham (Eds.), *Proceedings of 16th Annual Conference of International Group for the Psychology of Mathematics Education* (Vol. 3, pp. 235–261). Durham, NH: PME.

Goldin, G. (1998). The PME Working Group on Representation. *Journal of Mathematical Behavior, 17*(2), 283–301.

Goldin, G. (2000). Affective pathways and representation in mathematical problem solving. *Mathematical Thinking and Learning: An International Journal, 2*(3), 209–219.

Goldin, G., Epstein, Y., Schorr, R., & Warner, L. (2011). Beliefs and engagement structures: Behind the affective dimension of mathematical learning. *ZDM, 43*(4), 547–560.

Gomes, M. L. M., & D'Ambrosio, U. (2006). D'Ambrosio on ethnomathematics. *HPM Newsletter, 62*, 1–4.

González, N., Moll, L., & Amanti, C. (Eds.). (2005). *Funds of knowledge: Theorizing practice in households, communities, and classrooms.* Mahwah, NJ: Lawrence Erlbaum.

Goodwin, K., & Highfield, K. (2013). A framework for examining technologies and early mathematics learning. In L. D. English & J. T. Mulligan (Eds.), *Reconceptualising early mathematics learning* (pp. 205–226). New York: Springer.

Goos, M., Galbraith, P., Renshaw, P., & Geiger, V. (2003). Perspectives on technology mediated learning in secondary school mathematics classrooms. *Journal of Mathematical Behavior, 22*(1), 73. doi:10.1016/s0732-3123(03)00005-1.

Gorgorió, N., Planas, N., & Vilella, X. (2002). Immigrant children learning mathematics in mainstream schools. In G. de Abreu, A. Bishop, & N. Presmeg (Eds.), *Transitions between contexts of mathematical practices* (pp. 22–53). Dordrecht, The Netherlands: Kluwer.

Graham, B. (1988). Mathematical education and Aboriginal children. *Educational Studies in Mathematics, 19*(2), 119–136.

Grant, S., & Rudder, J. (Eds.). (2010). *A new Wiradjuri dictionary.* Wagga Wagga, NSW, Australia: Restoration House.

Gray, E., & Tall, D. (2007). Abstraction as a natural process of mental compression. *Mathematics Education Research Journal, 19*(2), 23–40.

Greene, M. (2001). *Variation on a blue guitar: The Lincoln Center Institute lectures on aesthetic education.* New York: Teachers College Press.

Greenes, C., Dacey, L., Cavanagh, M., Findell, C., Sheffield, L., & Small, M. (2003). *Navigating through measurement in prekindergarten-grade 2.* Reston, VA: National Council of Teachers of Mathematics.

Greeno, J. (2003). Situative research relevant to standards for school mathematics. In J. Kilpatrick, W. G. Martin, & D. Schifter (Eds.), *A research companion to Principles and Standards for School Mathematics* (pp. 304–332). Reston, VA: National Council of Teachers of Mathematics.

Greenspan, S. I. (2007). When a child has difficulty moving from place to place. *Early Childhood Today (J3), 21*(5), 16–19.

Gresalfi, M., & Barab, S. (2011). Learning for a reason: Supporting forms of engagement by designing tasks and orchestrating environments. *Theory Into Practice, 50*(4), 300–310.

Gruenewald, D. (2008). The best of both worlds: A critical pedagogy of place. *Environmental Education Research, 14*(3), 308–324.

Gruenewald, D., & Smith, G. (2007). *Place-based education in a global age: Local diversity.* New York: Lawrence Erlbaum.

Gunn, M. (1986). Rock art on Tabar, New Ireland Province, Papua New Guinea. *Anthropos, 81*(4–6), 455–467.

Gutiérrez, A. (1996). Visualization in 3-dimensional geometry: In search of a framework. In L. Puig & A. Gutiérrez (Eds.), *20th Conference of the International Group for Psychology of Mathematics Education* (Vol. 1, pp. 3–17). Valencia, Portugal: University of Valencia.

Gutiérrez, A., & Boero, P. (Eds.). (2006). *Handbook of research on the psychology of mathematics education: Past, present and future: PME 1976-2006*. Rotterdam, The Netherlands: Sense Publishers.

Gutstein, E. (2006). *Reading and writing the world with mathematics: Towards a pedagogy for social justice*. New York: Routledge.

Gutstein, E. (2007). "And that's just how it starts": Teaching mathematics and developing student agency. *Teachers College Record, 109*(2), 420–448.

Hagen, P. (2003). *Mathematics and imaginative education*. Vancouver, BC, Canada: Paper presented at the IERG conference.

Halat, E. (2007). Reform-based curriculum and acquisition of the levels. *Eurasia Journal of Mathematics, Science & Technology Education, 3*(1), 41–49.

Harris, P. (1989). *Mathematics in a cultural context: Aboriginal perspectives on space, time and money*. Geelong, Victoria: Deakin University Press.

Hauser-Schäublin, B. (1990). Settlement and house structures. In N. Lukehuas, W. Kaufmann, D. Michell, L. Newton, & L. Osmundsen (Eds.), *Sepik heritage: Tradition and change in Papua New Guinea. Proceedings of 1987 conference in Switzerland* (pp. 470–479). Durham, NC: Carolina Academic Press.

Hauser-Schäublin, B. (1996). The thrill of the line, the string, and the frond, or why the Abelam are a non-cloth culture. *Oceania, 67*(2), 81–106.

Healy, L., & Fernandes, S. (2011). The role of gestures in the mathematical practices of those who do not see with their eyes. *Educational Studies in Mathematics, 77*(2), 157–174. doi:10.1007/s10649-010-9290-1.

Healy, L., & Powell, A. (2013). Understanding and overcoming "disadvantage" in learning mathematics. In M. Clements, A. Bishop, C. Keitel, J. Kilpatrick, & F. Leung (Eds.), *Third international handbook of mathematics education* (Springer International Handbooks of Education). New York: Springer.

Hegarty, M., & Kozhevnikov, M. (1999). Types of visual–spatial representations and mathematical problem solving. *Journal of Educational Psychology, 91*(4), 684–689.

Hershkowitz, R. (1989). Visualization in geometry—Two sides of the coin. *Focus on Learning Problems in Mathematics, 11*(1), 61–76.

Hershkowitz, R. (1990). Psychological aspects of learning geometry. In P. Nesher & J. Kilpatrick (Eds.), *Mathematics and cognition: A research synthesis by the International Group for the Psychology of Mathematics Education* (pp. 70–95). Cambridge: Cambridge University Press.

Highfield, K. (2012). *Young children's mathematics learning with programmable toys*. Ph.D. thesis, Macquarie, Sydney, Australia.

Highfield, K., & Mulligan, J. (2007). The role of dynamic interactive technological tools in preschoolers' mathematical patterning. In J. Watson & K. Beswick (Eds.), *Mathematics: Essential research, essential practice. Proceedings of the 30th Annual conference of the Mathematics Education Research Group of Australasia* (Vol. 1, pp. 372–381). Adelaide: MERGA.

Highfield, K., & Mulligan, J. (2009). Young children's embodied action in problem-solving tasks using robotic toys. In M. Tzekaki, M. Kaldrimidou, & H. Sakonidis (Eds.), *Proceedings of the 33rd Conference of the International Group for the Psychology of Mathematics Education* (Vol. 2, pp. 273–280). Thessaloniki, Greece: PME.

Highfield, K., Mulligan, J., & Hedberg, J. (2008). Early mathematics learning through exploration with programmable toys. In O. Figueras, J. Cortina, S. Alatorre, T. Rojano, A. Sepulveda, & E. Rosch (Eds.), *32nd Conference of the International Group for the Psychology of Mathematics Education in Conjunction with PME-NAXXX* (Vol. 3, pp. 169–176). Morelia, Mexico: PME.

Highfield, K. (2010). Possibilities and pitfalls of techno-toys and digital play in mathematics learning. In M. Ebbeck & M. Waniganayake (Eds.), *Play in early childhood education: learning in diverse contexts* (pp. 177–196). South Melbourne, VIC: Oxford University.

Holton, D. L. (2010). How people learn with computer simulations. In H. Song & T. T. Kidd (Eds.), *Handbook of research on human performance and instructional technology* (pp. 485–504). Hershey, PA: IGI Global.

Hoyles, C., & Noss, R. (2003). What can digital technologies take from and bring to research in mathematics education? In A. Bishop, M. Clements, C. Keitel, J. Kilpatrick, & F. Leung (Eds.), *Second international handbook of research in mathematics education* (pp. 323–349). Dordrecht: Kluwer.

Hutchins, E. (1983). Understanding Micronesian navigation. In D. Gentner & A. Stevens (Eds.), *Mental models* (pp. 191–225). Hillsdale, NJ: Lawrence Erlbaum.

Hutchins, E. (1995). *Cognition in the wild*. Cambridge, MA: Massachusetts Institute of Technology.

Ison, B. (Ed.). (1986). *Lufakafai: Stories in design*. Aiyura, PNG: Aiyura National High School.

Jannok Nutti, Y. (2007). *Matematiskt tankesätt inom den samiska kulturen—Utifrån samiska slöjdares och renskötares berättelser, licentiate thesis [Mathematical thinking within the Sámi culture—On basis of Sámi handcrafters' and reindeer herders' stories]*. Masters thesis, Luleå University of Technology, Luleå, Sweden.

Jannok Nutti, Y. (2008). Sámi education in mathematics—A school development action research project. *Journal of Australian Indigenous Issues, 12*, 177–185.

Jannok Nutti, Y. (2010). *Ripsteg mot spetskunskap i samisk matematik: Lärares perspektive på transformeringsaktiviteter i samisk förskola ock sameskola*. Doktorsavhandling, Luleå Tekniska Universitet, Luleå, Sweden.

Jannok Nutti, Y. (2013). Indigenous teachers' experiences of the implementation of culture-based mathematics activities in Sámi school. *Mathematics Education Research Journal, 25*(1), 57–72. doi:10.1007/s13394-013-0067-6.

Jawahir, R. (2013). *Effective learning and teaching strategies of two-dimensional geometry at the upper primary grades in Mauritius*. Ph.D. thesis, University of Technology, Mauritius.

Jawahir, R., Owens, K., Sukon, K., & Sunhaloo, S. (2011). *An analysis of the contextual factors influencing performance in geometry at the upper primary level in Mauritius*. Paper presented at the 18th International Conference on Learning, Mauritius.

Jegede, P. J., & Aikenhead, G. S. (1999). Transcending cultural borders: Implications for science teaching. *Research in Science and Technological Education, 17*(1), 45–66.

Johansson, G. (2008). *Cultural knowledge in school curriculum in practice—Ecolonizing processes and school development at Sámi schools in Sweden*. Paper presented at the World Indigenous Peoples Conference on Education, Melbourne, Australia.

John, L. (2007). *Asaro mud mask*. Unpublished report. University of Goroka, Goroka, PNG.

Johnson, M. (1987). *The body and the mind: The bodily basis of meaning, imagination, and reason*. Chicago: University of Chicago Press.

Johnson, E., & Meade, A. (1985). *The JM battery of spatial tests: Lower battery*. Authors.

Johnston, E. (n.d.). *Aboriginal art—Australian Aboriginal art*. Retrieved from http://www.creativespirits.info/aboriginalculture/arts/

Johnston, W., & Heinz, S. (1978). Flexibility and capacity demands of attention. *Journal of Experimental Psychology, General, 107*, 420–435.

Johnston-Wilder, S., & Mason, J. E. (2005). *Developing thinking in geometry*. London: Open University and Paul Chapman.

Jonassen, D. (1999). *Welcome to the design of constructivist learning environments (CLEs)*. Retrieved from http://tiger.coe.mssouri.edu/~jonassen/courses/CLE/index.html

Jonassen, D., Peck, K., & Wilson, B. (1999). *Learning with technology: A constructivist perspective*. Upper Saddle River, NJ: Prentice Hall.

Jones, J. (1974). Quantitative concepts, vernaculars, and education in Papua New Guinea. *Education Research Unit Report 12*. University of Papua New Guinea.

Jones, K. (1998). *Theoretical frameworks for the learning of geometrical reasoning*. Paper presented at the Proceedings of the British Society for Research in to Learning Mathematics, King's College, University of London.

Jones, K. (2000). Providing a foundation for deductive reasoning: Students' interpretations when using dynamic geometry software and their evolving mathematical explanations. *Educational Studies in Mathematics, 44*(1), 55–85.

Jones, J. (2012). *Visualizing elementary and middle school mathematics methods*. Hoboken, NJ: Wiley.

Jones, K., & Mooney, C. (2003). Making space for geometry in primary mathematics. In I. Thompson (Ed.), *Enhancing primary mathematics teaching* (pp. 3–15). London, UK: Open University Press.

Joseph, G. (1991). Foundations of Eurocentrism in mathematics. In M. Harris (Ed.), *Schools, mathematics and work* (pp. 42–56). London: The Falmer Press.

Joseph, G. (2000). *The crest of the peacock: The non-European roots of mathematics*. Princeton: Princeton University Press.

Julius, A. (2007). Mathematics of traditional 'Tabare' of the Simbu Province. In W. Kaleva (Ed.), *Ethnomathematics reports 2007* (pp. 1–11). Goroka, PNG: University of Goroka.

Kahan, S. (2004). Engagement, identity and innovation: Etienne Wenger on communities of practice. *Journal of Association Leadership*. Retrieved from http://www.asaecenter.org/PublicationsResources/JALArticleDetail.cfm?ItemNumber=16217

Kahneman, D. (1973). *Attention and effort*. Englemwood Cliffs, NJ: Prentice-Hall.

Kaufmann, G. (1979). *Visual imagery and its relation to problem solving*. Bergen: Universitetsforlaget.

Khasawneh, A. A. (2009). Assesing Logo programming among Jordanian seventh grade students through turtle geometry. *International Journal of Mathematical Education in Science and Technology, 40*, 619–639.

Khisty, L. (1995). Making inequality: Issues of language and meanings in mathematics teaching with Hispanic students. In W. Secada, E. Fennema, & L. Adajian (Eds.), *New directions for equity in mathematics education* (pp. 279–297). New York: Cambridge University Press.

Kieran, C. (1986a). Logo and the notion of angle among fourth and sixth grade children. In Proceedings of Psychology in Mathematics Education 10 (pp. 99–104). London, England: City University.

Kidder, F. R. (1978). Conservation of length: A function of the mental operation involved. In R. Lesh (Ed.), *Recent research concerning the development of spatial and geometric concepts* (pp. 213–228). Columbus, OH: ERIC.

Kieras, D. (1978). Beyond pictures and words: Alternative information processing models for imagery effects in verbal memory. *Psychological Bulletin, 85*, 532–554.

Kilpatrick, J., Martin, W., & Schifter, D. (Eds.). (2003). *A research companion to Principles and Standards for School Mathematics*. Reston, VA: National Council of Teachers of Mathematics.

Kilpatrick, J., & Silver, E. (2000). Unfinished business: Challenges for mathematics educators in the next decades. In M. Burke & F. Curcio (Eds.), *Learning mathematics for a new century (2000 Yearbook of the National Council of Teachers of Mathematics)* (pp. 223–235). Reston, VA: National Council of Teachers of Mathematics.

Kim, M., Roth, W.-M., & Thom, J. (2011). Children's gestures and the embodied knowledge of geometry. *International Journal of Science and Mathematics Education, 9*, 207–238.

Kimball, S. T. (1974). *Culture and the educative process*. New York: Teachers College Press.

Knijnik, G. (2002). Curriculum, culture and ethnomathematics. *Journal of Intercultural Studies, 23*(2), 149–165.

Kono, S. (2007). Mathematics from bamboo wall patterns of the Simbu Province. In W. Kaleva (Ed.), *Ethnomathematics reports 2007* (pp. 113–116). Goroka, Papua New Guinea: University of Goroka.

Kosslyn, S. (1981). The medium and message in mental imagery: A theory. *Psychological Review, 88*(1), 46–66.

Kosslyn, S. (1983). *Ghosts in the mind's machine*. New York: W.W. Norton.

Kosslyn, S., & Pomerantz, J. (1977). Imagery, propositions and the form of internal representations. *Cognitive Psychology, 7*, 341–370.

Kouba, V., Brown, C., Carpenter, T., Lindquist, M., Silver, E., & Swafford, J. (1988). Results of the fourth NAEP assessment of mathematics: Measurement, geometry, data interpretation, attitudes, and other topics. *Arithmetic Teacher, 35*(9), 10–16.

Kritchevsky, M. (1988). The elementary spatial functions of the brain. In J. Stiles-Davis, M. Kritchevsky, & U. Bellugi (Eds.), *Spatial cognition: Brain bases and development* (pp. 111–139). Hillsdale, NJ: Lawrence Erlbaum Associates.

Krutetskii, V. (1976). The psychology of mathematical abilities in schoolchildren. In J. Kilpatrick & I. Wirszup (Eds.), *Soviet studies in the psychology of learning and teaching mathematics. Survey of recent East European mathematical literature* (Vol. II: The structure of mathematical abilities, pp. 5–58). Chicago: University of Chicago.

Kuchler, S. (1999). Binding in the Pacific: Between loops and knots. *Oceania, 69*(3), 145–156.

Kurina, F. (1992). *Geometry in the early childhood education*. Paper presented at the Subgroup 11.1: Geometry as a Part of Education in Early Childhood in Working Group 11: The Role of Geometry in General Education. International Congress on Mathematical Education, ICME 7, Quebec.

Kyllonen, P. C., Lohman, D. F., & Snow, R. E. (1984). Effects of aptitudes, strategy training, and task facets on spatial task performance. *Journal of Educational Psychology, 76*(1), 130–145.

Laborde, C., & Capponi, B. (1995). *Modelisation a double sens*. Paper presented at the Actes de la 8eme ecole d'ete de didactique des mathematiques, de Clermont-Ferrard.

Laborde, C., Kynoigos, C., Hollebrands, K., & Strässer, R. (2006). Teaching and learning geometry with technology. In A. Gutiérrez & P. Boera (Eds.), *Handbook of research on the psychology of mathematics education: Past, present and future*. Rotterdam: Sense Publishers.

Lakoff, G. (1987). *Women, fire, and dangerous things: What categories reveal about the mind*. Chicago: University of Chicago Press.

Landau, B. (1988). The construction and use of spatial knowledge in blind and sighted children. In J. Stiles-Davis, M. Kritchevsky, & U. Bellugi (Eds.), *Spatial cognition: Brain bases and development* (pp. 343–372). Hillsdale, NJ: Lawrence Erlbaum.

Landau, B., Gleitman, H., & Spelke, E. (1981). Spatial knowledge and geometric representation in a child blind from birth. *Science, 213*(4513), 1275–1278. doi:10.2307/1687359.

Lappan, G. (1999, December). Geometry: The forgotten strand. *National Council of Teachers of Mathematics News Bulletin*.

Lave, J. (1988). *Cognition in practice*. Cambridge, MA: Harvard University Press.

Lave, J., Smith, S., & Butler, M. (1989). Problem solving as an everyday practice. In R. Charles & E. Silver (Eds.), *The teaching and assessing of mathematical problem solving*. Hillsdale, NJ: Lawrence Erlbaum & National Council of Teachers of Mathematics.

Lean, G. (1984). *The conquest of space: A review of the research literatures pertaining to the development of spatial abilities underlying an understanding of 3-D geometry*. Paper presented at the Fifth International Congress on Mathematical Education, Adelaide, Australia.

Lean, G. (1992). *Counting systems of Papua New Guinea and Oceania*. Unpublished Ph.D. Thesis, University of Technology, Lae, Papua New Guinea. Retrieved from http://www.uog.ac.pg/glec/

Lean, G., & Clements, M. (1981). Spatial ability, visual imagery, and mathematical performance. *Educational Studies in Mathematics, 12*, 267–299.

Learmonth, A. E., Newcombe, N. S., Sheridan, N., & Jones, M. (2008). Why size counts: Children's spatial reorientation in large and small enclosures. *Developmental Science, 11*(3), 414–426.

Lefebvre, H. (1991). *The production of space (D. Nicholson-Smith, Trans.)*. Oxford: Blackwell.

Lehrer, R., & Chazan, D. (Eds.). (1998). *Designing learning environments for developing understanding of geometry and space*. Mahwah, NJ: Lawrence Erlbaum.

Lehrer, R., Jacobson, G., Thoyre, G., Kemeny, V., Strom, D., Horvath, J., et al. (1998). Developing understanding of geometry and space in the primary grades. In R. Lehrer & D. Chazan (Eds.), *Designing learning environments for developing understanding of geometry and space* (pp. 169–200). Mahwah, NJ: Lawrence Erlbaum.

Lehrer, R., Jenkins, M., & Osana, H. (1998). Longitudinal study of children's reasoning about space and geometry. In R. Lehrer & D. Chazan (Eds.), *Designing learning environments for developing understanding of geometry and space* (pp. 137–167). Mahwah, NJ: Lawrence Erlbaum.

Lehrer, R., & Littlefield, J. (1991). Misconceptions and errors in LOGO: the role of instruction. *Journal of Educational Psychology, 83*, 124–133.

Leinhardt, G., Zaslavsky, O., & Stein, M. (1990). Functions, graphs, and graphing: Tasks, learning and teaching. *Review of Educational Research, 60*(1–64).

Lerman, S. (2001). Cultural, discursive psychology: A sociocultural approach to studying the teaching and learning of mathematics. *Educational Studies in Mathematics, 46*, 87–113.

Lester, F. (Ed.). (1983). *Trends and issues in mathematical problem-solving research*. New York: Academic Press.

Lester, F. (Ed.). (2007). *Second handbook of research on mathematics teaching and learning*. Reston, VA: National Council of Teachers of Mathematics.

Lévi-Strauss, C. (1968). *The savage mind*. Chicago: University of Chicago Press.

Lewis, D. (1973). *We, the Navigators*. Honolulu: University Press of Hawaii.

Liben, L. S. (1988). Conceptual issue in the development of spatial cognition. In J. Stiles-Davis, M. Kritchevsky, & U. Bellugi (Eds.), *Spatial cognition: Brain bases and development* (pp. 167–194). Hillsdale, NJ: Lawrence Erlbaum.

Liben, L. S. (2006). Education for spatial thinking. In K. Renninger & I. Sigel (Eds.), *Handbook of child psychology* (Child psychology in practice, Vol. 4, pp. 197–247). Hoboken, NJ: Wiley (Reprinted from: 6th).

Lillo-Martin, D., & Tallal, P. (1988). Effects of different early experiences. In J. Stiles-Davis, M. Kritchevsky, & U. Bellugi (Eds.), *Spatial cognition: Brain bases and development* (pp. 433–441). Hillsdale, NJ: Lawrence Erlbaum.

Linn, M., & Hyde, J. (1989). Gender, mathematics, and science. *Educational Researcher, 18*(8), 17–19. 22–27.

Lipka, J., & Adams, B. L. (2004). Culturally based math education as a way to improve Alaska native students' math performance. Working Paper 20. Appalachian Collaborative Center for Learning, USA (ED484849).

Lipka, J., Mohatt, G., & The Ciulistet Group. (1998). *Transforming the culture of schools. Yup'ik Eskimo examples*. Mahwah, NJ: Lawrence Erlbaum Associates.

Lipka, J., Wildfeuer, S., Wahlberg, N., George, M., & Ezran, D. R. (2001). Elastic geometry and storyknifing: A Yup'ik Eskimo example. *Teaching Children Mathematics, 7*(6), 337–343.

Litteral, R. (2001). Language development in Papua New Guinea. *Radical Pedagogy, 3*(1). Retrieved from http://radicalpedagogy.icaap.org/content/issue3_1/01Litteral002.html.

Liu, Y., & Wickens, C. (1992). Visual scanning with or without spatial uncertainty and divided and selective attention. *Acta Psychologica, 79*, 139–153.

Llewellyn, G. (1991). *The teenage liberation handbook: How to quit school and get a real life and education*. Eugene, OR: Lowry House Publishers.

Lockheed, M., & Verspoor, A. (1991). *Improving primary education in developing countries*. Oxford: Oxford University Press.

Lohman, D. F. (1979). *Spatial ability: A review and re-analysis of the correlational literature (Technical Report No. 8)*. Stanford, CA: Aptitude Research Project, Stanford University School of Education.

Lohman, D. F., Pellegrino, J. W., Alderton, D. L., & Regian, J. W. (1987). Dimensions and components of individual differences in spatial abilities. In S. H. Irvine & S. E. Newstead (Eds.), *Intelligence and cognition* (pp. 253–312). Dordrecht, The Netherlands: Nijhoff Publishers.

Lossau, J. (2009). Pitfalls of (third) space: Rethinking the ambivalent logic of spatial semantics. In K. Ikas & G. Wagner (Eds.), *Communicating in the third space* (ebook, pp. 62–78). New York: Routledge.

Lovat, T., & Toomey, R. (Eds.). (2009). *Values education and quality teaching: Double helix effect*. Dordrecht, The Netherlands: Springer.

Love, E. (1996). Letting go: An approach to geometric problem solving. In L. Puig & A. Gutiérrez (Eds.), *Proceedings of 20th Annual Conference of the International Group for the Psychology of Mathematics Education, PME20* (Vol. 3, pp. 281–288). Valencia, Spain: IGPME.

Lowrie, T. (1992). *Developing talented children's mathematical ability through visual and spatial learning tasks*. Paper presented at the Annual Conference of the Australian Association for Research in Education, Deakin University, Geelong, VIC, Australia.

Lowrie, T., Diezmann, C., & Logan, T. (2012). A framework for mathematics graphical tasks: The influence of the graphic element on student sense making. *Mathematics Education Research Journal, 24*(1), 169–187. doi:10.1007/s13394-012-0036-5.

Lowrie, T., Logan, T., & Scriven, B. (2012). Perspectives on geometry and measurement in the Australian Curriculum: Mathematics. In B. Atweh, M. Goos, R. Jorgensen, & D. Siemon (Eds.), *Engaging the Australian National Curriculum: Mathematics—Perspectives from the field (Online Publication)* (pp. 71–88). Adelaide, SA, Australia: Mathematics Education Research Group of Australasia.

Luitel, B. C. (2009). *Culture, worldview and transformative philosophy of mathematics education in Nepal: A cultural-philosophical inquiry*. Perth: Curtin.

Mackenzie, M. A. (1991). *Androgynous objects: String bags and gender in central New Guinea*. Chur, Switzerland: Harwood Academic Publishers.

Macmillan, A. (1998). Investigating the mathematical thinking of young children: Some methodological and theoretical issues. In A. McIntosh & N. Ellerton (Eds.), *Research in mathematics education: A contemporary perspective*. Perth, WA: MASTEC, Edith Cowan University.

Macmillan, A. (2009). *Numeracy in early childhood*. Sydney, NSW: Oxford University Press.

Madang Teachers College. (1973). *Traditional arts and crafts* (Vol. 1). Madang, PNG: Kristen Press.

Magoon, R., & Garrison, K. (1976). *Educational psychology: An integrated view* (2nd ed.). Columbus, OH: Charles E. Merrill.

Malinowski, B. (1917–1920). Classificatory particles in the language of the Kiriwina. *Bulletin of the School of Oriental Studies, London Institution, 1*, 33–78.

Mammana, C., & Villani, V. (Eds.). (1998). *Perspectives on the teaching of geometry for the 21st century*. New York: Springer.

Mandler, J. (1988). The development of spatial cognition: On topological and Euclidean representation. In J. Stiles-Davis, M. Kritchevsky, & U. Bellugi (Eds.), *Spatial cognition: Brain bases and development* (pp. 423–432). Hillsdale, NJ: Lawrence Erlbaum Associates.

Mansfield, H., & Scott, J. (1990). Young children solving spatial problems. In G. Booker, P. Cobb, & T. de Mendicuti (Eds.), *Proceedings of the 14th PME conference* (Vol. II, pp. 275–282). Oaxlepec, Mexico: International Group for the Psychology of Mathematics Education.

Margetts, A. (2004). Spatial deictics in Saliba. In G. Senft (Ed.), *Deixis and demonstratives in Oceanic languages* (Vol. 562, pp. 37–58). Canberra, Australia: Pacific Linguistics.

Mariotti, M. A. (2006). Proof and proving in mathematics education. In A. Gutierrez & P. Boera (Eds.), *Handbook of research on the psychology of mathematics education: Past, present and future*. Rotterdam, The Netherlands: Sense Publishers.

Mark, K. (2006). *Ethno-mathematics: Is there any Mathematics in my culture?—Deriving Mathematics from culture*. Goroka, PNG: University of Goroka.

Martin, A. (2007). *Making fishing nets in Magi*. Goroka, Papua New Guinea: University of Goroka.

Martin, T., & Schwartz, D. (2005). Physically distributed learning: Adapting and reinterpreting physical environments in the development of fraction concepts. *Cognitive Science, 29*, 587–625.

Martlew, M., & Connolly, K. (1996). Human figure drawings by schooled and unschooled children in Papua New Guinea. *Child Development, 67*(6), 2743–2762. doi:10.1111/j.1467-8624.1996.tb01886.x.

Masingila, J. O. (1993). Learning from mathematics practice in out-of-school situations. *For the Learning of Mathematics, 13*(2), 4.

Mason, J. (2003). Structure of attention in the learning of mathematics. In J. Novotná (Ed.), *Proceedings of the International Symposium on Elementary Mathematics Teaching* (pp. 9–16). Prague: Charles University.

Mason, J., Burton, L., & Stacey, K. (1985). *Thinking mathematically* (revised ed.). Wokingham: Addison Wesley (Pearson).

Mason, J., Stephens, M., & Watson, A. (2009). Appreciating mathematical structure for all. *Mathematics Education Research Journal, 21*(2), 10–32. doi:10.1007/bf03217543.

Matang, R. (1996). *Towards an ethnomathematical approach to mathematics education in Papua New Guinea: An alternative proposal.* Unpublished QUT M.Ed. dissertation, Queensland University of Technology, Brisbane.

Matang, R. (1998). The role of ethnomathematics and reflective learning in mathematics education in Papua New Guinea. *Directions: Journal of Educational Studies, 20*(2), 22–29.

Matang, R. (2001). *Ethnomathematics and the age-old debate of conceptual knowledge versus procedural knowledge in mathematics: Implications for mathematics education in Papua New Guinea.* Paper presented at the Faculty of Science Conference, University of Goroka, Goroka, PNG.

Matang, R. (2003). The cultural context of mathematics learning and thinking in Papua New Guinea. In A. C. Maha & T. A. Flaherty (Eds.), *Education for 21st century in Papua New Guinea and the South Pacific* (pp. 161–168). Goroka, Papua New Guinea: University of Goroka.

Matang, R. (2008). *Enhancing children's formal learning of early number knowledge through Indigenous languages and ethnomathematics: The case of Papua New Guinea mathematics curriculum reform experience.* Paper presented at the 11th International Congress on Mathematics Education ICME 11, Monterray, Mexico. Retrieved from http://dg.icme11.org/document/get/322

Matang, R., & Owens, K. (2006). Rich transitions from Indigenous counting systems to English arithmetic strategies: Implications for mathematics education in Papua New Guinea. In F. Favilli (Ed.), *Ethnomathematics and mathematics education, Proceedings of the 10th International Congress on Mathematical Education Discussion Group 15 Ethnomathematics*. Pisa, Italy: Tipografia Editrice Pisana.

Matang, R., & Owens, K. (2014). The role of Indigenous traditional counting systems in children's development of numerical cognition: Results from a study in Papua New Guinea. *Mathematics Education Research Journal*. doi:10.1007/s13394-013-0115-2.

Matsuo, N. (1993). *Students' understanding of geometrical figures in transition from van Hiele level 1 to 2.* Paper presented at the 17th Conference of the International Group for the Psychology of Mathematics Education, Japan.

Maude, H., & Wedgwood, C. H. (1967). String figures from Northern New Guinea. *Oceania, 37*(3), 202–229. doi:10.2307/40329593.

McCosker, N., & Diezmann, C. M. (2009). Scaffolding students' thinking in mathematical investigations. *Australian Primary Mathematics Classroom, 14*, 27–32.

McGee, M. (1979). Human spatial abilities: Psychometric studies; Environmental, genetic, hormonal, and neurological influences. *Psychological Bulletin, 89*, 889–918.

McMurchy-Pilkington, C., & Bartholomew, H. (2009). Achieving in mathematics contested spaces and voices. In R. Hunter, B. Bicknell, & T. Burgess (Eds.), *32nd Annual Conference of the Mathematics Education Research Group of Australasia*. Palmerston North, New Zealand: MERGA.

Meaney, T., & Fairhall, U. (2003). Tensions and possibilities: Indigenous parents doing mathematics curriculum development. In L. Bragg, C. Campbell, G. Herbert, & J. Mousley (Eds.), *26th Annual Conference of Mathematics Education Research Group of Australasia* (Vol. 2, pp. 507–514). Geelong, VIC, Australia: MERGA.

Meaney, T., Trinick, T., & Fairhall, U. (2012). *Collaborating to meet language challenges in Indigenous mathematics classrooms.* Dordrecht, The Netherlands: Springer.

Meaney, T., Trinick, T., & Fairhall, U. (2013). One size does NOT fit all: Achieving equity in Maori mathematics classrooms. *Journal for Research in Mathematics Education, 44*(1), 235–263.

Meira, L. (1998). Making sense of instructional devices: The emergence of transparency in mathematical activity. *Journal for Research in Mathematics Education, 29*, 121–142.

Mellin-Olsen, S. (1987). *The politics of mathematics education*. Dordrecht, The Netherlands: D. Reidel Publishing.

Menghini, M. (2012). *From practical geometry to the laboratory method: The search for an alternative way to Euclid in the history of teaching geometry*. Paper presented at the International Congress on Mathematics Education ICME12, Korea.

Metzgar, E. (1991). *Traditional education in Micronesia: A case study of Lamotrek Atoll with comparative analysis of the literature on the Trukic continuum*. Ph.D. thesis, University of California, Los Angeles. Retrieved from http://www.tritonfilms.com/images/EMetzgar Dissertation.pdf

Metzgar, E. (2004). Sacred spaces, taboo places: Negotiating *Roang* on Lamotrek Atoll. *Micronesian Journal of the Humanities and Social Sciences, 3*(1–2).

Michael, W., Guilford, J., Fruchter, B., & Zimmerman, W. (1957). Description of spatial visualization abilities. *Educational and Psychological Measurement, 17*, 185–199.

Mildren, J. (1990). The elegant path to metacognition. In M. A. Clements (Ed.), *Whither Mathematics? Proceedings of the 27th Annual Conference of The Mathematical Association of Victoria* (pp. 373–379). Melbourne: The Mathematical Association of Victoria.

Millroy, W. (1992). *An ethnographic study of the mathematical ideas of a group of carpenters*. Reston, VA: National Council of Teachers of Mathematics.

Molnar, J., & Slezakova, J. (2012). Testing of geometrical imagination—Poster 16. *12th International Congress on Mathematics Education ICME12* (pp. 7516). Seoul, Korea: ICME12.

Montiel, C., & Managal, M. A. (2011). Commentary: Nuancing the meaning of cultural diversity. *Journal of Pacific Rim Psychology, 5*(2), 81–84.

Moschkovich, J. (1996). Learning in two languages. In L. Puig & A. Gutiérrez (Eds.), *20th Conference of the International Group for the Psychology of Mathematics Education PME20* (Vol. 4, pp. 27–34). Valencia, Spain: Universitat de Valencia and PME Organising Committee.

Moschkovich, J. (2002). A situated and sociocultural perspective on bilingual mathematics learners. *Mathematical Thinking and Learning, 4*(2 and 3), 189–212.

Mosel, U. (2004). Demonstratives in Samoan. In G. Senft (Ed.), *Deixis and demonstratives in Oceanic languages* (Vol. 562, pp. 141–174). Canberra, Australia: Pacific Linguistics.

Moses, B. (1977). *The nature of spatial ability and its relationship to mathematical problem solving*. Ph.D. thesis, Indiana University.

Moyer, J. C. (1978). The relationship between the mathematical structure of Euclidean transformations and the spontaneously developed cognitive structures of young children. *Journal for Research in Mathematics Education, 9*, 83–92.

Moyer, P., Niezgoda, D., & Stanley, M. (2005). Young children's use of virtual manipulatives and other forms of mathematical representation. In W. Masalski & P. Elliott (Eds.), *Technology-supported mathematics learning environments*. Reston, VA: National Council of Teachers of Mathematics.

Muke, C. (2000). *Ethnomathematics: Mid-Wahgi counting practices in Papua New Guinea*. Masters thesis, University of Waikato, Waikato, New Zealand.

Muke, C. (2012). *Role of local language in teaching mathematics in PNG*. Ph.D. thesis, Australian Catholic University, Melbourne, Australia.

Muke, C., & Clarkson, P. (2011). Teaching mathematics in the land of many languages. In M. Setati, T. Nkambule, & L. Goosen (Eds.), *Mathematics education and language diversity. Proceedings of ICMI Study Conference 21* (pp. 242–250). São Paolo, Brazil: International Committee on Mathematics Instruction.

Mullis, I. V. S., Martin, M. O., Kennedy, A. M., & Foy, P. (2007). *PIRLS 2006 International Report. IEA's progress in international reading literacy study in primary school in 40 countries*. Chestnut Hill, MA: TIMSS and PIRLS International Study Center, Boston College.

Munn, N. (1961). *Psychology: The fundamentals of human adjustment* (4th ed.). Boston: Houghton Mifflin.

Murdoch, K. (1998). *Classroom connections: Strategies for integrated learning*. Armadale, VIC: Eleanor Curtain.

Murray, M. (2011). *The language of geometry* (Vol. 4). Sydney: Mathematical Publications.

Nakata, M. (2011). Pathways for Indigenous education in the Australian Curriculum framework. *Australian Journal of Indigenous Education, 40*(1), 1–8.

Nasir, N., & de Royston, M. (2013). Power, identity, and mathematical practices outside and inside school. *Journal for Research in Mathematics Education, 44*(1), 264–287.

National Council of Teachers of Mathematics. (1989). *Curriculum and evaluation standards for school mathematics*. Reston, VA: National Council of Teachers of Mathematics.

National Department of Education PNG. (2003). *Cultural mathematics syllabus*. Retrieved from http://www.education.gov.pg/Teachers/elem-crip-materials/syllabus-elementary-cultural-mathematics.pdf

National Museum of Australia. (2013). *Old masters: Australia's great bark artists*. Canberra, Australia: Author.

National Research Council Committee on Geography. (2006). *Learning to think spatially: GIS as a support system in the K-12 curriculum*. Washington, DC: National Academies Press.

Ncedo, N., Pieres, M., & Morar, T. (2002). *Code- switching revisited: The use of languages in primary school science and mathematics classrooms*. Paper presented at the 10th Annual Conference of the Southern African Association for Research in Mathematics, Science and Technology Education, Durban.

Ness, D., & Farenga, S. (2007). *Knowledge under construction: The importance of play in developing children's spatial and geometric thinking*. Lanham, Maryland: Rowan & Littlefield.

Network, A. N. K. (1998). *Alaskan standards for culturally-responsive schools*. Fairbank, Alaska: Author.

Neville, H. (1988). Cerebral organization for spatial attention. In J. Stiles-Davis, M. Kritchevsky, & U. Bellugi (Eds.), *Spatial cognition: Brain bases and development* (pp. 327–342). Hillsdale, NJ: Lawrence Erlbaum Associates.

Newcombe, N., & Huttenlocher, J. (2000). *Making space: The development of spatial representation and reasoning*. Cambridge, MA: MIT Press.

Norris, R., & Norris, C. (2009). *Emu dreaming: An introduction to Australian Indigenous astronomy*. Sydney: Emu Dreaming.

Noss, R. (1987). Children's learning of geometrical concepts through Logo. *Journal for Research in Mathematics Education, 18*(5), 343–362.

NSW Department of Education and Training. (1998). *Count Me In Too*. Retrieved from http://www.curriculumsupport.education.nsw.gov.au/primary/mathematics/countmeintoo/index.htm

NSW Department of Education and Training Curriculum Support and Development. (2000). *Count Me Into Space. Resource for teachers with learning framework, exemplar lessons, assessment tasks, three videorecordings*. Sydney: Author.

Nunes, T. (1992). Ethnomathematics and everyday cognition. In D. A. Grouws (Ed.), *Handbook of research on mathematics teaching and learning* (pp. 557–574). New York: Macmillan.

O'Sullivan, D. (2008). The Treaty of Waitangi in contemporary New Zealand politics. *Australian Journal of Political Science, 43*(2), 317–331.

Odobu, D. (2007). Teaching area and surveying, circles and volume, ratio and rates by using examples from PNG culture. In W. Kaleva (Ed.), *Ethnomathematics projects 2007* (pp. 38–71). Goroka, Papua New Guinea: University of Goroka.

Onggi, C. (2005). *Deriving academic mathematics from Telefol traditional door board designs*. Unpublished report, University of Goroka, Goroka, PNG.

Opper, S. (1977). Concept development in Thai urban and rural children. In P. Dasen (Ed.), *Piagetian psychology: Cross-cultural contributions*. New York: Gardner Press.

Organisation for Economic Co-operation and Development. (2013). *PISA 2012 results: What students know and can do—Student performance in mathematics, reading, and science* (Vol. 1): PISA. OECD Publishing. Retrieved from http://www.oecd.org/pisa/keyfindings/pisa-2012-results-volume-I.pdf

Osborne, R., & Wittrock, M. (1983). Learning science: A generative process. *Science Education, 67*, 489–508.

Outhred, L. (1993). *The development in young children of concepts of rectangular area measurement*. Doctoral thesis, Macquarie University, Sydney.

Outhred, L., & Mitchelmore, M. (2004). *Students' structuring of rectangular arrays*. Paper presented at the 28th Conference of the International Group for the Psychology of Mathematics Education, Bergen, Norway.

Owens, K. (1990). *Getting inside the problem solver's head: Using retrospection and observation to access spatial and problem thinking processes*. Paper presented at the 13th Annual Conference of the Mathematics Education Research Group of Australasia MERGA13, Hobart.

Owens, K. (1992a). Spatial mathematics: A group test for primary school students. In K. Stephens & J. Izard (Eds.), *Reshaping assessment practices: Assessment in the mathematical sciences under challenge*. Melbourne: Australian Council for Education Research.

Owens, K. (1992b). Spatial thinking takes shape through primary school experiences. *Proceedings of the Sixteenth PME Conference* (Vol. 2, pp. 202–209). University of New Hampshire, Durham, NH. Durham, NH: Program Committee of the 16th PME Conference.

Owens, K. (1993). *Spatial thinking processes employed by primary school students engaged in mathematical problem solving*. Ph.D. thesis, Deakin University, Geelong, VIC, Australia. Retrieved from http://dro.deakin.edu.au/eserv/DU:30023339/owens-spatialthinking-1993.pdf informit database.

Owens, K. (1996a). Matematik-dold i ett pappersark (Mathematics wrapped up in a piece of paper). *NämNaren, 23*(3), 26–28.

Owens, K. (1996b). Recent research and a critique of theories of early geometry learning: The case of the angle concept. *Nordisk Matematikk Didaktikk-Nordic Studies in Mathematics Education, 4*(2/3), 85–106.

Owens, K. (1997a). Classroom views of space. In B. Doig & J. Lokan (Eds.), *Learning from children: Mathematics from a classroom perspective* (pp. 125–146). Melbourne: Australian Council for Educational Research.

Owens, K. (1997b). Visualisation and empowerment in mathematics education. In R. Ryding (Ed.), *Sveriges Matematiklararforening Arsbok (Swedish Mathematics Teaching and Learning Yearbook)* (pp. 36–46). Gothenburg, Sweden: National Center for Mathematics Education, Namnaren.

Owens, K. (1998a). Explaining spatial problem solving in terms of cognitive load or responsiveness and selective attention. In P. Jeffery (Ed.), *Annual Conference of Australian Association for Research in Education. File: Owe98243*. Melbourne: AARE.

Owens, K. (1998b). Investigating with paper. Reflections. *Reflections, 23*(1), 62–67.

Owens, K. (1999a). Investigating string and rope. *Square One, 9*(3), 2–5.

Owens, K. (1999b). The role of culture and mathematics in a creative design activity in Papua New Guinea. In E. Ogena & E. Golla (Eds.), *8th South-East Asia Conference on Mathematics Education: Technical papers* (pp. 289–302). Manila, The Philippines: Southeast Asian Mathematical Society.

Owens, K. (1999c). The role of visualisation in young students' learning. In O. Zaskavsky (Ed.), *23rd Annual Conference of International Group for Psychology of Mathematics Education, PME23, Haifa, Israel* (Vol. 1, pp. 220–234).

Owens, K. (2000a). Students' mapping and spatial knowledge. *Square One, 10*(1), 17–21.

Owens, K. (2000b). Traditional counting systems and their relevance for elementary schools in Papua New Guinea. *PNG Journal of Education, 36*(1 & 2), 62–72.

Owens, K. (2001a). *Development of the test: Thinking about 3D Shapes*. Sydney: NSW Department of Education and Training.

Owens, K. (2001b). Undersökning med snöre ock rep (Investigating with string and rope). *NämNaren, 28*(3), 30–31.

Owens, K. (2001c). The work of Glendon Lean on the counting systems of Papua New Guinea and Oceania. *Mathematics Education Research Journal, 13*(1), 47–71.

Owens, K. (2002a). *Count Me Into Space implementation over two years with consultancy support*. NSW Department of Education and Training Professional Support and Curriculum Directorate.

Owens, K. (2002b). *Final report on Count Me Into Space with school-based facilitators*. Sydney, Australia: NSW Department of Education and Training Professional Support and Curriculum Division.

References

Owens, K. (2002c). Report on Count Me Into Space implemented in 2001 by two groups of schools using facilitators from the schools. NSW Department of Education and Training Professional Support and Curriculum Directorate.

Owens, K. (2004a). Imagery and property noticing: Young students' perceptions of three-dimensional shapes. In P. Jeffery (Ed.), *Proceedings of the Annual Conference for the Australian Association for Research in Education*. AARE: Melbourne, Australia. Retrieved from http://www.aare.edu.au/conf04/. File: OWE04038.

Owens, K. (2004b). Improving the teaching and learning of space mathematics. In B. Clarke, D. Clarke, G. Emanuelsson, B. Johansson, D. Lambdin, F. Lester, A. Wallby, & K. Wallby (Eds.), *International perspectives on learning and teaching mathematics* (pp. 569–584). Gothenburg, Sweden: Göteborg University National Center for Mathematics Education.

Owens, K. (2005). Substantive communication of space mathematics in upper primary school. In H. Chick & J. Vincent (Eds.), *Proceedings of the 29th Annual Conference of the International Group for the Psychology of Mathematics Education* (Vol. 4, pp. 33–40). Melbourne: PME.

Owens, K. (2006a). *Creating space: Professional knowledge and spatial activities for teaching mathematics*. Retrieved from http://athene.riv.csu.edu.au/~kowens/creatingspaceit.

Owens, K. (2006b). Rethinking cultural research. *Proceedings of Third International Conference on Ethnomathematics*. Retrieved from www.math.auckland.ac.nz/Events/2006/ICEM-3/2.Prez%20Given/Prez%20given%20papers/Owens-paper.doc-2006–08–25

Owens, K. (2007a). Changing our perspective on measurement: A cultural case study. In J. Watson & K. Beswick (Eds.), *Proceedings of 30th Annual Conference of the Mathematics Education Research Group of Australasia* (pp. 563–573). Hobart: MERGA.

Owens, K. (2007/2008). Identity as a mathematical thinker. *Mathematics Teacher Education and Development, 9*, 36–50.

Owens, K. (2008). Diversity of approaches to mathematics education in a cultural context. *Proceedings of the Conference of the International Study Group on the Relations Between History and Pedagogy of Mathematics, HPM2008*. Mexico City: Organising Committee HPM2008.

Owens, K. (2012a). Identity and ethnomathematics projects in Papua New Guinea. In D. Jaguthsing, L. P. Cheng, & S. F. Ng (Eds.), *Mathematics Education: Expanding Horizons. Proceedings of 35th Annual Conference of Mathematics Education Research Group of Australasia*. Singapore: MERGA.

Owens, K. (2012b). Papua New Guinea Indigenous knowledges about mathematical concepts. *Journal of Mathematics and Culture (on-line), 6*(1), 15–50.

Owens, K. (2013a). Chapter Nine: The current status of ethnomathematics in Papua New Guinea: Its importance in education. In J. Pumwa (Ed.), *Mathematics digest: Contemporary discussions in various fields with some mathematics—ICPAM-Lae*. Lae, Papua New Guinea: PNG University of Technology.

Owens, K. (2013b). Diversifying our perspectives on mathematics about space and geometry: An ecocultural approach. *International Journal for Science and Mathematics Education*. doi:http://www.springerlink.com/openurl.asp?genre=article&id=doi:10.1007/s10763-013-9441-9

Owens, K. (2013c). *Improving the teaching of mathematics in elementary schools by using local languages and cultural practices (Papua New Guinea): Workshop materials*. Goroka, PNG: University of Goroka and Charles Sturt University.

Owens, K. (2013d). *Seli measures area*. Dubbo, NSW, Australia: OREC.

Owens, K. (2014). The impact of a teacher education culture-based project on identity as a mathematics learner. *Asia-Pacific Journal of Teacher Education, 42*(2), 186–207. doi:10.1080/1359866X.2014.892568.

Owens, K., & Clements, M. (1998). Representations used in spatial problem solving in the classroom. *Journal of Mathematical Behavior, 17*(2), 197–218.

Owens, K., Doolan, P., Bennet, M., Logan, P., Murray, L., McNair, M., et al. (2012). Continuities in education: Pedagogical perspectives and the role of Elders in education for Indigenous students. *Journal of Australian Indigenous Issues, 15*(1), 20–39.

Owens, K., Edmonds-Wathen, C., & Bino, V. (2014). *Bringing ethnomathematics to elementary school teachers in Papua New Guinea: A design-based research project*. Paper presented at the International Congress on Ethnomathematics 5, Maputo, Mozambique.

Owens, K., & Geoghegan, N. (1998). The use of children's worksamples for deciding learning outcomes and level of development in spatial thinking. In A. Olivier & K. Newstead (Eds.), *22nd Annual Conference of International Group for Psychology of Mathematics Education PME22* (Vol. 4, pp. 294). Stellenbosch: PME.

Owens, K., & Kaleva, W. (2008). Case studies of mathematical thinking about area in Papua New Guinea. In O. Figueras, J. Cortina, S. Alatorre, T. Rojano, & A. Sepúlveda (Eds.), *Annual Conference of the International Group for the Psychology of Mathematics Education (PME) and North America chapter of PME, PME32-PMENAXXX* (Vol. 4, pp. 73–80). Morelia, Mexico: PME.

Owens, K., & Kaleva, W. (2008b). *Indigenous Papua New Guinea knowledges related to volume and mass*. Paper presented at the International Congress on Mathematics Education ICME 11, Discussion Group 11 on The Role of Ethnomathematics in Mathematics Education, Monterray, Mexico. Retrieved from http://dg.icme11.org/document/get/311

Owens, K., McPhail, D., & Reddacliff, C. (2003). Facilitating the teaching of space mathematics: An evaluation. In N. Pateman, B. Dougherty, & J. Zilliox (Eds.), *Proceedings of 27th annual conference of the International Group for the Psychology of Mathematics Education* (Vol. 1, pp. 339–345). Hawaii: International Group for the Psychology of Mathematics Education.

Owens, K., & Outhred, L. (1996). Young children's understandings of tiling areas. *Reflections, 21*(3), 35–40.

Owens, K., & Outhred, L. (1997). Early representations of tiling areas. In E. Pehkonen (Ed.), *21st Annual Conference of International Group for Psychology of Mathematics Education* (Vol. 3, pp. 312–319). Lahti, Finland: Research and Training Institute & University of Helsinki.

Owens, K., & Outhred, L. (1998). Covering shapes with tiles: Primary students' visualisation and drawing. *Mathematics Education Research Journal, 10*(3), 28–41.

Owens, K., & Outhred, L. (2006). The complexity of learning geometry and measurement. In A. Gutiérrez & P. Boero (Eds.), *Handbook of research on the psychology of mathematics education: Past, present and future* (pp. 83–115). Rotterdam, The Netherlands: Sense Publishers.

Owens, K., Paraide, P., Jannok Nutti, Y., Johansson, G., Bennet, M., Doolan, P., et al. (2011). Cultural horizons for mathematics. *Mathematics Education Research Journal, 23*, 253–274. doi:10.1007/213394-011-0014-3.

Owens, K., & Perry, B. (1998). *Mathematics K-10 literature review*. Sydney: New South Wales Department of Education and Training.

Owens, K., Perry, B., Conroy, J., Geoghegan, N., & Howe, P. (1998). Responsiveness and affective processes in the interactive construction of understanding in mathematics. *Educational Studies in Mathematics, 35*(2), 105–127.

Owens, K., & Reddacliff, C. (2002). *Facilitating the teaching of space mathematics: An evaluation*. Paper presented at the Mathematics Education in the South Pacific, 25th Annual Conference of Mathematics Education Research Group of Australasia, Auckland.

Owens, K., & Students. (2007). *The reality of intellectual quality in the mathematics classroom*. Paper presented at the Australian Association of Mathematics Teachers, Hobart.

Ozanne-Rivierre, F. (2004). Spatial deixis in Iaai. In G. Senft (Ed.), *Deixis and demonstratives in Oceanic languages* (Vol. 562, pp. 129–140). Canberra, Australia: Pacific Linguistics.

Ozyürek, A. (1998). An analysis of the basic meaning of Turkish demonstratives in face-to-face conversational interaction. In S. Santi, I. Guaitella, C. Cave, & G. Konopczynski (Eds.), *Oralité et gestualite: Communication multimodale, interaction* (pp. 609–614). Paris: L'Harmattan.

Paivio, A. (1971). *Imagery and verbal processing*. New York: Holt, Reinhart, & Winston.

Paivio, A. (1986). *Mental representations: A dual coding approach*. New York: Oxford University Press.

Palais, R. (1999). The visualization of mathematics: Towards a mathematical Exploratorium. *Notices of the AMS, 46*(6), 647–658.

Pappas, S., Ginsburg, H. P., & Jiang, M. (2003). SES differences in young children's metacognition in the context of mathematical problem solving. *Cognitive Development, 18*(3), 431–450. doi:10.1016/S0885-2014(03)00043-1.

Paraide, P. (2003). *What skills have they mastered?* Paper presented at the The National Education Reform: Where Now, Where to?, Goroka, Papua New Guinea.

Paraide, P. (2008). *Counting and measurement systems of the Tolai.* Paper presented at the Connecting Indigenous Contexts for Continuities in Education, Dubbo, NSW, Australia.

Paraide, P. (2010). *Integrating Indigenous and Western mathematical knowledge in PNG early schooling.* Ph.D. thesis, Deakin University, Geelong, VIC, Australia.

Patronis, T. (1994). *On students' conceptions of axioms in school geometry.* Paper presented at the 18th Conference of the International Group for the Psychology of Mathematics Education, Lisbon, Portugal.

Pegg, J., & Baker, P. (1999). An exploration of the interface between Van Hiele's levels 1 and 2: Initial findings. In O. Zaslavsky (Ed.), *23rd Conference of the International Group for the Psychology of Mathematics Education* (Vol. 4, pp. 25–32). Haifa, Israel: PME.

Pegg, J., & Davey, G. (1998). Interpreting student understanding in geometry: A synthesis of two models. In R. Lehrer & D. Chazan (Eds.), *Designing learning environments for developing understanding of geometry and space* (pp. 109–136). Mahwah, NJ: Lawrence Erlbaum.

Peirce, C. (1998). *The essential Peirce* (Vol. 2, edited by the Peirce Edition Project). Bloomington, IN: Indiana University Press.

Pellegrino, J. W., & Hunt, E. B. (1991). Cognitive models for understanding and assessing spatial abilities. In H. Rowe (Ed.), *Intelligence: Reconceptualization and measurement*. Hillsdale, NJ: Lawrence Erlbaum and Australian Council for Educational Research.

Penn Museum. (1997). *Traditional navigation in the Western Pacific: A search for pattern.* Retrieved from http://www.penn.museum/sites/Navigation/Intro.html

Pepeta, P. (2007). Developing mathematics in the traditional/cultural activity. In W. Kaleva (Ed.). Goroka, Papua New Guinea: University of Goroka.

Perham, F. (1978). An investigation into the effect of instruction on the acquisition of transformation geometry concepts in first grade children and subsequent transfer to general spatial ability. In R. Lesh (Ed.), *Recent research concerning the development of spatial and geometric concepts* (pp. 229–242). Columbus, OH: ERIC.

Perry, B., Anthony, G., & Diezmann, C. (Eds.). (2004). *Research in mathematics education in Australasia 2000–2003*. Sydney: MERGA.

Piaget, J., & Inhelder, B. (1956). *The child's conception of space*. London: Routledge & Kegan Paul.

Piaget, J., & Inhelder, B. (1971). *Mental imagery in the child: A study of the development of imaginal representation*. London: Routledge & Kegan Paul.

Piaget, J., Inhelder, B., & Szeminska, A. (1960). *The child's conception of geometry*. New York: Basic Books.

Pickles, A. (2009). Part and whole numbers: An 'enumerative' reinterpretation of the Cambridge Anthropological Expedition to Torres Straits and its subjects. *Oceania, 79*(3), 293–315.

Pinxten, R. (1991). Geometry education and culture. *Learning and Instruction, 1*(3), 217–227.

Pinxten, R. (1997). Applications in the teaching of mathematics and the sciences. In A. Powell & M. Frankenstein (Eds.), *Ethnomathematics: Challenging Eurocentrism in mathematics education* (pp. 373–401). Albany, NY: State University of New York Press.

Pinxten, R., & François, K. (2011). Politics in an Indian canyon? Some thoughts on the implications of ethnomathematics. *Educational Studies in Mathematics, 78*(2), 261–273. doi:10.1007/s10649-011-9328-z.

Pinxten, R., & François, K. (2012a). Ethnomathematics: A challenge. *Journal of Mathematics and Culture (on-line), 6*(1), 76–87.

Pinxten, R., & François, K. (2012b). Ethnomathematics: A social and scientific choice. *Volkskunde, 113*(1), 72–91.

Pinxten, R., van Dooren, I., & Harvey, F. (1983). *The anthropology of space: Explorations into the natural philosophy and semantics of the Navajo*. Philadelphia: University of Pennsylvania Press.

Pinxten, R., van Dooren, I., & Soberon, E. (1987). *Towards a Navajo geometry*. Ghent, Belgium: KKI Books.

Pirie, S., & Kieren, T. (1991). Folding back: Dynamics in the growth of mathematical understanding. In F. Furinghetti (Ed.), *Proceedings of the 15th Annual Conference of the International Group for the Psychology of Mathematics Education* (Vol. 3, pp. 169–176). Italy: Program Committee for the International Group for the Psychology of Mathematics Education.

Pirie, S., & Kieren, T. (1994). Growth in mathematical understanding: How can we characterise it and how can we represent it? *Educational Studies in Mathematics, 26*(2 and 3), 165–190.

Piru, P. (2005). *Mathematics derived from Huli wig construction*. Unpublished report, University of Goroka, Goroka, PNG.

Planas, N., & Civil, M. (2009). Working with mathematics teachers and immigrant students: An empowerment perspective. *Journal of Mathematics Teacher Education, 12*(6), 391–409.

Poltrock, S., & Agnoli, F. (1986). Are spatial visualization ability and visual imagery ability equivalent? In R. J. Sternberg (Ed.), *Advances in the psychology of human intelligence* (Vol. 3, pp. 255–296). Hillsdale, NJ: Lawrence Erlbaum.

Poltrock, S., & Brown, P. (1984). Individual differences in visual imagery and spatial ability. *Intelligence, 8*, 93–138.

Polynesian Voyaging Society. (2003). Retrieved October 20, 2009, from http://pvs.kcc.hawaii.edu/index.html

Postman, N. (1996). *The end of education. Redefining the value of school*. London: Vintage.

Prediger, S., Clarkson, P., & Bose, A. (2012). *A way forward for teaching in multilingual contexts: Purposefully relating multilingual registers*. Paper presented at the Topic Group 30: Language and Communication in the Mathematics Classroom, International Congress on Mathematics Education ICME12, Seoul, Korea.

Presmeg, N. (1986). Visualisation in high school mathematics. *For the Learning of Mathematics, 6*(3), 42–46.

Presmeg, N. (1997). Reasoning with metaphors and metonymies in mathematics learning. In L. English (Ed.), *Mathematical reasoning: Analogies, metaphors, and images* (pp. 267–279). Mahwah, NJ: Lawrence Erlbaum.

Presmeg, N. (1998). Balancing complex human worlds: Mathematics education as an emergent discipline in its own right. In J. Kilpatrick & A. Sierpinska (Eds.), *Mathematics education as a research domain: A search for identity* (pp. 57–70). Dordrecht, The Netherlands: Kluwer.

Presmeg, N. (2002). Shifts in meaning during transitions. In G. De Abreu, A. Bishop, & N. Presmeg (Eds.), *Transitions between contexts of mathematical practices* (pp. 212–228). Dordrecht, The Netherlands: Kluwer.

Presmeg, N. (2006). Research on visualization in learning and teaching mathematics. In A. Gutiérrez & P. Boero (Eds.), *Handbook of research on the psychology of mathematics education* (pp. 205–304). Rotterdam: Sense Publishers.

Pylyshyn, Z. (1979). The rate of "mental rotation" of images: A test of a holistic analogue hypothesis. *Memory and Cognition, 7*(1), 19–28.

Pylyshyn, Z. (1981). The imagery debate: Analogue media versus tacit knowledge. *Psychological Review, 88*, 16–45.

Quinn, M. (1984). *Visualization in learning mathematics*. Ph.D. thesis, Monash University, Melbourne, Australia.

Radford, L. (2006). The anthropology of meaning. *Educational Studies in Mathematics, 61*, 39–65. doi:10.1007/s10649-006-7136-7.

RADMASTE Ethnomathematics Research Project. (1998). *RADMASTE Ethnomathematics Research Project*. Paper presented at the PME, Stellenbosch, South Africa.

Rahaman, J. (2012). An analysis of students' strategies for area measurement and its curriculum implications. In D. Jaguthsing, L. P. Cheng, & S. F. Ng (Eds.), *Mathematics education: Expanding horizons. Proceedings of 35th Annual Conference of Mathematics Education Research Group of Australasia* (p. 866). Singapore: MERGA.

Rawes, P. (2007). Reflective subjects in Kant and architectural design education. *Journal of Aesthetic Education, 41*(1), 74–89.

REHSEIS-UMR7219. (2005). *Eric Vandendriessche's work: Minutes of meeting 26/05/05*. Retrieved October 21, 2009, from http://www.rehseis.cnrs.fr/spip.php?article526&var_recherche=Vandendriessche

Reichard, G. (1933 (reprinted 1969)). *A study of Melanesian design* (Vol. 1). New York: Columbia University Press.

Reisman, F., & Kauffman, S. (1980). *Teaching mathematics to children with special needs*. Columbus, OH: Merrill.

Restivo, S., Van Bendegem, J., & Fischer, R. (1993). *Math worlds: Philosophical and social studies of mathematics and mathematics education*. Albany, NY: State University of New York.

Rivera, F. (2011). *Towards a visually-oriented school mathematics classrooms: Research, theory, practice, and issues*. New York: Springer.

Rivera, F., & Rossi Becker, J. (2007). Ethnomathematics in the global episteme: Quo vadis? In B. Atweh, A. Calabrese Barton, M. Borba, N. Gough, C. Keitel, C. Vistro-Yu, & R. Vithal (Eds.), *Internationalisation and globalisation in mathematics education* (pp. 209–226). Dohtrecht: Springer.

Rosa, M., & Orey, D. (2007). Pop: The ethnomathematics of globalization using the sacred Mayan mat pattern. In B. Atweh, A. Calabrese Barton, M. Borba, N. Gough, C. Keitel, C. Vistro-Yu, & R. Vithal (Eds.), *Internationalisation and globalisation in mathematics and science education* (pp. 227–246). Dordrecht: Springer.

Rosa, M., & Orey, D. (2012). *Ethnomodeling: A pedagogical action for uncovering ethnomathematical practices*. International Congress on Mathematics Education. Retrieved from http://www.icme12.org/upload/UpFile2/TSG/1799.pdf

Rosser, R., Lane, S., & Mazzeo, J. (1988). Order of acquisition of related geometric competencies in young children. *Child Study Journal, 18*(2), 75–89.

Roth, W.-M., & McGinn, M. (1998). Towards a theory of representing as social practice. *Review of Educational Research, 68*(1), 35–59. doi:10.3102/00346543068001035.

Rouse, J. (2007). Practice theory. *Division 1 Faculty Publications, 43*, Retrieved from http://wesscholar.wesleyan.edu/div1facpubs/43

Rowe, M. (1982). *Teaching in spatial skills requiring two- and three-dimensional thinking and different levels of internalization and the retention and transfer of these skills*. Melbourne, Australia: Monash University.

Royal Society, & Joint Mathematical Council. (2001). *Teaching and learning geometry Pre-19*. London: Author.

Rumsey, A., & Weiner, J. (Eds.). (2001). *Emplaced myth: Space, narrative, and knowledge in Aboriginal Australia and Papua New Guinea*. Hawaii: University of Hawaii.

Salzmann, Z. (2006). *Language, culture, and society: An introduction to linguistic anthropology* (4th ed.). Boulder, CO: Westview Press.

Sarama, J., & Clements, D. (2004). Building blocks for early childhood mathematics. *Early Childhood Research Quarterly, 19*(1), 181–189.

Sarama, J., & Clements, D. (2009). *Early childhood mathematics education research: Learning trajectories for young children*. Hoboken, NJ: Taylor & Francis.

Saunderson, A. (1973). The effect of a special training programme on spatial ability test performance. *New Guinea Psychologists, 5*, 13–23.

Saxe, G. (1985). The effects of schooling on arithmetical understandings: Studies with Oksapmin children in Papua New Guinea. *Journal of Educational Psychology, 77*(5), 503–513.

Saxe, G. (1991). *Culture and cognitive development: Studies in mathematical understanding*. Hillsdale, NJ: Lawrence Erlbaum Associates.

Saxe, G. (2012). *Cultural development of mathematical ideas: Papua New Guinea studies*. New York: Cambridge University Press.

Saxe, G. (n.d.). *Cultural development of mathematical ideas.* Retrieved from http://www.culturecognition.com/

Saxe, G., & Esmonde, I. (2005). Studying cognition in flux: A historical treatment of *fu* in the shifting structure of Oksapmin mathematics. *Mind, culture, and activity, 12*(3), 171–225.

Senechal, M. (1991). Visualisation and visual thinking. In J. Malkevitch (Ed.), *Geometry's future* (pp. 15–21). Arlington, MA: Community Map Analysis Project.

Senft, G. (Ed.). (1997). *Referring to space: Studies in Austronesian and Papuan languages.* Oxford: Oxford University Clarendon Press.

Senft, G. (2004a). Aspects of spatial deixis in Kilivila. In G. Senft (Ed.), *Deixis and demonstratives in Oceanic languages* (Vol. 562, pp. 59–80). Canberra, Australia: Pacific Linguists.

Senft, G. (2004b). *Deixis and demonstratives in Oceanic languages.* Canberra: Research School of Pacific and Asian Studies, The Australian National University.

Senft, G., & Senft, B. (1986). Ninikula fadenspiele auf den Trobriand Inseln Papua New Guinea. *Baessler-Archiv, 34*, 93–235.

Setati, M., & Adler, J. (2000). Between languages and discourses: Language practices in multilingual mathematics classrooms in South Africa. *Educational Studies in Mathematics, 43*(3), 243–269.

Sfard, A., & Prusak, A. (2005). Telling identities: In search of an analytic tool for investigating learning as a culturally shaped activity. *Educational Researcher, 34*(4), 14–22.

Shah, P., & Miyake, A. (2005). *The Cambridge handbook of visuospatial thinking.* New York: Cambridge University Press.

Shayer, M. (2003). Not just Piaget; not just Vygotsky, and certainly not Vygotsky as alternative to Piaget. *Learning and Instruction, 13*(5), 465–485. doi:http://dx.doi.org/10.1016/S0959-4752(03)00092-6.

Sheckels, M., & Eliot, J. (1983). Preference and solution patterns in mathematics performance. *Perceptual and Motor Skills, 57*, 811–816.

Shepard, R. (1971). Mental rotation of three dimensional objects. *Science, 171*, 701–703.

Shepard, R. (1975). Form, formation, and transformation of internal representations. In R. Solso (Ed.), *Information processing and cognition: The Loyola Symposium* (pp. 87–122). Hillsdale, NJ: Lawrence Erlbaum.

Shepard, R. (1988). The role of transformations in spatial cognition. In J. Stiles-Davis, M. Kritchevsky, & U. Bellugi (Eds.), *Spatial cognition: Brain bases and development* (pp. 81–110). Hillsdale, NJ: Lawrence Erlbaum Associates.

Shir, K., & Zaslavsky, O. (2001). *What constitutes a (good) definition? The case of a square.* Paper presented at the 25th Conference of International Group for the Psychology of Mathematics Education, Utrecht, The Netherlands.

Shir, K., & Zaslavsky, O. (2002). Students' conception of an acceptable geometric definition. In A. D. Cockburn & E. Nardi (Eds.), *26th Conference of the International Group for the Psychology of Mathematics Education* (Vol. 4, pp. 201–208). Norwich, UK: University of East Anglia.

Shirley, L. (1995). Using ethnomathematics to find multicultural mathematical connections. In P. House & A. Coxford (Eds.), *Connecting mathematics across the curriculum: National Council of Teachers of Mathematics Yearbook* (pp. 34–43). Reston, VA: National Council of Teachers of Mathematics.

Siegler, R. (1999). Strategic development. *Trends in Cognitive Sciences, 3*(11), 430–436. doi:10.1016/S1364-6613(99)01372-8.

Skemp, R. (1989). *Mathematics in the primary schools.* London: Routledge.

Smidt, D. (1990). Symbolic meaning in Kominimung masks. In N. Lukehuas, W. Kaufmann, D. Michell, L. Newton, & L. Osmundsen (Eds.), *Sepik heritage: Tradition and change in Papua New Guinea Proceedings of 1987 conference in Switzerland.* Durham, NC: Carolina Academic Press.

Smith, G. (1984). Morobe counting systems: An investigation into the numerals of the Morobe Province, Papua New Guinea. M.Phil. thesis, Papua New Guinea University of Technology, Lae, Papua New Guinea.

Sobel, D. (2008). *Childhood and nature: Design principles for educators*. Portland, ME: Stenhouse Publishers.

Soja, S. (2009). Thirdspace: Towards a new consciousness of space and spatiality. In K. Ikas & G. Wagner (Eds.), *Communicating in thirdspace* (pp. 49–61). New York: Routledge.

Somerville, M. (2007). Place literacies. *Australian Journal of Language & Literacy, 30*(2), 149–164.

Somerville, M., Power, K., & de Carteret, P. (Eds.). (2009). *Landscapes and learning: Place studies for a global world*. Rotterdam, The Netherlands: Sense Publishers.

Sophian, C. (1999). Children's ways of knowing: lessons from cognitive development research. In J. V. Copley (Ed.), *Mathematics in the early years* (pp. 11–20). Reston, VA: National Council of Teachers of Mathematics.

Spencer, C., & Darvizeh, Z. (1983). Young children's place-description, maps and route-findings: A comparison of nursery school children in Iran and Britain. *International Journal of Early Childhood, 15*, 26–31.

Spennemann, D. (1998). Essays on the Marshallese Past: Traditional Marshallese stickchart navigation. Retrieved in 2010, from http://marshall.csu.edu.au/html/essays/es-tmc-2.html

Steffe, L. (Ed.). (1991). *Epistemological foundations of mathematical experience*. New York: Springer.

Sternberg, R. J. (1987). Synopsis of a triarchic theory of human intelligence. In S. H. Irvine & S. E. Newstead (Eds.), *Intelligence and cognition* (pp. 141–176). Dordrecht: Martinus Nijoff Publishers.

Stigler, J., & Baranes, R. (1988). Culture and mathematics learning. *Review of Research in Education, 15*, 253–306.

Stiles-Davis, J., Kritchevsky, M., & Bellugi, U. (Eds.). (1988). *Spatial cognition: Brain bases and development*. Hillsdale, NJ: Lawrence Erlbaum Associates.

Stillman, G., & Balatti, J. (2001). Contributions of ethnomathematics to mainstream mathematics classroom practices: An international perspective. In B. Atweh, H. Forgasz, & B. Nebres (Eds.), *Sociocultural research on mathematics education* (pp. 313–328). Mahwah, NJ: Lawrence Erlbaum.

Stokes, J. (1982). A description of the mathematical concepts of Groote Eyandt Aborigines. *Work Papers of SIL-AAB Series B, B*(December), 33–152.

Sturrock, F., & May, S. (2002). *Programme for International Student Assessment (PISA 2000): The New Zealand Context. The reading, mathematical and scientific literacy of 15-year-olds*. Wellington, NZ: Ministry of Education.

Sukon, K., & Jawahir, R. (2005). Influence of home-related factors in numeracy performance of fourth-grade children in Mauritius. *International Journal of Education Development, 25*, 547–556.

Suwarsono, S. (1982). *Visual imagery in the mathematical thinking of seventh grade students*. Ph.D. thesis, Monash University, Melbourne, Victoria.

Sweller, J., & Chandler, P. (1991). Evidence for cognitive load theory. *Cognition and Instruction, 8*, 351–362.

Tartre, L. (1990a). Spatial skills, gender, and mathematics. In E. L. Fennema & G. Leder (Eds.), *Mathematics and gender*. New York: Teachers College Press.

Tartre, L. (1990b). Spatial orientation skill and mathematical problem solving. *Journal for Research in Mathematics Education, 21*, 216–229.

Téllez, K., Moschkovich, J., & Civil, M. (Eds.). (2011). *Latinos/as and mathematics education: Research on learning and teaching in classrooms and communities*. Charlotte, NC: Information Age Publishing.

Then, D. C.-O., & Ting, S.-H. (2011). Code-switching in English and science classrooms: More than translation. *International Journal of Multilingualism, 8*(4), 299–323. doi:10.1080/147907 18.2011.577777.

Thomas, D. (1978). Students' understanding of selected transformation geometry concepts. In R. Lesh (Ed.), *Recent research concerning the development of spatial and geometric concepts* (pp. 177–193). Columbus, OH: ERIC.

Thomas, N., Mulligan, J., & Goldin, G. (2002). Children's representation of numbers and structures 1–100. *Journal of Mathematical Behavior, 21*(1), 117–133. doi:10.1016/S0732-3123(02)00106-2.

Thornton, M., & Watson-Verran, H. (1996). *Living maths*. Abbotsford, Victoria: Yirrkala Community School and Boulder Valley Films.

Thune, C. (1978). Numbers and counting in Loboda: An example of a non-numerically oriented culture. *Papua New Guinea Journal of Education, 14*, 69–80.

Thurstone, L., & Thurstone, T. (1941). Factor studies of intelligence. *Psychological Monographs, 2*.

Tomasello, M., Carpenter, M., Call, J., Behne, T., & Moll, H. (2005). Understanding and sharing intentions: The origins of cultural cognition. *Behavioral and Brain Sciences, 28*(5), 675–691.

Treacy, K., & Frid, S. (2008). Recognising different starting points in Aboriginal students' learning of number. In M. Goos, R. Brown, & K. Maker (Eds.), *31st Annual Conference of the Mathematics Education Research Group of Australasia* (pp. 531–537). Brisbane: MERGA.

Treisman, A. (1988). Features and objects: The fourteenth Bartlett memorial lecture. *Quarterly Journal of Experimental Psychology, 40A*, 207–237.

Trninic, D., & Kim, H.-J. (2012). *Abstract, concrete, and embodied: An embodied cognition perspective of mathematics education*. Paper presented at the International Congress on Mathematics Education, Seoul, Korea.

Trudgen, R. (2000). *Why warriors lie down and die: Towards an understanding of why the Aboriginal people of Arnhem Land face the greatest crisis in health and education since European contact: Djambatj mala*. Darwin, Australia: Aboriginal Resource & Development Services.

Tuan, Y.-F. (1977). *Space and place: The perspective of experience*. London: Edward Arnold.

Tuinamuana, K. (2007). Reconstructing dominant paradigms of teacher education: Possibilities for pedagogical transformation in Fiji. *Asia-Pacific Journal of Teacher Education, 35*(2), 111–127. doi:10.1080/13598660701268544.

Tupper, I. (2007). *Emphatic pronouns in Namia*. Paper presented at the Papuan Linguists Association Conference, Madang, PNG.

Turkle, S., & Papert, S. (1990). Epistemological pluralism: Styles and voices. *Sign, 16*(1), 128–143.

Turner, E., Dominguez, H., Maldonado, L., & Empson, S. (2013). English learners' participation in mathematical discussion: Shifting positionings and dynamic identities. *Journal for Research in Mathematics Education, 44*(1), 199–234.

Tutak, F. A., Bondy, E., & Adams, T. L. (2011). Critical pedagogy for critical mathematics education. *International Journal of Mathematical Education in Science and Technology, 42*(1), 65–74. doi:10.1080/0020739X.2010.510221.

University of Goroka Students SMAC351. (1998–2007). *Culture and language in mathematics education*. Goroka, PNG: University of Goroka. Retrieved from http://www.uog.ac.pg/GLEC/teaching/teaching.html.

Uttal, D. H., Fisher, J. A., & Taylor, H. A. (2006). Words and maps: Developmental changes in mental models of spatial information acquired from descriptions and depictions. *Developmental Science, 9*(2), 221–235.

Vakalahi, H. (2011). Commentary: Embracing culture as essential to Pacific people. *Journal of Pacific Rim Psychology, 5*(2), 85–89.

Valdés, G. (1998). The world outside and inside schools: Language and immigrant children. *Educational Researcher, 27*(6), 4–18.

Valero, P., & Zevenbergen, R. (Eds.). (2004). *Researching the socio-political dimensions of mathematics education: Issues of power in theory and methodology*. New York: Kluwer.

van der Heijden, A. H. C. (1992). *Selective attention in vision*. London: Routledge.

van Hiele, P. (1986). *Structure and insight: A theory of mathematics education*. New York: Academic Press.

Van Moer, A. (2007). Logic and intuition in mathematics and mathematical education. In K. Francois & J. Bendegem (Eds.), *Philosophical dimensions in mathematics education*. New York: Springer Science+Business Media.

Vandendriessche, E. (2007). Les jeux de ficelle: Une activité mathématique dans certainess sociétés traditionnelles (String figures: A mathematical activity in some traditional societies). *Revue d'histoire des mathématiques, 13*(1), 7–84.

Verlot, M., & Pinxten, R. (2000). Intercultural education and complex instruction. Some remarks and questions from an anthropological perspective on learning. *Intercultural Education, 11*, 7–14. doi:10.1080/14675980020010818.

Verner, I., Massarwe, K., & Bshouty, D. (2013). Constructs of engagement emerging in an ethnomathematically-based teacher education course. *Journal of Mathematical Behavior, 32*(3), 494–507.

Voigt, J. (1985). Patterns and routines in classroom interaction. *Recherches en Didactique des Mathematiques, 6*(1), 69–118.

Voigt, J. (1994). Negotiation of mathematical meaning and learning mathematics. *Educational Studies in Mathematics, 26*(2 and 3), 275–298.

von Glasersfeld, E. (1991). Abstraction, re-presentation, and reflection: An interpretation of experience and Piaget's approach. In L. Steffe (Ed.), *Epistemological foundations of mathematical experience* (pp. 46–65). New York: Springer.

Vurpillot, E. (1976). *The visual world of the child*. New York: International Universities Press.

Walden, R., & Walkerdine, V. (1982). *Girls and mathematics: The early years (Bedford Way Papers 8)*. London: University of London, Institute of Education.

Walkerdine, V. (1988). *The mastery of reason*. Cambridge: Routledge.

Ward, A., & Wong, L. (1996). Equity, education and design in New Zealand: The Whare Wananga Project. *Journal of Architectural Education, 49*(3), 136–155.

Warner, L. (1937). *A black civilisation*. New York: Harper.

Warren, E., Cole, A., & Devries, E. (2009). Closing the gap: Myths and truths behind subitisation. *Australasian Journal of Early Childhood, 34*(4), 46–53.

Wassmann, J. (1994). The Yupno as post-Newtonian scientists: The question of what is 'natural' in spatial description. *MAN, 29*(3), 645–666.

Wassmann, J. (1997). Finding the right path. The route knowledge of the Yupno of Papua New Guinea. In G. Senft (Ed.), *Referring to space: Studies in Austronesian and Papuan languages*. Oxford, UK: Oxford University Press.

Wassmann, J., & Dasen, P. (1994a). 'Hot' and 'cold': Classification and sorting among the Yupno of Papua New Guinea. *International Journal of Psychology, 29*(1), 19–28.

Wassmann, J., & Dasen, P. (1994b). Yupno number system and counting. *Journal of Cross-Cultural Psychology, 25*, 78–94.

Watson, R. (1965). *Psychology of the child* (2nd ed.). New York: Wiley.

Watson-Verran, H., & Turnbull, D. (1995). Science and other indigenous knowledge systems. In S. Jasanoff, G. Markle, J. Perersen, & T. Pinch (Eds.), *Handbook of science and technology studies* (pp. 115–139). London: Sage.

Webb, N. (1979). Processes, conceptual knowledge and mathematical problem-solving ability. *Journal for Research in Mathematics Education, 10*, 83–93.

Wenger, E. (1998). *Communities of practice: Learning, meaning, and identity*. New York: Cambridge University Press.

Were, G. (2003). Objects of learning: An anthropological approach to mathematics education. *Journal of Material Culture, 8*(25), 25–44. doi:10.1177/1359183503008001761.

Were, G. (2010). *Lines that connect: Rethinking pattern and mind in the Pacific*. Honolulu, HA: University of Hawai'i Press.

Werner, H. (1964). *Comparative psychology of mental development* (rev ed.). London: International University Press.

Wessels, D., & Van Niekerk, R. (1998). Semiotic models and the development of secondary school spatial knowledge. Short oral communication. Paper presented at the 22nd Conference of the International Group for the Psychology of Mathematics PME22, Stellenbosch, South Africa.

Wheatley, G., & Cobb, P. (1990). Analysis of young children's spatial constructions. In L. Steffe & T. Wood (Eds.), *Transforming children's mathematics education* (pp. 161–173). Hillsdale, NJ: Lawrence Erlbaum.

White, P., & Mitchelmore, M. (2010). Teaching for abstraction: A model. *Mathematical Thinking and Learning: An International Journal, 12*(3), 205–226.

Wickens, C., & Prevett, T. (1995). Exploring the dimensions of egocentricity in aircraft navigation displays. *Journal of Experimental Psychology: Applied, 1*(2), 110–135.

Wickler, S. (2002). Oral traditions and archeology: Modeling village settlement in Palau, Micronesia. *Micronesian Journal of the Humanities and Social Sciences, 1*(1–2), 39–47.

Williams, L. (2008). Tiering and scaffolding: Two strategies for providing access to important mathematics. *Teaching Children Mathematics, 14*, 324–330.

Willis, S. (2000). *Strengthening numeracy: Reducing risk*. Paper presented at the ACER Research Conference. Improving Numeracy Learning: What Does the Research Tell Us?, Brisbane, Australia.

Wilson, J. (1997). *Self regulated learners and distance education theory*. Retrieved from http://www.usask.ca/education/coursework/802papers/wilson/wilson.html

Wilson, D. (2007). Beyond puzzles: Young children's shape-composition abilities. In W. G. Martin, M. Strutchens, & P. Elliott (Eds.), *The learning of mathematics (NCTM 69th Yearbook)* (pp. 239–255). Reston, VA: National Council of Teachers of Mathematics.

Witelson, S., & Swallow, J. (1988). Neuropsychological study of the development of spatial cognition. In J. Stiles-Davis, M. Kritchevsky, & U. Bellugi (Eds.), *Spatial cognition: Brain bases and development* (pp. 373–409). Hillsdale, NJ: Lawrence Erlbaum.

Wood, T. (2003). Complexity in teaching and children's mathematical thinking. In N. Pateman, B. Doherty, & J. Zilliox (Eds.), *27th Annual Conference of the International Group for the Psychology in Mathematics Education* (Vol. 4, pp. 435–442). Honolulu, Hawaii: PME.

Woodhouse, J., & Knapp, C. (2000). *Place-based curriculum and instruction*. ERIC Document Reproduction Service No. EDO-RC- 00-6.

Woodward, E., Gibbs, V., & Shoulders, M. (1992). Similarity unit. *Arithmetic Teacher, 39*(8), 22–25.

Worsley, P. (1997). *Knowledges: Culture, counterculture, subculture*. New York: The New Press.

Wright, R., Stranger, G., Stafford, A. K., & Martland, J. (2006). *Teaching number: Advancing children's skills and strategies* (2nd ed.). London, UK and Thousand Oaks, CA: Paul Chapman and Sage.

Yakimanskaya, I. S. (1991). Volume 3. In R. Silverman, Trans. P. Wilson, & E. Davis (Eds.), *The development of spatial thinking in school children*. Reston, VA: National Council of Teachers of Mathematics.

Yambi, K. (2004). *Moka gift system and cane bridge making*. Unpublished report, University of Goroka. Goroka, PNG.

Yamu, A. (2000). *Examples of mathematics in PNG cultures*. Goroka, Papua New Guinea: University of Goroka.

Yelland, N. J., & Masters, J. (2007). Rethinking scaffolding in the information age. *Computers & Education, 48*(3), 362–382.

Yunkaporta, T., & McGinty, S. (2009). Reclaiming aboriginal knowledge at the cultural interface. *The Australian Educational Researcher, 36*(2), 55–72.

Zimmerman, B. (1990). Self-regulated learning and academic achievement: An overview. *Educational Psychologist, 25*, 3–17.

Zimmermann, W., & Cunningham, S. (Eds.). (1991). *Visualisation in teaching and learning mathematics*. Washington, DC: Committee on Computers in Mathematics Education of the Mathematical Association of America.

Author Index

A
Abe, K., 34
Adam, N.A., 202, 219, 289
Adam, S., 308
Adams, B.L., 237, 247
Adams, T.L., 106, 111
Adendorff, R., 119
Adler, J., 13, 116, 118, 119, 224, 255
Agnoli, F., 23, 31
Aikenhead, G., 93, 94, 224
Aikenhead, G.S., 99, 145
Akerblom, K., 208
Alangui, W., 308
Alderton, D.L., 3, 19, 21–24
Allport, D., 76, 78, 80
Amanti, C., 111, 231, 246, 295, 307
Amos, S., 225, 321
Anderson, D., 118, 235
Anderson, R., 25
Andrade, R., 275
Anthony, G., 5
Apple, M., 17, 111
Ascher, M., 123, 207, 210, 214–216
Atkin, T., 201
Atran, S., 121
Atweh, B., 13, 232, 246
Aust, R., 22
Averill, R., 118, 235
Avosa, M., 248

B
Baddeley, A., 75
Baenninger, M., 33
Baillargeon, R., 108
Baker, P., 232
Balatti, J., 216
Barab, S., 293
Baranes, R., 13
Barnett, R., 113
Barnhardt, R., 117, 118
Barratt, E.S., 31, 32
Bartholomew, H., 247
Barton, A.C., 13, 232, 246
Barton, B., 13, 118, 142, 210, 224, 308
Battista, M., 6, 108, 136, 137, 232, 234, 267, 282
Bauersfeld, H., 24
Behne, T., 300
Bellugi, U., 24, 25
Bellwood, P., 221
Ben-Chaim, D., 33
Bennet, M., 13, 93, 118, 235, 246, 304
Berry, J., 206, 299
Bhabha, H.K., 99
Bino, V., 221, 248
Bishop, A., 5, 13, 20, 23, 33, 34, 70, 75, 132, 184, 235
Boero, P., 6
Bondy, E., 111
Borba, M., 13, 232, 246, 276
Bose, A., 116, 117
Bowden, R., 169
Bowers, C.A., 112
Brayboy, B., 272
Brofenbrenner, U., 11, 93
Brown, A., 4
Brown, C., 3
Brown, K., 287
Brown, P., 22, 23

Bruner, J., 22
Bryan, E., 107, 208, 209
Bshouty, D., 307
Burden, L., 23, 31
Burton, L., 291
Bush, W., 111
Butler, M., 307
Butterworth, B., 107

C
Call, J., 300
Callingham, R., 234
Cameron, J., 135, 306
Campbell-Jones, S., 40
Campbell, K., 5
Capell, A., 122, 128, 263
Capponi, B., 234
Carpenter, M., 300
Carpenter, P., 23, 24, 31, 32, 56, 142
Carpenter, T., 3
Carraher, T., 142, 224
Casey, B., 4
Castagno, A., 272
Ceci, S., 11, 93
César, M., 307
Chandler, P., 75
Chazan, D., 84, 282, 296, 303
Cheeseman, J., 106
Cheng, K., 109
Cherinda, M., 218, 219, 234, 245–273, 300
Christou, C., 275
Civil, M., 89, 223, 247, 275
Clark, L., 4
Clarkson, P., 116, 117, 224, 234, 247, 255
Clements, D., 5, 7, 14, 29, 31, 32, 35, 65, 74, 76–78, 80, 108, 109, 199, 203, 230, 232, 234, 267, 268, 281–282, 297, 298, 300–301, 304, 361
Clements, M., 5, 14, 15, 32, 34, 35, 61, 74, 80, 90, 141, 199, 203, 230, 268, 297, 298, 301
Cobb, P., 28, 268
Codrington, R.H., 125
Coiffier, C., 201
Cole, A., 81, 245
Collis, K., 5
Connolly, K., 27, 143
Conroy, J., 132
Cook, E., 123
Cooper, L., 23
Corte, E.d., 252
Costigan, K., 13, 201
Coulson, S., 23, 31

Cox, M., 34
Critchlow, K., 234
Cunningham, S., 5, 24
Cuoco, A., 3, 5, 7, 294
Currie, P., 232

D
Dalin, J., 277
D'Ambrosio, U., 111, 142, 238
Darvizeh, Z., 110
Dasen, P., 92, 108, 141, 223
Davenport, W.H., 208, 209
Davey, G., 11, 232, 304
David, D., 250, 251
David, M., 266
Davis, B., 14–16, 114, 301
Davis, R., 96
Davis, W., 208, 249
de Abreu, G., 100, 203, 230
de Carteret, P., 111
de Castello Branco Fantinato, M.C., 221, 223
de la Fuente Arias, J., 278
de Mello, C., 266
de Ribaupierre, A., 108, 141
de Royston, M., 203
De Villiers, M., 232
Deacon, A.B., 214, 215
Dearing, E., 4
Dehaene, S., 107, 235
Del Campo, G., 74
Del Grande, J., 21, 34
Dennett, H., 159
Deregowski, J.B., 24, 34
Derewianka, B., 112
Devries, E., 81, 245
Díaz, A.L., 278
Diezmann, C., 4, 5, 272
Do, C.S., 142
Doig, B., 106
Dominguez, H., 117
Domite, M., 142
Doolan, P., 13, 93, 118, 235, 246, 304
Dörfler, W., 63, 89
Dreyfus, T., 65, 232

E
Easton, H., 118, 235
Edmonds-Wathen, C., 123–125, 221
Egan, D.E., 23, 31
Egan, K., 75
Eglash, R., 13, 111, 221, 223, 247, 276, 287, 288

Ehrich, V., 122
Eisenberg, T., 232
Eisenman, P.
Eliot, J., 3, 20, 31, 32
Empson, S., 117
Enderson, M., 305
English, L., 75
Epstein, Y., 307
Esmonde, I., 13, 92, 99, 142, 223, 305
Eynde, P.O.t., 252
Ezran, D.R., 237

F

Fairhall, U., 119, 120, 130, 237, 239, 247
Falcade, R., 6, 279
Falcao, J., 224, 237
Falsetti, M., 287
Farenga, S., 6, 108–110, 142, 243, 300
Farnham-Diggory, S., 26
Fassler, R., 248
Favilli, F., 223
Fennema, E., 33
Fernandes, S., 26
Ferrare, J., 17, 111
Finau, K., 216
Fiorentino, G., 223
Fischer, K., 110
Fischer, R., 224
Fisher, J., 179
Fisher, J.A., 131
Flavell, J., 67, 75
Flemming, D., 266
Flemming Luz, E., 266
Flores, A., 34
Foy, P., 4
Frade, C., 224, 237
François, K., 78, 118, 120, 121, 136, 206, 207, 263, 306
Freire, P., 111
Frid, S., 81
Frostig, M., 34
Fruchter, B., 20
Furinghetti, F., 233
Fuson, K., 66

G

Gagné, R., 22, 52
Galbraith, P., 146, 276, 279, 280
Gammage, B., 343
Ganley, C., 4
Garbutt, R., 93
Garrison, K., 24

Gee, J., 180
Geiger, V., 146, 276, 279, 280
Gell, A., 75, 145
Genkins, E., 34
Geoghegan, N., 80, 132, 199
George, M., 237
Gerdes, P., 108, 118, 123, 218, 246
Gervasoni, A., 14
Giaquinto, M., 64–66, 81, 82
Gibbs, V., 61
Gibson, J., 77
Gilsdorf, T.E., 217
Ginsburg, H.P., 108
Giroux, H., 111
Gleitman, H., 109
Godfrey, C., 268
Goetzfridt, N., 211, 214
Goldberg, M., 91
Goldenberg, P., 3, 5, 7, 90, 278, 294
Goldin, G., 15, 20, 24, 66, 82, 301, 307
Gomes, M.L.M., 111
González, N., 111, 231, 246, 295, 307
Goos, M., 146, 276, 279, 280
Gorgorió, N., 247
Graham, B., 125
Grant, S., 94, 124
Gray, E., 89
Greene, M., 271, 281
Greeno, J., 226
Greenspan, S.I., 109
Gresalfi, M., 293
Gruenewald, D., 17, 92, 111–113, 294, 306
Guilford, J., 20
Gunn, M., 144
Gutiérrez, A., 6, 8, 235
Gutstein, E., 89, 112

H

Halat, E., 232
Hamadi, T.,
Hannibal, M.,
Harrell, G., 106
Harris, P., 110, 126, 135
Harvey, F., 78, 108, 113, 118, 121, 136, 206, 247, 263, 303
Hauser-Schäublin, B., 168, 170, 201
Healy, L., 26
Hedberg, J., 132, 275, 278
Hegarty, M., 33, 63, 89
Heinz, S., 75
Hershkowitz, R., 24, 59
Highfield, K., 132, 275, 278, 280, 282, 284–286, 301

Hollebrands, K., 81, 278
Holton, D.L., 287
Horne, D., 34
Horvath, J., 84, 234, 235, 245, 246, 282, 303
Houang, R., 33
Howe, P., 132
Hunt, E.B., 21, 22
Hutchins, E., 8, 105, 131, 208, 209, 300
Huttenlocher, J., 77, 109
Hyde, J., 73
Hynds, A., 118, 235

I

Inhelder, B., 3, 26
Ison, B., 159, 169
Izard, V., 107, 235

J

Jacobson, G., 84, 234, 235, 245, 246, 282, 303
Jannok Nutti, Y., 118, 228, 235–237, 247, 304
Jawahir, R., 108, 116, 120, 245–273, 300
Jegede, P.J., 99, 145
Jenkins, M., 30, 303
Jiang, M., 108
Johansson, G., 118, 235, 248, 304
John, L., 165, 252, 253
Johnson, E., 21
Johnson, M., 41, 65
Johnston, W., 75
Johnston-Wilder, S., 12
Jonassen, D., 15, 275, 288
Jones, J., 5, 118, 127
Jones, K., 268, 275
Jones, M., 109
Jones, P., 112
Joseph, G., 111, 223
Julius, A., 252, 254
Just, M., 23, 24, 31, 32, 56, 142

K

Kahan, S., 16
Kahneman, D., 75
Kaleva, W., 118, 127, 224, 228, 247, 248, 250
Kauffman, S., 26
Kaufmann, G., 41, 73
Keitel, C., 5
Kemeny, V., 84, 234, 235, 245, 246, 282, 303
Kennedy, A.M., 4
Khisty, L., 269
Kieras, D., 22

Kieren, T., 5, 63, 231, 312
Kilpatrick, J., 5, 5, 245
Kim, H.-J., 297
Kim, M., 8, 47, 89, 297
Kimball, S.T., 13
Knapp, C., 112
Knijnik, G., 223, 266
Kono, S., 176, 241, 242, 252
Kosslyn, S., 22–24, 31
Kouba, V., 3
Kozhevnikov, M., 33, 63, 89
Krinsky, C., 201
Kritchevsky, M., 24, 25
Krutetskii, V., 20, 24, 31, 61
Kuchler, S., 168
Kull, M., 248
Kurina, F., 34
Kyllonen, P.C., 33
Kynoigos, C., 81, 278

L

Laborde, C., 6, 81, 234, 278, 279
Lakoff, G., 41, 65, 122
Landau, B., 24, 109
Lane, S., 26
Lappan, G., 4, 33
Lave, J., 15, 203, 224, 307
Lean, G., 20, 32–35, 61, 94, 101, 141,
   177–179, 268
Learmonth, A.E., 109
Lefebvre, H., 98
Lehrer, R., 30, 84, 234, 235, 245, 246, 282,
   296, 303
Leinhardt, G., 64, 297
Lerman, S., 96
Lester, F., 6, 67, 311
Leung, F., 5
Lévi-Strauss, C., 75, 276
Lewis, D., 208
Liben, L.S., 8, 132
Lillo-Martin, D., 25
Lin, C.-l., 108
Lindquist, M., 3
Lindsay, J., 106
Linn, M., 73
Lipka, J., 237, 247, 264
Litteral, R., 247
Liu, Y., 9, 75
Llewellyn, G., 1
Lockheed, M., 119
Logan, P., 13, 93, 246, 304
Logan, T., 6, 90, 272

Lohman, D.F., 3, 19, 21–24, 31, 33
Lossau, J., 99
Lovat, T., 9
Love, E., 278
Lowrie, T., 4, 6, 31, 90, 272
Luitel, B.C., 295

M
Mackenzie, M.A., 168, 170, 171
Macmillan, A., 15, 92, 99
Magoon, R., 24
Maher, C., 96
Maldonado, L., 117
Malinowski, B., 178, 179, 319
Mammana, C., 10
Managal, M.A., 104
Mandler, J., 25
Margetts, A., 125
Mariotti, M., 6, 279
Mariotti, M.A., 232
Mark, J., 3, 5, 7, 294
Mark, K., 191
Martin, A., 251
Martin, M.O., 4
Martin, T., 106
Martin, W., 5
Martland, J., 245
Martlew, M., 27, 143
Masingila, J.O., 203, 239
Mason, J., 8, 90, 272, 278, 291, 300
Mason, J.E., 12
Massarwe, K., 307
Matang, R., 13, 125, 224, 238, 239, 247, 261, 263
Matsuo, N., 232
Maude, H., 174
May, S., 4
Mazzeo, J., 26
McFarlane-Smith, I., 32
McGee, M., 20, 31
McGinn, M., 203
McGinty, S., 247
McMillen, S., 282
McMurchy-Pilkington, C., 247
McNair, M., 13, 93, 246, 304
McPhail, D., 5, 235
Meade, A., 21
Meaney, T., 119, 120, 130, 237, 239, 247
Meira, L., 232
Mellin-Olsen, S.
Menghini, M., 224, 276
Mesquita, M., 120, 121

Metzgar, E., 220
Michael, W., 20
Mildren, J., 67
Millroy, W., 142, 199
Mitchelmore, M., 234, 240
Miyake, A., 6, 8, 11, 232
Mohatt, G., 264
Moll, H., 300
Moll, L., 111, 231, 246, 295, 307
Molnar, J., 8
Montiel, C., 104
Mooney, C., 268
Morar, T., 119
Moschkovich, J., 89, 119, 258
Mosel, U., 124
Moses, B., 27, 33, 34
Mousoulides, N., 275
Moyer, P., 280
Muke, C., 14, 137, 255
Mulligan, J., 82, 132, 275, 278, 280, 284–286
Mullis, I.V.S., 4
Munn, N., 2
Murdoch, K., 257, 307
Murray, C., 66
Murray, L., 13, 93, 246, 304
Murray, M., 250

N
Nakata, M., 202
Nasir, N., 203
Ncedo, N., 119
Ness, D., 6, 108–110, 142, 243, 300
Network, A.N.K., 273
Neville, H., 25
Newcombe, N., 77, 109
Newcombe, N.S., 33, 109
Niezgoda, D., 280
Noddings, N., 96
Norris, C., 211, 212
Norris, R., 211, 212
Nunes, T., 13

O
Odobu, D., 228, 229, 258
Onggi, C., 187
Opper, S., 108
Orey, D., 13, 212, 220, 221, 223, 225, 266, 276
Osana, H., 30, 303
Osborne, R., 19, 20, 97
O'Sullivan, D., 13

Outhred, L., 8, 27, 54, 57, 62, 63, 68, 89, 232, 234
Owens, K., 4, 5, 8, 13–16, 22, 24, 27–30, 35, 47, 49, 54, 57, 60–65, 67, 68, 70, 72–74, 78–82, 84, 89, 93, 94, 101, 118, 120, 126, 127, 132, 133, 180, 188, 191, 199, 200, 202, 203, 219, 221, 223, 224, 228, 230–232, 234, 235, 244, 246–250, 252, 261–263, 266, 285, 297, 298, 300, 301, 303, 304, 313, 323, 324, 327, 328
Ozanne-Rivierre, F., 124
Ozyürek, A., 122

P

Paivio, A., 22, 23, 31
Paola, D., 233
Papert, S., 276, 277
Pappas, S., 108
Paraide, P., 94, 118, 148, 177, 178, 219, 235, 247, 250, 304
Patronis, T., 234
Peck, K., 15
Pegg, J., 11, 232, 304
Peirce, C., 96, 97
Pellegrino, J.W., 3, 19, 21–24
Pepeta, P., 241, 243
Perham, F., 34
Perry, B., 5, 80, 132, 232, 328
Piaget, J., 2, 26
Pica, P., 107, 235
Pickles, A., 146
Pieres, M., 119
Pinxten, R., 78, 108, 113, 118, 120, 121, 136, 206, 207, 247, 263, 303, 307
Pirie, S., 5, 63, 231, 312
Piru, P., 130, 250, 251
Pittalis, M., 275
Planas, N., 223, 247
Poisard, C., 142
Poltrock, S., 22, 23, 31
Pomerantz, J., 22
Powell, A., 26
Power, K., 111
Prediger, S., 116, 117
Presmeg, N., 5, 9, 43, 45, 47, 51, 63, 81, 115, 224
Prevett, T., 7–9, 47, 107
Prusak, A., 301
Pylyshyn, Z., 22, 31

Q

Quinn, M., 20, 32, 34

R

Radford, L., 92, 96–98
Rahaman, J., 106
Rawes, P., 297
Reddacliff, C., 5, 47, 82, 235
Regian, J.W., 3, 19, 21–24
Reichard, G., 145
Reisman, F., 26
Renshaw, P., 146, 276, 279, 280
Restivo, S., 224
Rivera, F., 63, 64, 89, 118, 119, 225, 226, 279, 280, 297, 299, 302, 308
Rodríguez, M., 287
Rosa, M., 13, 212, 220, 221, 223, 225, 266, 276
Rosser, R., 26
Rossi Becker, J., 118, 119, 308
Roth, W.-M., 8, 47, 89, 203, 297
Rouse, J., 203
Rowe, M., 34
Rudder, J., 94, 124
Rumsey, A., 135, 214

S

Sabo, H., 24, 25
Sakopa, P., 221
Salzmann, Z., 130
Sarama, J., 29, 76, 108, 110, 267, 281, 282, 301
Saunderson, A., 34
Saxe, G., 13, 81, 92, 100, 101, 117, 142, 177, 223, 300, 305
Schifter, D., 5
Schorr, R., 307
Schwartz, D., 106
Scriven, B., 6, 90
Senechal, M., 7
Senft, B., 173
Senft, G., 118, 121–123, 125, 131, 173, 182
Seo, K-H., 108
Setati, M., 116, 224, 255
Sfard, A., 301
Shah, P., 6, 8, 11, 232
Shayer, M., 141
Sheckels, M., 32
Shepard, R., 3, 23, 25, 31, 41, 141
Sheridan, N., 109
Shir, K., 232
Shirley, L., 223
Shoulders, M., 61
Silver, E., 3, 245
Silvern, L., 110
Skemp, R., 19, 40

Slezakova, J., 8
Smidt, D., 169
Smith, D., 118, 235
Smith, G., 12, 92, 111, 239, 281
Smith, S., 307
Snow, R.E., 33
Sobel, D., 112, 113
Soja, S., 98, 136, 295
Somerville, M., 92, 111, 112
Spelke, E., 107, 109, 235
Spencer, C., 110
Spennemann, D., 208
Stacey, K., 291
Stafford, A.K., 245
Stanley, M., 280
Steffe, L., 19
Stein, M., 64, 297
Stephens, M., 300
Sternberg, R.J., 277
Stigler, J., 13
Stiles-Davis, J., 24, 25
Stillman, G., 216
Stokes, J., 125–127
Stranger, G., 245
Strässer, R., 81, 278
Strom, D., 84, 234, 235, 245, 246, 282, 303
Sturrock, F., 4
Sukon, K., 108, 116, 120
Sunhaloo, S., 120
Suwarsono, S., 20, 32
Swafford, J., 3
Swallow, J., 24
Swaminathan, S., 282
Sweller, J., 75
Szeminska, A., 3

T
Tall, D., 89
Tallal, P., 25
Tartre, L., 20, 31, 32, 41, 47, 72
Tau, K., 248
Taylor, H.A., 131
Te Maro, P.N., 118, 235
Téllez, K., 89
Then, D.C.-O., 120
Thom, J., 8, 47, 89, 297
Thomas, D., 45
Thomas, N., 82
Thornton, M., 9, 13, 133, 223, 304
Thoyre, G., 84, 234, 235, 245, 246, 282, 303
Thune, C., 177
Thurstone, L., 21
Thurstone, T., 21

Tine, M., 4
Ting, S.-H., 120
Tomasello, M., 300
Tomaz, V., 266
Toomey, R., 9
Treacy, K., 81
Treisman, A., 76
Trinick, T., 119, 120, 130, 237, 239
Trninic, D., 297
Trudgen, R., 231, 247
Tuan, Y.-F., 8, 17, 112, 113, 131, 201
Tuinamuana, K., 307
Tupper, I., 122, 130
Turkle, S., 276, 277
Turnbull, D., 133, 134, 210, 291
Turner, E., 117
Tutak, F.A., 111

U
Uttal, D.H., 131

V
Vaid, J., 24, 25
Vakalahi, H., 96
Valdés, G., 234
Valero, P., 13
Van Bendegem, J., 224
van der Heijden, A.H.C., 76
van Dooren, I., 78, 108, 113, 118, 121, 136, 206, 247, 263, 303
van Hiele, P., 5, 9, 27, 34, 61, 92, 120, 232
Van Moer, A., 302
Van Niekerk, R., 88
Vandendriessche, E., 65, 173, 250
Vasilyeva, M., 4
Verlot, M., 307
Verner, I., 307
Verschaffel, L., 252
Verspoor, A., 119
Vilella, X., 247
Villani, V., 10
Villarrea, M., 276
Voigt, J., 63, 74, 224
von Glasersfeld, E., 96
Vurpillot, E., 64, 65, 75

W
Wahlberg, N., 237
Walden, R., 234
Walkerdine, V., 9, 234
Ward, A., 199, 200

Warner, L., 211, 307
Warren, E., 81, 245
Wassmann, J., 92, 135, 223, 276
Watson, A., 300
Watson, J., 5
Watson, R., 2, 5
Watson-Verran, H., 9, 13, 133, 135, 210, 223, 291, 304
Webb, N., 33
Wedgwood, C.H., 174, 214, 215
Weiner, J., 135, 214
Wenger, E., 16, 203, 301, 304
Were, G., 141, 145, 146, 245
Werner, H., 77
Wessels, D., 88
Wheatley, G., 28, 268
White, P., 240
White, R., 22, 52
Wickens, C., 7–9, 47, 75, 107
Wickler, S., 220
Wildfeuer, S., 237
Willis, S., 81
Wilson, B., 15
Wilson, D., 29
Wilson, J., 243
Witelson, S., 24
Wittrock, M., 19, 20, 97
Wong, L., 199, 200
Wood, T., 74
Woodhouse, J., 112
Woodward, E., 61
Worsley, P., 8, 107, 126
Wright, R., 245

Y

Yakimanskaya, I.S., 108
Yambi, K., 250, 251
Yamu, A., 226
Yunkaporta, T., 247

Z

Zaslavsky, O., 64, 232, 297
Zevenbergen, R., 13
Zimmerman, B., 301
Zimmerman, W., 20
Zimmermann, W., 5, 24

Subject Index

A
Affect, 271, 301
Africa
  Mozambique weaving, 263
  Temne, Sierra Leone, 206
  Tshokwe, Bantu, Angola, 215
America
  Hñähñu, the Otomies, 217
  Inuit, Alaska, 206
  Latinas/os, 111
  Mayan cosmology, 220
  Navajo, 136, 207
  Yup'ik, 177
Architecture, 193, 200
  Africa, settlement patterns, 221
  PNG architecture student project, 193
  power and relationships, 201
  settlement patterns, 222
  Trobriands, settlement patterns, 220
  Yap, settlement patterns, 220
Asia
  Malaysia, food covers, 219
  Tamil Nadu, 214
  Timor Leste, ikat weaving, 219
Astronomy
  Australian Aboriginal and Torres Strait Islanders, 212
  stone arrangements, 212
Attention, 4, 77, 78, 80, 81
  cultural, 82
Australian Aboriginal and Torres Strait Islanders
  Anindilyakwa, 125, 126
  Iwaidja, 125
  Nganguraku, Murray River, 212
  Walpiri, 125
  Wathaurung, Victoria, 212
  Wiradjuri, 124
  Yolngu, 133

B
Bilum, 172
Block and construction play, 109
Bridges, 190–191, 250

C
Carving, 169
Community of practice, 246
Concept development, 61
Count me into space, 83
Continuity between home and school mathematics, 247
Culturally responsive curriculum, 273

D
Deixis, 125
  Bali, Indonesia, 110
Diagrams, 1, 3–5, 7, 8, 56, 57, 63, 64, 68, 74, 80, 85, 88, 96, 97, 110, 115, 117, 135, 139, 210, 217, 224, 242, 252, 272
Dimensions, 121, 126, 131
Drawing
  children, PNG, 143
  PNG rock art, curves, 144

E

Early childhood education, 99, 233, 255
  ecocultural perspective, 108
  spatial referencing, 109
Ecocultural, 292
  Brazil examples, 221
  context, 297
  definition, 12, 113, 211, 231
  examples, ecology, 94, 207
  examples, PNG, 250
Embodiment, 229, 297
Embodiment, gestures, kinaesthetic, physical, 107
Ethnomathematics, 13, 17, 142, 224
Ethnomodelling, 266

G

Genetic development, 102
Geography, 6, 7, 9, 11, 232, 293
Geometry, 10, 88, 167, 205, 212, 233, 253, 267
  angles of polygons, 242
  current education, 232
  geometry education theory, 83
  juxtaposed symmetry, 216
  position, frames of reference, 291
  reflective and slide symmetries, 217
  right angles, PNG, 226
  rotational and reflective symmetry, 215
  sine wave, 250
  skills or processes, 271
  symmetry, 242
  symmetry, litima designs, Africa, 218
  symmetry, stylised tapa, 216
  symmetry, weaving board, 263
  tessellations, tiling, 214
  topology, 167
  topology, curves around arrays of points, 215
  triangular pyramid, tipi, 221

H

House construction, 189–190

I

Identity, 99, 104
  definition, 14, 301
  ecocultural, 16
  mathematical, 15–16, 197
  mathematical and cultural, 241, 297
Information processing, 32

Intention, 108
Intention and attention, 230
Intuition, 92, 272

K

Kapkap, 145

L

Language, 224, 231, 240, 260
  classification, 121, 122
  code switching, 116, 120, 248
  Creole, 120
  Creole, Mauritius, 269
  further language issues, 248
  identity, 119
  metaphor and imagery, 249
  prepositions, prefix, suffix, infix, 125
  registers, 117
  Learning, information processing, 23
Learning theory
  disability studies, 26
  ecological, 94
  folding back and visual, 5
  piagetian, 27
  signification, meaning, practice, 98
  visual imagery as central, 5
  visualising, 65
Locating, position, 109
Location
  western language, 121

M

Malaysia, food covers, 201
Maori
  architecture, 201
  language, 119
Mapping, 132, 133
  Australian Aboriginal, 135
  context, 132
  Navajo, 136, 264
  Papua New Guinea, 135
  reference point, stars, sea roads, 208
  young children, 132
Material culture PNG, 167
Mauritius, 120
Measurement
  area, 106, 185
  area and length lesson, 258
  estimation, 106
  length, 182
  length composite units, 182

Papua New Guinea languages, 126
practical, 276
visuospatial ecocultural units, 184
visuospatial reckoners, 228
volume, 188
volume, Yup'ik, 264
Measurement study, 180–181
ecocultural variance, 180
Middle East and Meditterenean, La Alhambra, Spain, 212
Myth, 135, 144, 174, 176

N

Navigation, 227
Pacific cultures, Marshall Islands, 211
USA Navy, 105
Number
Australian, 5-cycle, Wiradjuri, 94
classifiers, 179
classifiers, PNG, 178
by display, PNG, 177
fish sharing, Motu, PNG, 258
groupings, PNG counting systems, 178
Manus type base 10 counting systems, 179
Mayan number pairs, meanings, 220
PNG body-part counting, fu, 103
PNG counting systems, 104
PNG 5-cycle Gahuku-Asaro, 94
visualising irrational numbers, 225

P

Pacific
Fiji tapa, 216
Palau, settlement patterns, 220
Tonga, 217
Vanuatu, sand drawings, 215
Pacific cultures/languages (Oceania), Iaai, New Caledonia, 124
Papua New Guinea, cultural activity as a whole context, 191
Papua New Guinea cultures/languages
Abelam, Middle Sepik river, East Sepik Province, 168
Asaro mudmen, 252
Baining. East New Britain, 122
Buang, Morobe Province, 144
coastal, fishing, 250
Finschaffen, Morobe Province, 250
Gahuku-Asaro Alekano, Eastern Highlands Province, 124
Gahuku-Asaro/Alekano-Kaveve, Goroka, EHP, 95
Hagen, Melpa, 191
Hela, 250
Kagua/Erave, 174
Kamano-Kafe, Eastern Highlands Province, 186
Kâte, 182
Kewapi, Southern Highlands, 187
Kilivila, Trobriands, Milne Bay, 126, 173
Kilivila, Trobriands, Milne Bay Provicne, 178
Kiwai, Western Province, 122
Kominimung, Ramu,river, Madang Province, 170
Kuruti, Ere, Lele, Geke, Manus Province, 179
Malalamai, Madang Province, 226
Manumanu, Central Province, 228
Mekeo, Central Province, 176
Motu, Central Province, 176, 178, 258
Narak, Jimi valley, Jiwaka, 123
New Ireland, 168
Panim, Madang Province, 187
Rai coast, Madang, 260
Tabare. Sinasina, Simbu Province, 254
Tabar, New Island, 144
Tinatatuna, Tolai, Kuanua, 177
Wantoat, Morobe Province, 170
Wosera, East Sepik Province, 169, 182
Yupna, 135
Patterns, 251
Perception, 77–78
Place, 201
Place-based, 12
Place based education, 111, 272, 306
Position, language patterns, 123
Pottery, 169
Problem solving, 63, 79, 231, 297
metaphor, metonomy, imagery, 65–67
robotics, 284
Psychology, 19–20

R

Ratio, 192–193
Representations, 9
Mayan diamonds, number, 220
number, 180
Responsiveness, 12, 14, 20, 45, 78–82, 108, 114, 138, 199, 263, 292–294, 298–299

S

Sámi, 237
Self regulation, 14–16, 60, 132, 194, 198, 199, 219, 237, 244, 252, 260, 292, 294, 295, 297–299, 301, 302, 306, 308
  computing, 278
  strategies, 281
Shapes, 232
  action visuospatial reasoning, 227
  definitions, 136
  language classifiers in counting, 179
  woven, Sinasina, PNG, 254
Situated cognition, 230
Social justice, 12, 14, 17, 92, 111, 112, 116, 118, 142, 220, 239, 244, 272, 308
  Alaskan Universities Council Policy, 222
  border crossings, continuities, 295
  decolonising thinking, 113
Spatial abilities, 2–3, 20–22
Spatial imagery, 7, 8
Spatiality, 306
  thirdspace, 99, 136
String figures (cat's cradles), 172

T

Tapa, 159, 170, 216
Tattoo, 176, 258
Teaching
  Alaskan curriculu, 273
  classrexamples, 271
  classroom context, 75, 271
  computer use in Singapore, 280
  computing and culture, 288
  Count Me Into Space, NSW, 235
  culturally responsive, 111, 236
  cultural mathematics lessons, PNG, 260
  curriculum, 5, 7, 11
  curriculum geometry general, 235
  dynamic geometry software, 278
  elementary school project, PNG, 256
  ethnomathematics curriculum on processes, 266
  examples, 308
  general geometry curriculum, 235
  geometry Australia, 73
  importance of stories, 247
  inquiry method, 257, 307
  investigative activities, 270
  key study on geometry education, 38
  Latinos/as, 112
  logo, 282
  mathematical processes, 264
  Mauritius language and investigation study, 268
  Navajo mapping, 264
  robotics, 282
  robots and measurement, 286
  robots and position, 286
  role of teacher in digital age, 285
  South African curriculum, 235
  teacher education, 304
  teachers' beliefs, 247
  technology principles, 285
  transcultural principles, 248
  virtual manipulatives, 280
  weaving boards, 263
  Yup'ik, 264
Testing, 1, 3–6, 20, 22, 33, 54, 67, 81, 90, 110, 119, 210, 231, 268, 294
Thirdspace, 295
Time
  Melpa, PNG, 192

V

Values, 96, 230, 234, 294
van Hiele, 92, 120, 232
van Hiele theory, 27
Visuality, 32, 33
Visuospatial reasoning, 227
  action, 49, 51
  action-school, community, 234
  assessment, 38, 87
  computer graphics, 276
  definition, 8, 9, 11, 24, 88, 294
  drawings, 282
  3D shapes, 72
  dynamic, 47, 227
  ecocultural contexts, 303
  example, 29, 90, 302
  holistic, 45
  indigenous cultures, 304
  logical examples, 230
  material objects, 242
  number, 258
  objects' role, 146
  pattern, 55
  pattern and dynamic, 300
  pattern-area, 57
  physical/mental, 198
  procedure, 59

Subject Index

ratio in visuospatial reckoners, 202
synergy of culture and school, 296
tacit knowledge, 238
tasks, 31, 35
training, 35
Volume
pig measuring, 186
ratio small volumes, 187

W
Weaving
  Hñähñu, the Otomies, 217
  Mozambique, Africa, 219
  Papua New Guinea, 3D objects, 219
  Sinasina, PNG, 254

Y
Yolngu, 9, 13, 99, 134, 135, 212, 220, 306

CPSIA information can be obtained at www.ICGtesting.com
Printed in the USA
LVOW01*1959200115

423614LV00001B/14/P

9 783319 024622